SYNER-G: Systemic Seismic Vulnerability and Risk Assessment of Complex Urban, Utility, Lifeline Systems and Critical Facilities

GEOTECHNICAL, GEOLOGICAL AND EARTHQUAKE ENGINEERING

Volume 31

For further volumes:
http://www.springer.com/series/6011

K. Pitilakis • P. Franchin • B. Khazai • H. Wenzel
Editors

SYNER-G: Systemic Seismic Vulnerability and Risk Assessment of Complex Urban, Utility, Lifeline Systems and Critical Facilities

Methodology and Applications

Springer

Editors
K. Pitilakis
Department of Civil Engineering
Aristotle University
Thessaloniki, Greece

P. Franchin
Department of Structural and Geotechnical Engineering
Sapienza University of Rome
Rome, Italy

B. Khazai
Geophysical Institute
Karlsruhe Institute of Technology
Karlsruhe, Germany

H. Wenzel
VCE - Vienna Consulting Engineers
Vienna, Austria

ISSN 1573-6059 ISSN 1872-4671 (electronic)
ISBN 978-94-024-0113-4 ISBN 978-94-017-8835-9 (eBook)
DOI 10.1007/978-94-017-8835-9
Springer Dordrecht Heidelberg New York London

Printed on acid-free paper

Springer is part of Springer Science+Business Media (www.springer.com)

Foreword

When I first reviewed the proposal of the project SYNER-G, I thought that it was a 10 years effort funded at a few tens of million euros. Even if it was a 10 year, tens of millions of euros, it would have been a milestone in geographically-broad multidisciplinary earthquake impact assessment. That the project was funded at a fraction of what I anticipated and had a life span of only 4 years adds enormously to the credit that the research team, and the project leadership, deserve. The first publication of the SYNER-G project (Pitilakis et al. 2014) presented an impressive array of methods and models for fragilities that surpass those existing in the literature and in modeling software worldwide. As such, the first publications pushed the envelope to the limit of our knowledge and capabilities, and beyond. I want though to step back and remind us of the components of broad geography earthquake impact assessment. What is it for? Why is it so critically important? What does it require? Who are the end users? I attempt to answer these questions below, and will answer the last first. Truly, 'Who is it for' conditions the response to the other questions. Engineering communities have traditionally owned quantitative disaster impact modeling; engineers know the mathematics and physics, know the earth sciences, they know all that is there to be known for the estimation of the physical impact. Unfortunately, this is not sufficient, as necessary as it is. In my early days of regional impact assessment studies around 1999–2000, I was hit by the gap between us engineers and those who will use our data, and even more those who will make decisions. Presenting our 'interstorey drift maps' to the clients did not go down well. We tried to explain the great significance of interstorey drift! But this too did not go down well. It took effort from both sides, the engineers and let us say the non-engineers (who include the extremely important communities of social scientists, economists, emergency managers, to name a few), to develop a common vocabulary, and to articulate the transition from physical and social impact. It took my team and I at the Mid-America Earthquake (MAE) Center (a US National Science Foundation Engineering Research Center) years to understand, make the case for, and apply an interdisciplinary approach to earthquake risk management. We even developed a framework, the Consequence-Based Risk Management approach, summarized in the figure shown, courtesy of the MAE Center researcher Professor Steven

French, which was a further development of an earlier construct of the MAE Center referred to as Consequence-based Earthquake Engineering. When we understood that engineering develops physical impact models, passes the impact assessment to social and economic sciences, where societal impact models are used to estimate impact, that various scenarios have to be investigated to aid in decision-making, that emergency shelter and temporary housing are part of our responsibility, and many other nuances that eluded us for decades, we were able to truly support society's goals of managing risk in a holistic and hence effective manner.

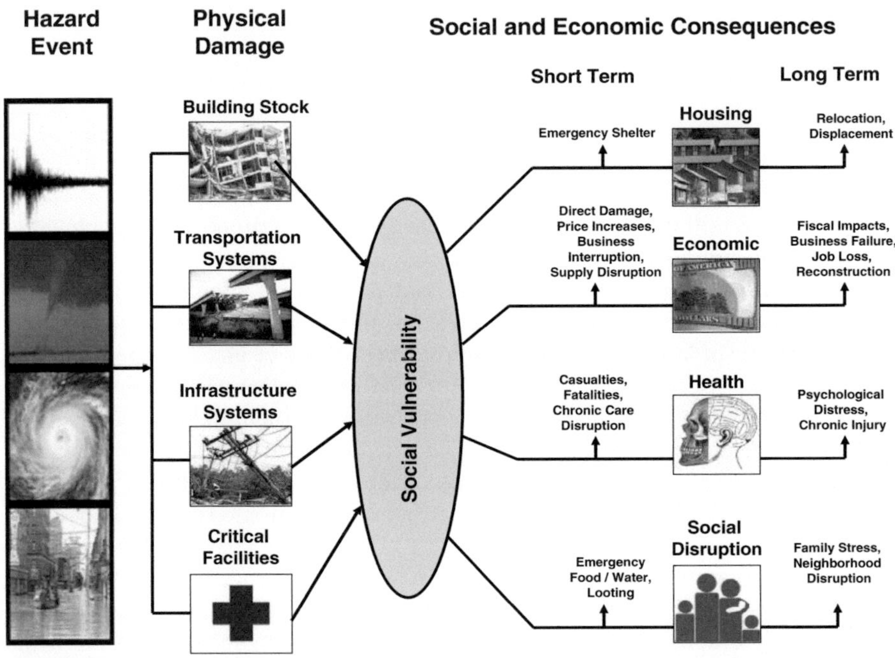

Framework for consequence-based risk management in the Mid-America Earthquake Center

I wish that we had the current SYNER-G book then! Chapters 1 and 2 in this book set the scene and provide the context. The challenging task of estimating the consequences on large spatially distributed systems is addressed; such systems have different levels of demand-capacity ratios in space and time and thus require intricate simulation approaches at the forefront of the state-of-the-art of dynamic network analysis. Chapters 4 and 5 address physical and social vulnerability in quite a novel manner, especially the social vulnerability part that has broken new ground. In my view, Chapters 1, 2, 3, 4, and 5 are the intellectual core of the book. Chapters 6, 7, 8, 9, 10, 11, and 12 though are where SYNER-G comes into its own, with a set of practical applications that are eye-opening and exceptionally insightful. These are the chapters that I would have wished were there when the

MAE Center was wrestling with the challenges of regional impact of earthquakes on the eight central US states, the so-called Mid-America. That SYNER-G not only developed component and system assessment tools, and integration frameworks, but also applied its research developments to the wide array of cities (Thessaloniki, L'Aquila, Vienna) is impressive and very useful to many technical and social communities. Whereas I was involved on the fringes of SYNER-G, I am a relative outsider, with good experience in the challenges of developing engineering, social and economic earthquake impact models, information technology implementation and practical application, in passing judgment on the present SYNER-G book. My verdict is that this is a top tier book that has breached many research boundaries and took the research findings right to the doorstep of emergency managers and policy- and decision-makers. I recommend the book to all communities interested in protecting communities from the perils of earthquake, and my recommendation comes with no reservation.

Harold and Inge Marcus Dean of Engineering
The Pennsylvania State University
University Park, PA, USA

Amr S. Elnashai, FREng

Reference

Pitilakis K, Crowley H, Kaynia A (eds) (2014) SYNER-G: typology definition and fragility functions for physical elements at seismic risk, Geotechnical, geological and earthquake engineering, vol 27. Springer, Dordrecht. ISBN 978-94-007-7871-9

Preface

Modern societies and economies become more complex and sophisticated and at the same time more vulnerable to multiple hazards. Although design provisions against seismic hazards have been considerably improved over the last two decades, the urban built environment and the various infrastructures serving it are still very vulnerable. The main reason is that they have been built with lower code or no code provisions and in some cases, especially in Europe, several centuries ago. The resilience of the modern cities and infrastructures built in prone seismic areas relies on an *integrated seismic risk approach*, where physical damages and socio-economic impact are studied as a whole considering the interactions among different systems and infrastructures at city or regional scale.

The European funded research project SYNER-G consists a major step forward in this challenging task, at least in Europe, giving special emphasis to the interactions among systems at urban or regional scale, interactions that generally increase the overall impact of a catastrophic earthquake.

The present volume presents an integrated framework and tools that have been developed for the systemic seismic vulnerability and risk analysis of complex systems exposed to earthquake hazard, like buildings and aggregates in urban scale, lifelines, transportation and utility networks, gas and electric power systems, critical facilities, and infrastructures. The core novelty of the work is the integration of interactions between different components and systems that may increase considerably the global vulnerability and impact in case of strong seismic events. SYNER-G methodology encompasses in an integrated way all aspects in the chain, from hazard to the physical vulnerability and loss assessment of components and systems and to the socio-economic impacts of earthquakes, accounting for all relevant uncertainties within an efficient quantitative simulation scheme, modeling interactions between the multiple components and systems.

With its companion book *SYNER-G: Typology Definition and Fragility Functions for Physical Elements at Seismic Risk*, also published in same series of Springer editions, this book provides the state of the art on this topic at least within the European context. The ambition is to offer to the European and international scientific and engineering community a standard reference book for the systemic seismic vulnerability and risk analysis of complex systems exposed to earthquake hazard.

The Editor would like to express his acknowledgement to all partners and researchers who contributed to the successful realization of the ambitious work undertaken. Most of them are contributing in the different chapters of this volume. Finally, the support of the three co-editors, Paolo Franchin, Bijan Khazai and Helmut Wenzel, and above all the devotion and hard work of Dr. Sotiris Argyroudis in the preparation of this volume is gratefully acknowledged. Special acknowledgement to Dr. Denis Peter, project officer of the European Commission, for supporting the successful accomplishment of SYNER-G.

Professor in Aristotle University
Thessaloniki, Greece

Kyriazis Pitilakis

Contents

About the Editors

Kyriazis Pitilakis
Professor in Earthquake Geotechnical Engineering
Department of Civil Engineering, Aristotle University
54124 Thessaloniki, Greece, Tel +30-2310-995693
e-mail: kpitilak@civil.auth.gr
http://users.auth.gr/~kpitilak/

Professor **Kyriazis Pitilakis** graduated from Aristotle University of Thessaloniki, Greece, and took his Ph.D. in Ecole Centrale Paris. He has more than 30 years of intensive academic, research and professional experience in civil, earthquake and geotechnical engineering. He is Chairman of the Technical Committee "Geotechnical Earthquake Engineering and Associated Problems" (TC4), of the International Society of Soil Mechanics and Geotechnical Engineering (ISSMGE) and President of the Greek Society of Earthquake Engineering. He has been Head of the Civil Engineering Department of Aristotle University (1997–2001) and Chairman of

the Institute of Earthquake Engineering and Engineering Seismology in Greece (ITSAK) and coordinator of several important EU research projects namely EURO-SEISTEST (http://euroseis.civil.auth.gr) and SYNER-G (www.syner-g.eu). His main fields of interest are in earthquake, geotechnical and lifeline earthquake engineering, vulnerability and risk assessment, soil dynamics, site effects and microzonation. He has been state-of-the-art and keynote lecturer in numerous international conferences, author of more than 400 scientific papers, and reviewer in numerous important scientific journals and international research projects. He is member of the editorial advisory board in Springer (series of Geotechnical, Geological and Earthquake Engineering), member of many international societies in earthquake and geotechnical engineering and member of national and international committees for seismic standards.
Honors: Chevalier des Palmes Académiques, République Française.

Paolo Franchin
Assistant Professor in Department of Structural and Geotechnical Engineering
Sapienza University of Rome
via Gramsci, 53, 00197 Rome, Italy, Tel: +39 0649919-189
e-mail: paolo.franchin@uniroma1.it
http://w3.uniroma1.it/franchin/Paolo_Franchin/MAIN.html

Paolo Franchin graduated with honors in Civil Engineering from Sapienza University of Rome in 1997, received an M.Sc. in Structural Engineering from University of California, Berkeley in 2000, and a doctoral degree from Sapienza in 2001. Post-doc research fellow at Sapienza from 2001 to 2006, when he became Assistant Professor in Structural Design. Since 2001 he is a faculty of the School for Advanced Studies in Earthquake Engineering (ROSE School) in Pavia. His teaching activity covers structural concrete design, earthquake engineering and structural reliability. His current research interests are performance-based seismic design and assessment of buildings and bridges, and risk analysis of structural and infrastructural systems. He serves as a reviewer for most international journals in the fields of his research, and has taken part in many national and international research projects funded by the European Union (SAFERR (FP5), SPEAR (FP5), LESSLOSS (FP6) and SYNER-G(FP7)), by the Italian Ministry for Higher Education and Research (PRIN2002, PRIN2004, PRIN2007), by the Italian National Research Council (VIA, MIUR-CNR) and by the Italian Department of Civil Protection (Progetto Esecutivo DPC-Reluis 2005–2008, Progetto Esecutivo DPC-Eucentre 2009–2012). He is a member of Commission 7 "Seismic design", and of Task Groups 7.5 "High-performance materials and structural systems" and 7.7 "Performance-based Seismic Design", of the International Federation of Structural Concrete (fib); he is also a member of Sub-Committee 3 "Structural Reliability and Optimization" of the International Association of Structural Safety and Reliability (IASSAR).

Dr. Bijan Khazai
Karlsruhe Institute of Technology (KIT)
Geophysical Institute (GPI)
D-76187 Karlsruhe, Germany, Tel: +49 721 608 44442
e-mail: khazai@kit.edu

Bijan Khazai is a senior research scientist at Karlsruhe University's Center for Disaster Management and Risk Reduction Technology (CEDIM) where he leads the Risk and Vulnerability research group. He holds masters and doctoral degrees in Geotechnical Earthquake Engineering from the University of California at Berkeley. He was a post-doctoral research fellow at Columbia University's Earth Institute and Kyoto University's Disaster Prevention Research Institute (DPRI), where he was involved with interdisciplinary research investigating post-disaster recovery and reconstruction processes in Iran (Bam), Sri Lanka, Pakistan (Northwest Frontier Province), and New Orleans. The basis of his research work is in applying spatial analysis, statistical and probabilistic methods, as well as indicator-based frameworks to address various issues of systemic vulnerability and resilience in urban environments. Over the last 7 years he has engaged closely with stakeholders as a project specialist for the Earthquake and Megacities Initiative in a number of projects in Istanbul, Mumbai, Amman, Metro Manila, Kathmandu and Dhaka to address cross-sectoral issues in urban risk reduction and disaster response. He is Principal Investigator of the Social Vulnerability and Integrated Risk Project of the Global Earthquake Model (GEM) and also leads a collaborative research program on risk and resilience between KIT and Heidelberg University (HEIKA).

Prof. Dr. Helmut Wenzel
President, VCE Vienna Consulting
Engincers ZT GmbH, Hadikgasse 60, A-1140 Wien, Austria
Tel: +43 (1) 897 53 39 – 1116
e-mail: wenzel@vce.at

Dr. Helmut Wenzel has worked for large infrastructure projects in Europe and overseas for the first 15 years of his career. Driven by the demand for new technologies and practical applications in the construction industry he recognised the value of research and development and devoted his attention to national and international research projects. Since 1995 (in the BRITE-EURAM program of the European Commission) he took over the responsibility of coordinating large collaborative research projects (HARIS, CasCo, IMAC, OASYS, MOBILE, SAFEPIPES, IRIS, NERA) and organised dedicated international networks (SAMCO and I-SAMCO).

Research and development has created a number of very successful applications grouped around the global brand **BRIMOS** (Bridge Monitoring System). This development brought him international recognition in the structural health monitoring community (two books). The main objectives of recent projects concentrate on life cycle engineering issues which are supported by the demand of safe life extension of our European critical infrastructures. Dr. Wenzel is the author of three internationally well recognised books and has contributed with keynotes and papers to numerous initiatives of the scientific society.

In his capacity as internationally well recognised expert, he is performing evaluations of research programs, contributes to strategic research agenda and road maps and is engaged as technical and scientific advisor to major agencies and projects.

Long-standing international collaboration has been established with leading Universities and Research Institutes worldwide (i.e. University of Tokyo, Chinese Academy of Science, Stanford University, JRC, NCREE, NCSA, Drexel University, Rutgers University, BARC Mumbai, University of Ottawa, ETHZ, AUTH, Sheffield, KU Leuven, Porto).

Contributors

Sotiris Argyroudis Department of Civil Engineering, Aristotle University, Thessaloniki, Greece

Anna Bosi Vienna Consulting Engineers (VCE), Vienna, Austria

Francesco Cavalieri Department of Structural and Geotechnical Engineering, Sapienza University of Rome, Rome, Italy

James E. Daniell Karlsruhe Institute of Technology, Karlsruhe, Germany

Nicolas Desramaut Bureau de Recherches Géologiques et Minières (BRGM), Orléans Cedex 2, France

Şebnem Düzgün Middle East Technical University, Çankaya, Ankara, Turkey

Simona Esposito Dipartimento di Strutture per l'Ingegneria e l'Architettura, Università degli Studi di Napoli Federico II, Naples, Italy

Paolo Franchin Department of Structural and Geotechnical Engineering, Sapienza University of Rome, Rome, Italy

Pierre Gehl Formerly at Bureau de Recherches Géologiques et Minières (BRGM), Orléans Cedex 2, France

Iunio Iervolino Dipartimento di Strutture per l'Ingegneria e l'Architettura, Università degli Studi di Napoli Federico II, Naples, Italy

Kalliopi Kakderi Department of Civil Engineering, Aristotle University, Thessaloniki, Greece

Bijan Khazai Geophysical Institute, Karlsruhe Institute of Technology, Karlsruhe, Germany

Tina Kunz-Plapp Karlsruhe Institute of Technology, Karlsruhe, Germany

Alessio Lupoi Department of Structural and Geotechnical Engineering, Sapienza University of Rome, Rome, Italy

Hormoz Modaressi Formerly at Bureau de Recherches Géologiques et Minières (BRGM), Orléans Cedex 2, France

Paolo Emilio Pinto Department of Structural and Geotechnical Engineering, Sapienza University of Rome, Rome, Italy

Kyriazis Pitilakis Department of Civil Engineering, Aristotle University, Thessaloniki, Greece

David Schäfer Vienna Consulting Engineers (VCE), Vienna, Austria

Jacopo Selva Istituto Nazionale di Geofisica e Vulcanologia, Bologna, Italy

Graeme Weatherill European Centre for Training & Research in Earthquake Engineering (EUCENTRE), Pavia, Italy

Friedemann Wenzel Karlsruhe Institute of Technology, Karlsruhe, Germany

Helmut Wenzel Vienna Consulting Engineers (VCE), Vienna, Austria

List of Symbols

A	matrix of connectivity-related coefficients in the flow equations of a water network
AD	average displacement of a rupture
AF(T)	soil amplification factor at spectral period T
A_i	random amplification factor at the i-th site
B_i	Boolean random variable modeling the state of complete damage (i.e. rupture) of the i-th linear (e.g. pipe) component
CC_i	customer connectivity at demand node i
CR	casualty ratio
D	damage
D_i	random state of physical damage of the i-th component
E[ ]	expect value operator
f(x)	probability density function of random vector x
h	vector of water heads in the nodes of a water network
h(x)	sampling probability density function of vector x
H_{0i}	water head in non-seismic, normal operation conditions at node i
HR_i	head ratio at node i
H_{si}	water head in seismically damaged network at node i
$I_E(x)$	indicator function that takes the value one when x belongs to Ω_E, zero otherwise
L	geographic location of the seismic event generated on source S
M	earthquake magnitude
N_0	number of connected nodes in non-seismic conditions
N_{Li}	random number of leaks along the i-th pipe segment
NO_t	number of occupants (at the time of the event) by building type
N_{si}	number of connected nodes in seismic conditions
$N_{t,i}$	number of buildings of type t having damage level j
$N^i_{customer,0}$	number of customer at node i in the normal condition
$N^i_{customer,s}$	number of customer at node i in the seismic condition
$N^i_{source,0}$	number of sources connected to the sink node i in the normal condition

$N^{i}_{source,s}$	number of sources connected to the sink node i in the seismic condition
p	probability
P_{0i}	pressure in non-seismic, normal operation conditions at node i
PR_i	pressure ratio at node i
P_{si}	pressure in seismically damaged network at node i
Q	vector of boundary conditions in the flow equations of a water network
q	vector of water flows in the links of a water network
$QD_{t,i}$	proportion of deaths by building type and damage level
$QI_{t,i}$	proportion of injured by building type and damage level
r	crane productivity
R	epicentral distance (source to site)
$R = [u_i L_i]$	diagonal matrix of resistances in the flow equations of a water network, with u resistance per unit length and L the length
S	seismic or seismogenic source
$S_a(T)_{SURFACE}$	spectral acceleration at the engineering bedrock
$S_a(T)_{SURFACE}$	spectral acceleration at the ground surface
S_{gi}	random geotechnical hazard at the i-th site of interest
S_{gir}	random seismic intensity at the i-th grid point, on rock/stiff-soil
S_{ri}	random seismic intensity at the i-th site of interest, on rock/stiff-soil
S_{si}	random seismic intensity at the i-th site of interest, at the surface
TCaH	total cargo handled (loaded and unloaded) per day
TCaM	total cargo movements per day
TCoH	total number of containers handled (loaded and unloaded) per day
TCoM	total number of containers' movements per day
T_R	return period
u	vector of standard normal random variables
$V_{i,0}$	voltage magnitude at node i in the non-seismic, normal conditions
$V_{i,s}$	voltage magnitude at node i in the seismically damaged network
VR_i	voltage ratio at node i
V_{s30}	average shear-wave velocity in the upper 30 m
x	vector collecting random variables x
$\alpha(x) = f(x)/h(x)$	importance sampling ratio
α, β	parameters of the Gutenberg-Richter magnitude recurrence law
ε	vector collecting random variables ε modeling intra-event variability in ground-motion prediction equations
ε_S	transient ground strain
η	random variable modeling inter-event variability in ground-motion prediction equations

$\boldsymbol{\theta}$	vector collecting random variables θ modeling epistemic uncertainty in model parameters
λ	mean annual frequency of exceedance
μ	moving average
σ	moving standard deviation
Ω_E	domain in the sample space of x where the event E occurs

List of Acronyms

AHR	Average head ratio
APR	Average pressure ratio
ATC	Applied Technology Council
BC	Building Census
BDG	Buildings
BHI	Building Habitability Index
BIP	Building identification procedure
CEM	Casualty estimation models
COM	Mixed Operations Centers
DCI	Damage Consequence Index
DPI	Displaced Population Index
EAU	European Urban Audit
ECL	Electric power connectivity loss
ELE	Earthquake loss estimation
EMS	European Macroseismic Scale
EPN	Electric power network
FEMA	Federal Emergency Management Agency
GAS	Gas network
GIS	Geographical information systems
GMPE	Ground motion prediction equation
HBR	Harbor
HII	Health Impact Index
ICT	Information and communications technology
ID	Identification number
IDU	Impianto di Derivazione Utenza (in Italian): end-user node in the case of low-pressure gas networks
IM	Intensity measure
LOA	berth (ship) length (ship length overall)
LRE	Lack of resistance to evacuation
LUP	Land use plan
MAF	Mean annual frequency

MAVT	Multi-Criteria Value Theory
MC	Monte Carlo
MCDA	Multi-criteria decision analysis
MCS	Monte Carlo simulation
NEHRP	National Earthquake Hazard Reduction Program
PFDHA	Probabilistic fault displacement hazard analysis
PGA	Peak ground acceleration
PGD	Permanent ground deformation
PGV	Peak ground velocity
PI	Performance indicator
PSA	Pseudo-spectral acceleration
PVC	Polyvinyl chloride
RC, R/C	Reinforced concrete
RDN	Roadway network
SCD	Sub-city district
SCL	Connectivity loss
SI	Serviceability Index
SR	Serviceability ratio
SSI	Shelter Seeking Index (Chaps. 4 and 7)
SSI	System Serviceability Index
TAZ	Traffic analysis zones
TEU	Twenty-foot equivalent units
UBI	Uninhabitable Building Index (Chaps. 4 and 7)
UBI	Upgrade Benefit Index
UCD	Utility customer density
UL	Utility loss
UML	Unified Modeling Language
UR	Usability ratio
URM	Unreinforced masonry
WCL	Water connectivity loss
WSS	Water supply system

Chapter 1
Introduction

Kyriazis Pitilakis and Paolo Franchin

Abstract This chapter outlines the SYNER-G project, its objectives and structure. A short literature review of the vulnerability and risk assessment of infrastructural systems and their components highlights the framework of the past works and the challenges anticipated. The main issues for the systemic risk analysis are shortly described including the SYNER-G taxonomy, the seismic hazard estimates, the intensity measures and fragility curves, the systemic analysis methods and performance indicators, the treatment of uncertainties and socio-economic issues of the analysis. Finally, the applications that have been performed to test the SYNER-G methodology and tools are also outlined.

1.1 Background: Scope and Aim of the Book

The book presents the results of the work carried out within the SYNER-G project (see Sect. 1.3) on the physical modelling of the systems made up of several components, of their interactions, of the seismic hazard acting upon them and of all the relevant uncertainties that affect the evaluation of the systemic vulnerability. The book is closely related to a previous one in the same series of Springer editions, entitled *"SYNER-G: Typology definition and fragility functions for physical elements at seismic risk"* (Pitilakis et al. 2014). The later is devoted to the characterization of components' fragility.

K. Pitilakis (✉)
Department of Civil Engineering, Aristotle University, 54124 Thessaloniki, Greece
e-mail: kpitilak@civil.auth.gr

P. Franchin
Department of Structural and Geotechnical Engineering, Sapienza University of Rome, Via Gramsci 53, 00197 Rome, Italy
e-mail: paolo.franchin@uniroma1.it

K. Pitilakis et al. (eds.), *SYNER-G: Systemic Seismic Vulnerability and Risk Assessment of Complex Urban, Utility, Lifeline Systems and Critical Facilities*, Geotechnical, Geological and Earthquake Engineering 31, DOI 10.1007/978-94-017-8835-9_1,

The present book is comprised of two parts: Part I collects Chaps. 2, 3, 4, and 5 and presents models and methods for systemic analysis, while Part II (Chaps. 6, 7, 8, 9, 10, 11, and 12) illustrates their application to a number of case studies employed during the project as test beds.

Chapter 2 focuses on the methodological framework developed to bind together all the necessary models and describes the probabilistic assessment procedure used to evaluate performance indicators.

Chapter 3 discusses the distributed seismic hazard model employed to predict probabilistically and physically consistent vector fields of intensity ("shake fields"), to be fed to all components in order to evaluate their state of physical damage.

Chapter 4 introduces the methodological advancements made in the modelling of the social consequences/impact of the earthquake within the framework of multi-criteria decision analysis.

Chapter 5, finally, describes the specification of the general methodology (Chap. 2) to all the systems considered in the detailed taxonomy drawn within SYNER-G, and reported later in Sect. 1.4.

The chapters in Part II illustrate the applications of the SYNER-G methodology and tools for the analysis to selected systems, as a gas distribution network, a road network, an electric power network, a regional health care system, a district in Vienna, the city of Thessaloniki and the harbour of the latter.

1.2 Literature Review

The degree to which our society depends upon the reliable functioning of infrastructural systems and more in general of the built environment is underlined by the ubiquitous term *critical infrastructures* (CI) with which this set of interconnected systems is indicated (PCCIP 1997).

This extreme dependence and the increased vulnerability of CI, due to ageing but also and more importantly to the ever deeper interdependence, are somewhat ironically paralleled by very high expectations on their performance held by the general public: CI tends to be given for granted most of the time, until of course spectacular and unexpected failures occur (Macaulay 2008). These failures, however, are not unexpected at all to emergency managers and researchers in the field.

The literature on vulnerability of infrastructural systems and their components to natural disasters as well as to targeted malevolent actions is vast. It must be recognized, however, that the largest proportion of these studies focuses on single systems, without considering interactions, cascading failures, and complex impacts. These studies, which in some cases started very early, have covered[1] buildings

[1]The references cited in this section by no means intend to be an exhaustive review of the relevant literature, and they represent only a subjective selection for illustrative purposes.

(Rossetto and Elnashai 2003; Spence et al. 2007; Goda and Hong 2008; Bal et al. 2010; Parodi et al. 2010) but also utilities, with a fairly large number of contributions on water supply networks (Isoyama and Katayama 1981; Shinozuka et al. 1981, 1992; O'Rourke et al. 1985; Ballantyne et al. 1990; Kawakami 1990; Taylor 1991; Awumah et al. 1991; ATC-25 1992; Markov et al. 1994; Hwang et al. 1998; Chang et al. 2002; Hoshiya and Yamamoto 2002; Kalungi and Tanyimboh 2003; Hoshiya et al. 2004; Adachi and Ellingwood 2006; Scawthorn et al. 2006; Javanbarg et al. 2006; Li et al. 2006; Shi et al. 2006; Javanbarg and Takada 2009; Wang et al. 2010), on electric power grids (Matsuda et al. 1991; Pires et al. 1996; Vanzi 1996, 2000; Giannini et al. 1999; Xingbin and Singh 2004; Helseth and Holen 2006; Shumuta 2007; Nuti et al. 2007; Schläpfer et al. 2008; Arianos et al. 2009; Ma et al. 2010; Bompard et al. 2011; Buritica et al. 2012), a relatively minor number of works on gas distribution networks (O'Rourke and Palmer 1996; Helseth and Holen 2006; Chang and Song 2007; Kim and Kang 2013). Transportation systems have also been the object of several studies (Shinozuka et al. 2003a, b; Zhou et al. 2004; Franchin et al. 2006; Shiraki et al. 2007; Kiremidjian et al. 2007; Kang et al. 2008; Chang et al. 2011).

The importance of the interconnection between different systems is a more recent acquisition (PCCIP 1997; Kameda 2000; Rinaldi et al. 2001; Peerenboom et al. 2001; Little 2002; Menoni et al. 2002; Li and He (2002), Bush et al. 2003; Benoît et al. 2003; Yao et al. 2004; Rinaldi 2004; Karaca 2005; Dudenhoeffer and Permann 2006; Leung et al. 2007; Laprie et al. 2007; Cardellini et al. 2007; Dueñas-Osorio et al. 2007a, b; Tang and Wen 2008; Rosato et al. 2008; Adachi and Ellingwood 2008; Dueñas-Osorio and Vemuru 2009; Ouyang et al. 2009; Shizuma et al. 2009; Nojima 2010; Johansson and Hassel 2010; Zhang and Peeta 2011; Hernandez-Fajardo and Dueñas-Osorio 2011; Dueñas-Osorio and Kwasinski 2012) and studies that target two or, rarely, more systems are relatively few (Kim et al. 2007; Cagno et al. 2011; Ouyang and Dueñas-Osorio 2011; Poljanšek et al. 2012; Hernandez-Fajardo and Dueñas-Osorio 2013).

In parallel with the above studies some large concerted efforts to come up with frameworks and tools for carrying out vulnerability and loss assessment at the regional or urban scale, have been funded in the US. These are the HAZUS (FEMA 1999) and the MAEviz (MAE 2013) initiatives. Other initiatives aimed at developing tools for regional risk/loss estimation include e.g. Rt (Mahsuli and Haukaas 2013) and CAPRA (Cardona et al. 2012). Finally, the most ambitious current project to develop a globally applicable consistent and extensible regional loss estimation methodology is the Global Earthquake Model (GEM 2013).

The National Institute for Building Sciences (NIBS) originally developed HAZUS (Hazard U.S.) on behalf of the Federal Emergency Management Agency (FEMA) back in the 1990 as a closed system, limited to seismic hazard and to U.S.A. scenarios. The current version, called HAZUS-MH (MR4) includes multiple hazards (earthquakes, hurricanes and floods), up to date inventory data and hazard characterization, and efforts have been made to develop an internationally applicable version, which has results so far in HAZ-TAIWAN, a country-specific release for Taiwan (Yeh et al. 2006). The main merit of the HAZUS platform is that of having provided for the first time an unparalleled set of fragility models for basically every component in every system in which the built environment can be subdivided. It must be recognized, however, that many of these models have been derived based solely on expert judgment and overall the consistency of derivation is limited. One effect of the sheer size of the HAZUS framework and set of tools is that it established itself very soon as the reference for all studies in the sector. By so doing, some of the basic choices made during its development have had a very important influence in the following research. For instance, many researchers have adopted as a default choice, somewhat uncritically, the five damage states/levels introduced by HAZUS. Most fragility studies published after its appearance employed this discretization of damage that, in many cases, can be too refined for the considered component. Also, HAZUS has basically introduced the lognormal distribution for fragility functions, rapidly become the de facto standard.

The development of MAEviz (later re-branded as the Earthquake module of Multi-Hazard Assessment, Response, and Planning, mHARP-EQ, and recently renamed ERGO-EQ) started somewhat later than HAZUS and was the product of the research efforts carried out at the Mid-America Earthquake Centre in collaboration with the National Center for Supercomputing Applications' (NCSA). In particular, MAEviz is an open-source and incorporates many of the design concepts and capabilities motivated by NCSA efforts to develop "Cyberenvironments" that span scientific disciplines and that can rapidly evolve to incorporate new research results (Elnashai et al. 2008). An important aspect of MAEviz is its extensibility, both in terms of analysis/features modules, and of visualization/representation (GIS) modules. The framework has been designed to implement the Consequence-based Risk Management (CRM) paradigm supported by the MAE center.

A different view characterizes the software 'Rt', the outcome of continuous development started with 'InRisk', a the 3 years research project on Infrastructure Risk funded by the Natural Sciences and Engineering Research Council of Canada (NSERC) started in 2006. Rt is a computer program for reliability and optimization analysis with multiple probabilistic models. To orchestrate the multi-model analyses, Rt has an object-oriented architecture. Rt is also fully parameterized, with individual objects for random variables, design variables, and model responses. The main emphasis of the project and of the developed software is on the adoption of proper probabilistic models, i.e. models that provide a deterministic output when fed with a deterministic input. In this respect, Rt is probably a unicum in the current landscape of framework for infrastructure risk assessment, in that it does not make use of almost ubiquitous fragility functions.

The Central American Probabilistic Risk Assessment (CAPRA) platform was developed in partnership with Central American governments, the support of the Central American Coordination Centre for Disaster Prevention (CEPREDENAC), the Inter-American Development Bank (IDB) and the International Strategy of United Nations for Disaster Reduction (UN-ISDR) and the World Bank. It is a free, modular, extensible platform aimed at risk analysis and decision making. Modularity means that hazard information is combined with exposure and physical vulnerability data, allowing the user to determine conjoint or cascade risk on an inter-related multi-hazard basis, distinguishing the platform from previous single hazard analyses. The CAPRA suite of software includes hazard mapping, risk assessment and cost-benefit analysis tools to support pro-active risk management. CAPRA can also be used to design risk-financing strategies.

The GEM initiative aims to build state-of-the-art, widely accepted basic datasets, models, best-practice and software/tools for the assessment of seismic risk on a global scale (Crowley et al. 2013). The ambitious project has started in response to the fact that while vulnerability to earthquakes is increasing, reliable risk assessment tools and data are most often still out of reach in many areas of the world. The non-profit and independent GEM Foundation drives the effort, and receives funding and support from both the public and private sector.

1.3 The SYNER-G Project: Short Description

SYNER-G is a collaborative integrated research project funded (2009–2013) by the European Commission Directorate-General for Research within the so-called Framework Programme 7. The 14 Consortium partners include representative institutions from many European countries, as well as non-funded international partners and industry representatives, as shown in Fig. 1.1.

The project proposal started from acknowledging that: (a) previous research on the seismic risk and vulnerability assessment of urban systems (buildings, building aggregates, lifeline networks and critical infrastructures), at international, European and national levels, were focused on the vulnerability of individual elements at risk, and there was a need for constraining the uncertainty associated with the employed

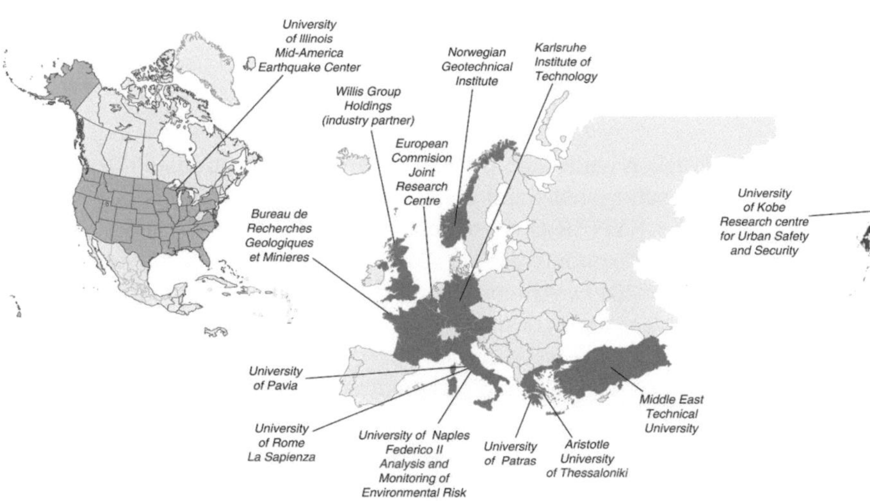

Fig. 1.1 Partners in the SYNER-G consortium

fragility and loss models; (b) most fragility models were developed outside Europe and their applicability for elements at risk in the European context was to be assessed; (c) systemic vulnerability and the associated increased impact had not been considered so far in a rigorous and unified way for all kind of systems; (d) the ability to model damage to non-structural systems and social and economic consequences was in need of significant improvement.

Therefore the project goal was to revise and when necessary propose fragility models to be applied in the European context, and to develop a systemic and holistic approach to loss estimation able to capture final loss estimates at the global level (i.e. socio-economic impacts), accounting for their dependence on the vulnerability and interactions of the whole system.

The work was organized into packages (WPs) as shown in Fig. 1.2, with the core technical WPs being numbered 2–6. One of the first tasks to be completed within WP2, and one that was instrumental to set all other WPs in motion, was the definition and preliminary description of the domain to be studied. This resulted in a detailed taxonomy of the system of systems that makes up the "Infrastructure", which is briefly outlined in Sect. 1.4. Work then started in parallel on the components' and the systemic lines, with WP3 focusing on the collection, review and proposal of fragility models for all elements in the taxonomy, and WP2 aimed at developing a framework for systemic analysis in close interaction with WP4 (socio-economic impacts) and WP5 (specification of the general methodology to each system in the taxonomy).

Shortly after the initial phase, data collection on the case studies in WP6 was started, especially for the two main applications to the cities of Thessaloniki and Vienna. The unfortunate occurrence of the April 6th 2009 earthquake in L'Aquila after the project submission provided another important case study, the gas network of L'Aquila that was included in the work programme after the project kick-off.

1.4 Elements at Risk and Taxonomy

The first task undertaken within the project was the identification and description of a set of systems, sub-systems and components to focus on. This has resulted in what is called the SYNER-G taxonomy, described in this section. A more detailed version of this taxonomy can be found in the SYNER-G reference report 2 (Hancilar and Taucer 2013). All considered systems and their components have been assigned unique tags used consistently throughout the project. This taxonomy has been the guidance for the work carried out within work packages 3 (physical vulnerability and losses) and 5 (socio-economic vulnerability and losses), where typology fragility models have been revised and/or developed for each component, with a focus on European distinctive features, and systems have been modelled, respectively (Fig. 1.3).

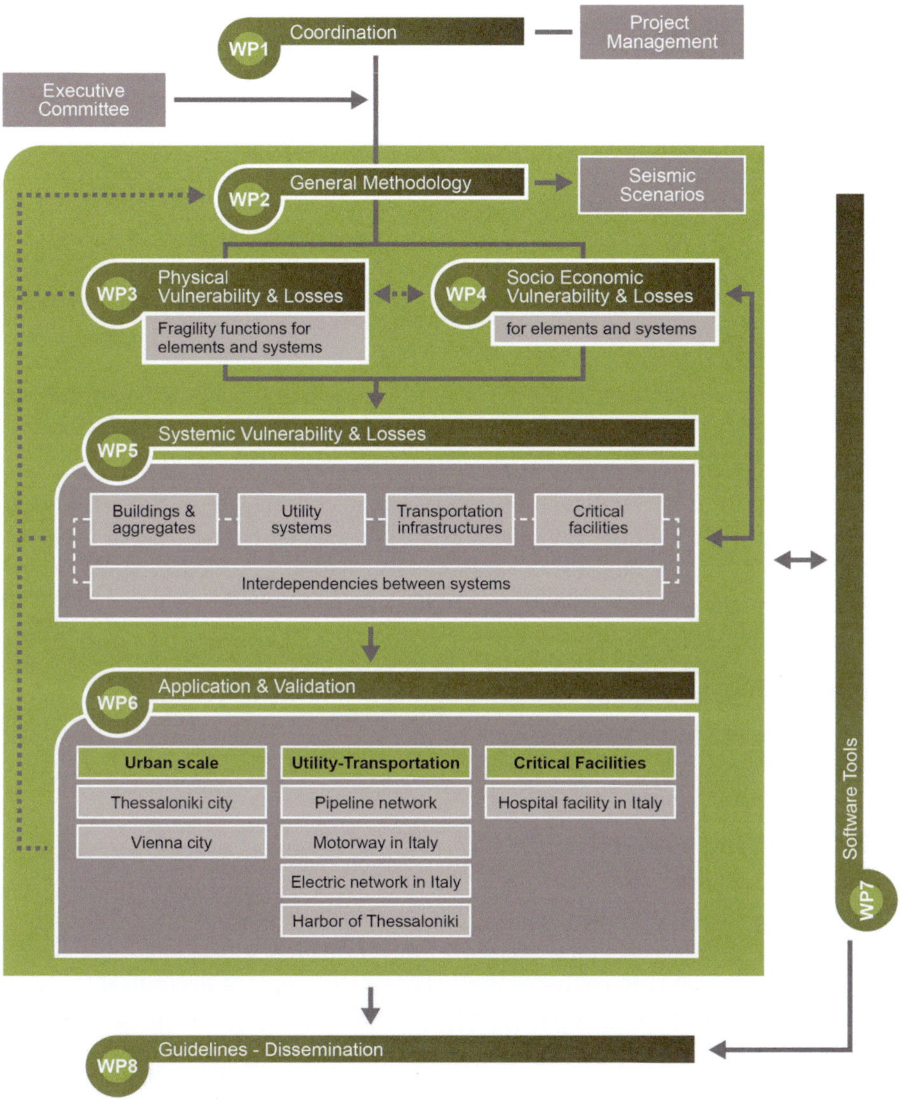

Fig. 1.2 The project workflow and subdivision into work-packages

1.4.1 Building Aggregates (BDG)

Buildings are the basic point-like component of building aggregates/agglomerates/ blocks (where buildings may or may not be in contact, with the ensuing interactions), which are delimited by roads and served by all other utility systems.

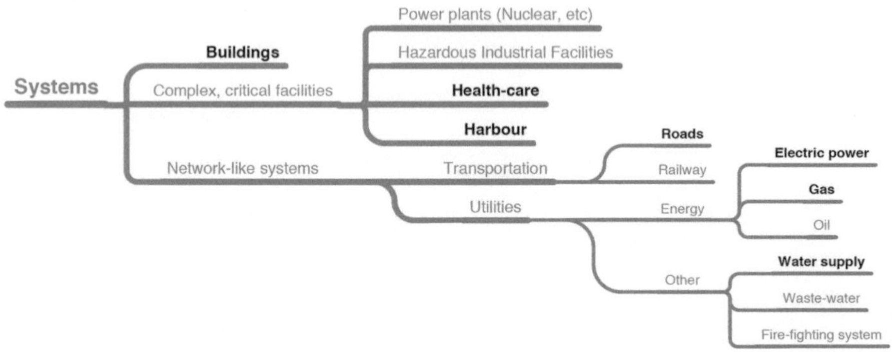

Fig. 1.3 Systems considered in the SYNER-G taxonomy: systems that have been developed and implemented in the model are in *bold black* typeface

The description of the vulnerability of an urbanized area (e.g. a census tract, where several such building agglomerates are present) for the purpose of a system study requires fragility analysis of representative buildings for each typology, and statistical data on the incidence of each typology in the building population and services. Buildings are mainly described and classified by the following parameters: Force Resisting Mechanism (FRM1), FRM Material (FRMM1), Plan (P), Elevation (E), Cladding (C), Detailing (D), Floor System (FS), Roof System (RS), Height Level (HL), Code Level (CL).

1.4.2 Electric Power Network (EPN)

The electric-power system as a whole is composed of a number of point-like critical facilities (i.e. power generation facilities, transformation substations) and of the electric power transmission network itself. The internal logic of the critical facilities and their function in the management of the whole system should be modelled explicitly. The network portion of the system can be subdivided into four major parts: Generation, Transformation, Transmission and Distribution and Loads.

The identified main system components are:

- EPN01: Electric power grid
- EPN02: Generation plant
- EPN03: Substation (distribution, transformation-distribution)
- EPN04: Distribution circuits
- EPN05-09: Substation macro-components
- EPN10-23: Substation micro-components
- EPN24: Transmission or distribution line

1.4.3 Natural Gas System (GAS) and Oil System (OIL)

The natural gas or oil system as a whole is composed of a number of point-like critical facilities (i.e. production and gathering facilities, treatment plants, storage facilities, intermediate stations where gas is pressurized/depressurized or simply metered) and of the transmission/distribution network itself. The internal logic of the critical facilities and their function in the management of the whole system should be modelled explicitly. The network portion of the system is made of pipelines and of the supervisory control and data acquisition (SCADA) sub-system.

The identified components for GAS system are:

- GAS01: Production and gathering facility (Onshore, Offshore)
- GAS02: Treatment plant
- GAS03: Storage tank farm
- GAS04: Station (Compression, Metering Compression/metering, Regulator/metering)
- GAS05: Pipe
- GAS06: SCADA

The identified components for OIL system are:

- OIL01: Production and gathering facility (onshore, offshore)
- OIL02: Refinery
- OIL03: Storage tank farm
- OIL04: Pumping plant
- OIL05: Pipe
- OIL06: SCADA

1.4.4 Water Supply System (WSS)

The water-supply system as a whole is composed of a number of point-like critical facilities (i.e. water sources, treatment plants, pumping stations, storage tanks) and the water distribution network itself. The internal logic of the critical facilities and their function in the management of the whole system should be modelled explicitly. The network portion of the system is made of pipelines, tunnels and canals and the supervisory control and data acquisition – SCADA – sub-system.

The identified system components are:

- WSS01: Source (springs, rivers, natural lakes, impounding reservoirs, shallow or deep wells)
- WSS02: Treatment plant
- WSS03: Pumping station
- WSS04: Storage tank
- WSS05: Pipe

- WSS06: Tunnel
- WSS07: Canal
- WSS08: SCADA system

1.4.5 *Waste Water Network (WWN)*

The waste water system as a whole is composed of a number of point-like critical facilities (i.e. treatment plants, pumping stations) and of the distribution network itself. The internal logic of the critical facilities and their function in the management of the whole system should be modelled explicitly. The network portion of the system is made of pipelines, tunnels.

The identified system components are:

- WWN01: Waste-water treatment plant
- WWN02: Pumping (lift) station
- WWN03: Pipe
- WWN04: Tunnel
- WWN05: SCADA system

1.4.6 *Road Network (RDN)*

The road network is composed of a number of nodes and edges. It is a transportation network where edges can be directed (one-way) or undirected (two-way). All edges are in general vulnerable to seismic shaking or geotechnical hazards, with pavements that can rupture due to surface ground deformation. Some types of edges or road segments, like those identified below have specific types of response to seismic action and associated vulnerability.

The main identified system components are:

- RDN01: Bridge
- RDN02: Tunnel
- RDN03: Embankment (road on)
- RDN04: Trench (road in)
- RDN05: Unstable slope (road on, or running along)
- RDN06: Road pavement (ground failure)
- RDN07: Bridge abutment

1.4.7 *Railway Network (RWN)*

The railway system as a whole is composed of a number of point-like critical facilities (stations) and of the railway network itself. The internal logic of the stations and their function in the traffic management of the whole system should be modelled explicitly. The network portion of the system has the same components as a road network, plus a supervisory control and data acquisition – SCADA – sub-system. The difference is in the fragility models: the underlying limit-state relative to continued traffic over railway bridges, embankments, etc. must consider the limitation and tolerances associated with the tracks. This will lead in general to limitations to relative, maximum and residual, displacements stricter than for roadway bridges.

The identified system components are:

- RWN01: Bridge
- RWN02: Tunnel
- RWN03: Embankment (track on)
- RWN04: Trench (track in a)
- RWN05: Unstable slope (track on, or running along)
- RWN06: Track
- RWN07: Bridge abutment
- RWN08: Station

1.4.8 Harbour (HBR)

A harbour is a complex system comprising all the activities related to the transfer of goods/passengers between the maritime transportation and the earth-bound transportation systems. It is serviced by a number of other systems including: EPN, WSN, WWN, FFS, GAS, RDN, RWN. The identified system components are:

- HBR01: Waterfront components (wharves, breakwaters, etc.)
- HBR02: Earthen embankments (hydraulic fills and native soil material)
- HBR03: Cargo handling and storage components (cranes, tanks, etc.)
- HBR04: Buildings (sheds, warehouse, offices, etc.)
- HBR05: Liquid fuel system (components as per the OIL system)
- Utility systems and transportation networks

1.4.9 Health-Care System (HCS)

The health-care system is made up of health-care facilities (HCF), or hospitals. Hospitals are systems whose function is to deliver medical services. From a social point of view, hospitals provide a fundamental assistance to citizens in every-day life; their function becomes of paramount importance in the case of an earthquake event. This is the reason for including them among the critical facilities group.

Medical services, which consist of standardized procedures to guarantee an adequate treatment of patients, are delivered to patients by a joint contribution of the three "active" components of the system:

- The *operators* (human component): doctors, nurses and in general whoever plays an active role in providing medical care;
- The *facility* (physical component): where medical services are delivered;
- The *organisation* (organizational component): hospital management, responsible of setting up adequate conditions (standardized procedures for ordinary and emergency conditions) so that the medical services can be delivered.

The identified system components are:

- HCS01: Organisational component
- HCS02: Human component
- HCS03: Physical component
    - HCS03-1: Structural elements (of the buildings within the complex/facility)
    - HCS03-2: Non-structural elements
    - HCS03-3: Architectural (walls, ceilings, windows etc.)
    - HCS03-4: Basic installations (generation/distribution)
    - HCS03-5: Basic installations/medical gases
    - HCS03-6: Basic installations/power system
    - IICS03-7: Basic installations/water system
    - HCS03-8: Basic installations/conveying system
    - HCS03-9: Building contents

1.4.10 Fire-Fighting System (FFS)

The fire-fighting system as a whole can be a separate system or part of the WSS. In case it is a separate system, it is composed of a number of point-like facilities (i.e. fire-fighters stations, pumping stations, storage tanks, fire-hydrant) and of the distribution network itself. The internal logic of the critical facilities and their function in the management of the whole system should be modelled explicitly. The network portion of the system is made of pipelines.

The identified system components are:

- FFS01: Fire-fighters station
- FFS02: Pumping station
- FFS03: Storage tank
- FFS04: Fire-hydrant
- FFS05: Pipe

1.5 Important Issues in the Systemic Seismic Risk Analysis

1.5.1 Seismic Hazard Assessment

The seismic hazard assessment of spatially distributed systems with various typologies differs from the point like hazard assessment. In Chap. 3 an innovative comprehensive approach is presented, summarized herein by making reference to the abstract of the relevant Chapter. "*The analysis of seismic risk to multiple systems of spatially distributed infrastructures presents new challenges in the characterisation of the seismic hazard input. A general procedure entitled "Shakefield" is established within SYNER-G, which allows for the generation of samples of ground motion fields for both single scenario events, and for stochastically generated sets of events needed for probabilistic seismic risk analysis. For a spatially distributed infrastructure of vulnerable elements, the spatial correlation of the ground motion fields for different measures of the ground motion intensity is incorporated into the simulation procedure. This is extended further to consider spatial cross-correlation between different measures of ground motion intensity. In addition to the characterisation of the seismic hazard from transient ground motion, the simulation procedure is extended to consider secondary geotechnical effects from earthquake shaking. Thus the Shakefield procedure can also characterise the site effects, site amplification and transient strain, and also provide estimates of permanent ground displacement due to liquefaction, slope displacement and coseismic fault rupture*".

1.5.2 Intensity Measures

A main issue related to the fragility curves is the selection of an appropriate earthquake Intensity Measure (IM) for each Infrastructure class and component that characterizes the strong ground motion and best correlates with the response of each element, for example, building, pipeline or harbour facilities like cranes. Examples of IMs include the peak ground acceleration/velocity/displacement or the spectral acceleration/velocity/displacement. Each intensity measure may describe different characteristics of the motion, some of which may be more adverse for the structure or system under consideration. SYNER-G encompasses an extensive review of common IMs for each element at risk.

1.5.3 Fragility Curves

Fragility curves constitute one of the key elements of seismic risk assessment. They relate the seismic intensity to the probability of reaching or exceeding a level of damage (e.g. minor, moderate, extensive, collapse) for the elements at risk. Several methods are available in the literature to derive fragility functions for different elements exposed to seismic and geotechnical hazard. Conventionally, they are classified into four categories: empirical, expert elicitation, analytical and hybrid. In the framework of SYNER-G a comprehensive review of fragility functions for most important elements at risk has been carried out. Moreover, new fragility curves have been developed where necessary, considering the distinctive features of European elements. The result of these studies is presented in a joint volume also published in Springer (Pitilakis et al. 2014).

1.5.4 Systemic Analysis and Performance Indicators

The quantitative measure of the performance of the whole system and its elements when subjected to a seismic hazard is given by Performance Indicators (PI's). They express numerically either the comparison of a demand with a capacity quantity, or the consequence of a mitigation action, or the assembled consequences of all damages (the "impact"). Performance indicators, at the component or the system level, depend on the type of analysis that is performed. Four main types of system evaluations are considered in the SYNER-G approach (Chap. 5):

- **Vulnerability analysis:** This level considers only the potential physical damages of the components of the systems, with no consideration of functionality of either the elements or the whole system.
- **Connectivity analysis:** Here the probability of the demand nodes to be connected to functioning supply nodes through undamaged paths is analyzed. In this approach the damaged components are removed from the network and the adjacency matrix is updated accordingly, thus pointing out the nodes or areas that are disconnected from the rest of the system. This qualitative approach is used for all utility networks (water, electricity, gas) and the road transportation system. Connectivity analysis gives access to indices such as the connectivity loss (measure of the reduction of the number of possible paths from sources to sinks).
- **Capacity analysis**: The ability of the system to provide to the users the required functionality is quantified. For utility networks, graph algorithms and flow equations can be used to estimate capacitive flows from sources (e.g. generators, reservoirs) to sinks (i.e. distribution nodes), based on the damages sustained by the network components (from total destruction to slight damages reducing the capacity). Capacitive modelling yields more elaborate performance indicators at the distribution nodes (e.g. head ratio for water system, voltage ratio for electric buses) or for the whole system (e.g. system serviceability index comparing the customer demand satisfaction before and after the seismic event).
- **Fault-tree analysis**: It concerns critical infrastructures, where multiple conditions are necessary for the systems to ensure its function. This approach aims to

evaluate the remaining operating capacity of objects such as health-care facilities. The system is broken down into structural, non-structural or human components, each one of them being connected with logic operators. It is generally used for the derivation of fragility curves for specific components that comprise a set of sub-components (e.g. health care facilities, water treatment plants).

1.5.5 Treatment of Uncertainties

Several sources of uncertainties are inherent in the analysis, which are related among others to the seismic hazard and spatial correlation models, the fragility and loss assessment or the functionality thresholds of each component, the methods to estimate adequate fragility curves, and the data available for the different infrastructures and systems. The SYNER-G methodology incorporates a rather comprehensive representation of uncertainty in the problem, with a refined and effective seismic hazard model (Chap. 3) and vulnerability model (Chap. 5), including epistemic modelling of the uncertainty in a hierarchical fashion.

1.5.6 Socioeconomic Analysis

An important issue in the seismic risk analysis of urban systems is to compute the expected social losses such as displaced population, shelter needs or health impacts. Economic losses are by themselves another important issue which is not treated explicitly in this volume. This way of conceptualizing integrated risk emphasizes the importance of understanding the interrelations between physical and social systems. In other words, the goal of the present effort is to provide a methodology and a tool on how direct physical losses can potentially aggravate existing vulnerabilities in society and how these vulnerabilities can ultimately lead to greater impacts from physical damage and losses.

A unified approach for modelling shelter needs and health impacts caused by earthquake damage has been developed in SYNER-G. In particular, the proposed models bring together the state-of-the-art casualty and displaced population estimation models into a comprehensive modelling approach based on multi-criteria decision support, which provides decision makers with a dynamic platform to capture post-disaster emergency shelter demand and health impact decisions. The focus in the shelter needs model is to obtain shelter demand as a consequence of building usability, building habitability and social vulnerability of the affected population rather than building damage alone. The shelter model simulates households' decision-making and considers physical, socio-economic, climatic, spatial and temporal factors in addition to modelled building damage states. The health impact model combines a new semi-empirical methodology for casualty estimation with models of health impact vulnerability, and transportation accessibility to obtain a holistic assessment of health impacts in the emergency period after earthquakes.

The models for shelter needs proposed in the present state of SYNER-G could be expanded to cover other post-earthquake needs in the frame of seismic risk management, mitigation and preparedness.

1.6 Applications

The applicability of the SYNER-G methodology and tools is tested through several case studies at urban and regional level as well as at complex infrastructure level. In particular, the following case studies are presented in Part II:

The **city of Thessaloniki** in Northern Greece (Chap. 7). The study area covers the municipality of Thessaloniki, which is divided in 20 Sub City Districts. It includes the building stock (BDG), road network (RDN), water supply system (WSS) and electric power network (EPN), considering specific interdependencies between systems. The purpose of this application is to study the systemic risk in a large urban area of high seismicity and to investigate the effect of interactions between systems in terms of network connectivity loss or displaced people. Furthermore, an accessibility analysis to hospital facilities considering the damages in RDN is performed (Chap. 4) and a shelter demand analysis based on a multi-criteria approach is applied (Chap. 7). Through the latter application, the districts with higher needs for shelters are identified, supporting in this way an efficient planning of shelter allocation.

The **Brigittenau district in Vienna**, Austria (Chap. 8). It is a heavily populated urban area with many residential buildings and several networks and infrastructures exposed to relatively low seismic risk. This test case is mainly an attempt to look at SYNER-G methods at the building level, using high-resolution data in a small area.

The medium-pressure **gas distribution system of L'Aquila** in Italy (Chap. 9). The functionality of the network is examined through a connectivity analysis considering the pipelines and the Reduction Groups (M/R stations). A probabilistic seismic and geotechnical (landslide) hazard analysis is performed based on characteristic earthquakes of moment magnitude $M_w = 6.3$, generated by the Paganica fault.

The **road network of Calabria region** in Southern Italy (Chap. 10). A pure connectivity analysis is performed and specific performance indicators that describe the loss of connectivity between traffic analysis zones and minimum travel time to reach a hospital are applied. The seismic hazard is modeled through 20 faults of the broader area.

The **electric power network of Sicily** in Italy (Chap. 10). The study here is carried out at the capacitive level, i.e. computing the actual power flows, voltages and currents in the network, both in the undamaged or reference state and in the damaged one. The seismic hazard is modeled through 18 faults of the broader area.

A **regional health care system** (Chap. 11). The earthquake effects both on hospitals and on the RDN, connecting towns to hospitals, are evaluated and the interaction among them is accounted. The estimated risk is described through several indicators such as the un-hospitalized victims, the inability of hospitals to provide medical care, the demand of medical care on hospitals or the hospitalization travel time.

The **harbour of Thessaloniki** in Greece (Chap. 12). The performance of the harbour is measured with the total cargo/containers handled and/or delivered (to the port's gate) in a pre-defined time frame per terminal and for the whole port system, considering the seismic damages as well as specific interdependencies. In particular, the effect of disruption of electric power supply to cranes and road closures due to building collapses is analyzed.

References

Adachi T, Ellingwood BR (2006) Serviceability of earthquake-damaged water systems: effects of electrical power availability and power backup systems on system vulnerability. Reliab Eng Syst Saf 93:78–88

Adachi T, Ellingwood BR (2008) Service ability of earthquake-damaged water systems: effects of electrical power availability and power back up systems on system vulnerability. Reliab Eng Syst Saf 93(1):78–88

Arianos S, Bompard E, Carbone A, Xue F (2009) Power grid vulnerability: a complex network approach. Chaos Interdiscip J Nonlin Sci 19(1):013119

ATC-25 (Applied Technology Council) (1992) A model methodology for assessment of seismic vulnerability and impact distribution of water supply systems. Report no: ATC-25-1, Applied Technology Council, Redwood City, CA

Awumah K, Goulter I, Bhatt S (1991) Entropy-based redundancy measures in water distribution network design. J Hydraul Eng 117(3):595–614

Bal IE, Bommer JJ, Stafford PJ, Crowley H, Pinho R (2010) The influence of geographical resolution of urban exposure data in an earthquake loss model for Istanbul. Earthq Spectra 26(3):619–634

Ballantyne DB, Berg E, Kennedy J, Reneau R, Wu D (1990) Earthquake loss estimation modelling of Seattle water system. Technical report. Kennedy/Jenks/Chilton, Federal Way, 139p

Benoît R, Sabourin J-P, Glaus M, Petit F, Senay M-H (2003) Chapter 17: A new structural approach for the study of domino effects between life support networks. In: Kreimer A, Arnold M, Carlin A (eds) Building safer cities. The future of disaster risk. The World Bank, Disaster Management Facility, Washington, DC

Bompard E, Wu D, Xue F (2011) Structural vulnerability of power systems: a topological approach. Electr Power Syst Res 81(7):1334–1340

Buritica JA, Tesfamariam S, Sánchez-Silva M (2012) Seismic vulnerability assessment of power transmission networks using complex-systems based methodologies. In: Proceedings of the 15th world conference on earthquake engineering, Lisbon, Portugal

Bush B, Giguere P, Holland J, Linger S, McCown A, Salazar M, Unal C, Visarraga D, Werley K, Fisher R, Folga S, Jusko M, Kavicky J, McLamore M, Portante E, Shamsuddin S (2003) Interdependent Energy Infrastructure Simulation System (IEISS) technical reference manual, Version 1.0. Los Alamos National Laboratory, LA-UR-03-1318

Cagno E, De Ambroggi M, Grande O, Trucco P (2011) Risk analysis of underground infrastructures in urban areas. Reliab Eng Syst Saf 96(1):139–148

Cardellini V, Casalicchio E, Galli E (2007) Agent-based modeling of interdependencies in critical infrastructures through UML. In: Proceedings of Agent-Directed Simulation Symposium (ADS'07) of spring simulation multiconference, Norfolk, VA, March 2007

Cardona OD, Ordaz MG, Reinoso E, Yamín LE, Barbar AH (2012) CAPRA – comprehensive approach to probabilistic risk assessment: international initiative for risk management effectiveness. In: Proceedings of the 15th world conference of earthquake engineering, Lisbon, Portugal, 24–28 Sept 2012

Chang L, Song J (2007) Matrix-based system reliability analysis of urban infrastructure networks: a case study of MLGW natural gas network. In: Proceedings of the 5th China-Japan-US trilateral symposium on lifeline earthquake engineering, Haikou, China, 26–28 Nov

Chang SE, Svekla WD, Shinozuka M (2002) Linking infrastructure and urban economy: simulation of water disruption impacts in earthquakes. Environ Plan B 29(2):281–301

Chang L, Elnashai AS, Spencer BF Jr, Song J, Ouyang Y (2011) Transportation system modelling and applications in earthquake engineering. Mid-America Earthquake (MAE) Center, Report 10-03

Crowley H, Pinho R, Pagani M, Keller N (2013) Chapter 30: Assessing global earthquake risks: the Global Earthquake Model (GEM) initiative. In: Tesfamariam S, Goda K (eds) Handbook of seismic risk analysis and management of civil infrastructure systems. Woodhead Publishing, Philadelphia, pp 815–838

Dudenhoeffer P, Permann D (2006) Critical infrastructure interdependency modelling: a survey of US and international research. Report INL/EXT-06-11464, Idaho National Laboratory

Dueñas-Osorio L, Kwasinski A (2012) Quantification of lifeline system interdependencies after the 27 February 2010 Mw 8.8 offshore Maule, Chile, earthquake. Earthq Spectra 28(S1):S581–S603

Dueñas-Osorio L, Vemuru SM (2009) Cascading failures in complex infrastructure systems. Struct Saf 31(2):157–167

Dueñas-Osorio L, Craig JI, Goodno BJ (2007a) Seismic response of critical interdependent networks. Earthq Eng Struct Dyn 36:285–306

Dueñas-Osorio L, Craig JI, Goodno BJ, Bostrom A (2007b) Interdependent response of networked systems. ASCE J Infrastruct Syst 13:185–194

Elnashai A, Hampton S, Lee JS, McLaren T, Myers JD, Navarro C, Spencer B, Tolbert N (2008) Architectural overview of MAEviz-HAZTURK. J Earthq Eng 12(S2):92–99. doi:10.1080/13632460802013610

FEMA (1999) HAZUS-99: earthquake loss estimation methodology, developed by the Federal Emergency Management Agency with the National Institute of Building Sciences

Franchin P, Lupoi A, Pinto PE (2006) On the role of road networks in reducing human losses after earthquakes. J Earthq Eng 10(2):195–206

GEM (2013) Global earthquake model. URL: http://www.globalquakemodel.org/. Last accessed 3 Sept 2013

Giannini R, Pinto PE, Vanzi I (1999) Earthquake hazard analysis software: ASKxELP. Assessment of Seismic risK for Electric Power networks and interaction with other indicators of seismic damage. National Information Service for Earthquake Engineering, University of California, Berkeley, CA

Goda K, Hong HP (2008) Estimation of seismic loss for spatially distributed buildings. Earthq Spectra 24(4):889–910

Hancilar U, Taucer F (eds) (2013) Guidelines for typology definition of European physical assets for earthquake risk assessment. SYNER-G reference report 2, Publications Office of the European Union. ISBN 978-92-79-28973-6

Helseth A, Holen AT (2006) Reliability modelling of gas and electric power distribution systems; similarities and differences. In: Proceedings of the 9th international conference on probabilistic methods applied to power systems, Stockholm, Sweden, 11–15 June

Hernandez-Fajardo I, Dueñas-Osorio L (2011) Sequential propagation of seismic fragility across interdependent lifeline systems. Earthq Spectra 27(1):23–43

Hernandez-Fajardo I, Dueñas-Osorio L (2013) Probabilistic study of cascading failures in complex interdependent lifeline systems. Reliab Eng Syst Saf 111:260–272

Hoshiya M, Yamamoto K (2002) Redundancy index of lifeline systems. J Eng Mech ASCE 128(9):961–968

Hoshiya M, Yamamoto K, Ohno H (2004) Redundancy index of lifeline for mitigation measures against seismic risk. Probab Eng Mech 19:205–210
Hwang H, Lin H, Shinozuka M (1998) Seismic performance assessment of water distribution systems. ASCE J Infrastruct Syst 4(3):118–125
Isoyama R, Katayama T (1981) Reliability evaluation of water supply systems during earth-quakes. Report of Institute of Industrial Science University at Tokyo, Feb, 30(1)
Javanbarg MB, Takada S (2009) Seismic reliability assessment of water supply systems. In: Proceedings 10th international conference on structural safety and reliability, Osaka, 13–17 Sept 2009
Javanbarg MB, Takada S, Kuwata Y (2006) Seismic vulnerability evaluation of water delivery system. In: Proceedings of the 12th Japan earthquake engineering symposium, Tokyo, Japan
Johansson J, Hassel H (2010) An approach for modelling interdependent infrastructures in the context of vulnerability analysis. Reliab Eng Syst Saf 95(120):1335–1344
Kalungi P, Tanyimboh TT (2003) Redundancy model for water distribution system. Reliab Eng Syst Saf 82:275–286
Kameda H (2000) Engineering management of lifeline systems under earthquake risk. In: 12th world conference on earthquake engineering, New Zealand society for earthquake engineering, Upper Hutt, NZ
Kang WH, Song J, Gardoni P (2008) Matrix-based system reliability method and applications to bridge networks. Reliab Eng Syst Saf 93:1584–1593
Karaca E (2005) Regional earthquake loss estimation: role of transportation network, sensitivity and uncertainty, and risk mitigation. PhD thesis, MIT, Cambridge, MA
Kawakami H (1990) Earthquake physical damage and functional serviceability of lifeline network models. Earthq Eng Struct Dyn 19:1153–1165
Kim Y, Kang W-H (2013) Network reliability analysis of complex systems using a non-simulation-based method. Reliab Eng Syst Saf 110:80–88
Kim Y-S, Spencer BF Jr, Song J, Elnashai AS, Stokes T (2007) Seismic performance assessment of interdependent lifeline systems. Mid-America Earthquake Center report TR 07-16
Kiremidjian AS, Stergiou E, Lee R (2007) Chapter 19: Issues in seismic risk assessment of transportation networks. In: Earthquake geotechnical engineering. Springer, Berlin, pp 939–964
Laprie J-C, Kanoun K, Kaâniche M (2007) Modelling interdependencies between the electricity and information infrastructure. In: 26th international conference on computer safety, reliability and security, SAFECOMP-2007, Nurenberg
Leung M, Haimes YY, Santos JR (2007) Supply- and output-side extensions to the inoperability input-output model for interdependent infrastructures. J Infrastruct Syst ASCE 13(4):299–310
Li J, He J (2002) A recursive decomposition algorithm for network seismic reliability evaluation. Earthq Eng Struct Dyn 31(8):1525–1539
Li J, Shulin W, Wei L (2006) Seismic reliability analysis of urban water distribution network. Earthq Eng Eng Vib 5(1):71–77
Little RG (2002) Controlling cascading failure: understanding the vulnerabilities of interconnected infrastructures. J Urb Technol 9(1):109–123
Ma J, Huang Z, Wong PC, Ferryman T, Northwest P (2010) Probabilistic vulnerability assessment based on power flow and voltage distribution. In: Proceedings of the transmission and distribution conference and exposition, 2010 IEEE PES, pp 1–8
Macaulay T (2008) Critical infrastructure: understanding its component parts, vulnerabilities, operating risks, and interdependencies. CRC Press/Taylor & Francis Group, Boca Raton
MAE (2013) MAEviz, developed by the Mid-America Earthquake Center and the National Centre for Supercomputing Applications. URL: http://mharp.ncsa.illinois.edu/. Last accessed 3 Sept 2013
Mahsuli M, Haukaas T (2013) Computer program for multimodel reliability and optimization analysis. J Comput Civ Eng 27(1):87–98
Markov IJ, Grigoriu M, O'Rourke TD (1994) An evaluation of seismic serviceability of water supply networks with application to San Francisco auxiliary water supply system, Technical report, NCEER-94-0001. Multidisciplinary Center for Earthquake Engineering Research, Buffalo

Matsuda EN, Savage WU, Williams KK, Laguens GC (1991) Earthquake evaluation of a substation network. In: Proceedings of 3rd U.S. conference technical council on lifeline earthquake engineering, ASCE, pp 295–317

Menoni S, Pergalani F, Boni MP, Petrini V (2002) Lifelines earthquake vulnerability assessment: a systemic approach. Soil Dyn Earthq Eng 22(12):1199–1208

Nojima N (2010) Earthquake damage assessment of multiple lifelines for system interaction analysis. In: 7th international conference on urban earthquake engineering (7CUEE) & 5th international conference on earthquake engineering (5ICEE), Tokyo Institute of Technology, Tokyo, Japan

Nuti C, Rasulo A, Vanzi I (2007) Seismic safety evaluation of electric power supply at urban level. Earthq Eng Struct Dyn 36(2):245–263

O'Rourke TD, Palmer MC (1996) Earthquake performance of gas transmission pipelines. Earthq Spectra 20(3):493–527

O'Rourke TD, Grigoriu MD, Khater MM (1985) Seismic response of buried pipes. In: Sundrarajan C (ed) Pressure vessel and piping technology – a decade of progress. ASME, New York, pp 281–323

Ouyang M, Dueñas-Osorio L (2011) Efficient approach to compute generalized interdependent effects between infrastructure systems. J Comput Civ Eng 25(5):394–406

Ouyang M, Hong L, Mao Z-J, Yu M-H, Qi F (2009) A methodological approach to analyze vulnerability of interdependent infrastructures. Simul Model Pract Theory 17:817–828

Parodi S, Cattari S, Lagomarsino S, Pagnini C (2010) Probabilistic seismic damage scenario by mechanical models: the case study of Sulmona (Italy), 14ECEE, MAEE

PCCIP (1997) Critical foundations: protecting America's infrastructures, report of the President's commission on critical infrastructure protection. Available from http://www.fas.org/sgp/library/pccip.pdf. Accessed 7 Sept 2011

Peerenboom J, Fisher R, Whitfield R (2001) Recovering from disruptions of interdependent critical infrastructures. Workshop on mitigating the vulnerability of critical infrastructures to catastrophic failures, Lyceum, Alexandria, Virginia

Pires JA, Ang AH-S, Villaverde R (1996) Seismic reliability of electrical power transmission systems. Nucl Eng Des 160(3):427–439

Pitilakis K, Crowley H, Kaynia A (eds) (2014) SYNER-G: typology definition and fragility functions for physical elements at seismic risk, vol 27, Geotechnical, geological and earthquake engineering. Springer, Dordrecht. ISBN 978-94-007-7871-9

Poljanšek K, Bono F, Gutiérrez E (2012) Seismic risk assessment of interdependent critical infrastructure systems: the case of European gas and electricity networks. Earthq Eng Struct Dyn 41(1):61–79

Rinaldi SM (2004) Modeling and simulating critical infrastructures and their interdependencies. In: Proceedings of the thirty-seventh annual Hawaii international conference on system sciences. doi:10.1109/HICSS.2004.1265180

Rinaldi SM, Peerenboom JP, Kelly TK (2001) Identifying, understanding, and analyzing critical infrastructure interdependencies. IEEE Control Syst Mag 21(6):11–25

Rosato V, Issaacharoff L, Tiriticco F, Meloni S, De Porcellinis S, Setola R (2008) Modelling interdependent infrastructures using interacting dynamical models. Int J Crit Infrastruct 4(1/2):63–79

Rossetto T, Elnashai A (2003) Derivation of vulnerability functions for European-type RC structures based on observational data. Eng Struct 25(10):1241–1263

Scawthorn C, O'Rourke TD, Balckburn FT (2006) The San Francisco earthquake and fire of 1906 – enduring lesson for fire protection and water supply. Earthq Spectra 22(S2):S135–S158

Schläpfer M, Kessler T, Kröger W (2008) Reliability analysis of electric power systems using an object-oriented hybrid modeling approach. In: Proceedings of the 16th power systems computation conference, Glasgow, Scotland

Shi P, O'Rourke TD, Wang Y (2006) Simulation of earthquake water supply performance. In: Proceedings of 8th U.S. national conference on earthquake engineering, San Francisco, CA, Paper no. 1285

Shinozuka M, Tan RY, Koike T (1981) Serviceability of water transmission systems under seismic risk. The current state of knowledge. In: Proceedings of the ASCE specialty conference on lifeline earthquake engineering, Oakland, CA, pp 97–110
Shinozuka M, Hwang H, Murata M (1992) Impact on water supply of a seismically damaged water delivery system. In: Ballantyne DB (ed) Lifeline earthquake engineering in the central and eastern US, Technical council on lifeline earthquake engineering monograph, no. 5. ASCE, Reston, pp 43–57
Shinozuka M, Murachi Y, Dong X, Zhou Y, Orlikowski M (2003a) Effect of seismic retrofit of bridges on transportation networks. Research progress and accomplishments 2001–2003. Multidisciplinary Center for Earthquake Engineering Research, pp 35–49
Shinozuka M, Murachi Y, Dong X, Zhu Y, Orlikowski MJ (2003b) Seismic performance of highway transportation networks. In: Proceedings of China-US workshop on protection of urban infra-structure and public buildings against earthquakes and man-made disasters, Beijing, China
Shiraki N, Shinozuka M, Moore JE II, Chang SE, Kameda H, Tanaka S (2007) System risk curves: probabilistic performance scenarios for highway networks subject to earthquake dam-age. J Infrastruct Syst 213(1):43–54
Shizuma T, Nakamura T, Yoshikawa H (2009) Evaluation of outage time for a system consisting of distributed facilities considering seismic damage correlation. In: 10th international conference on structural safety and reliability, ICOSSAR, Osaka, Japan
Shumuta Y (2007) Practical seismic upgrade strategy for substation equipment based on performance indices. Earthq Eng Struct Dyn 36(2):209–226
Spence et al (2007) LESSLOSS report 2007/07. Earthquake disaster scenarios prediction and loss modelling for urban areas. (Note: risk scenarios in Lisbon, Istanbul and Thessaloniki)
Tang A, Wen A (2008) An intelligent simulation system for earthquake disaster assessment. Comput Geosci 35(5):871–879
Taylor CE (1991) Seismic loss estimation for a hypothetical water system, Technical council on lifeline earthquake engineering monograph no. 2. ASCE, Reston
Vanzi I (1996) Seismic reliability of electric power networks: methodology and application. Struct Saf 18(4):311–327
Vanzi I (2000) Structural upgrading strategy for electric power networks under seismic action. Earthq Eng Struct Dyn 29(7):1053–1073
Wang Y, Siu-Kui A, Qiang F (2010) Seismic risk assessment and mitigation of water supply systems. Earthq Spectra 26(1):257–274
Xingbin Y, Singh C (2004) A practical approach for integrated power system vulnerability analysis with protection failures. Power Syst IEEE Trans 19(4):1811–1820
Yao B, Xie L, Huo E (2004) Study effect on lifeline interaction under seismic conditions. In: 13th world conference on earthquake engineering, Vancouver, BC, Canada
Yeh CH, Loh CH, Tsai KC (2006) Overview of Taiwan earthquake loss estimation system. Nat Hazards 37(1–2):23–37
Zhang P, Peeta S (2011) A generalized modeling framework to analyze interdependencies among infrastructure systems. Transp Res B Methodol 45(3):553–579
Zhou Y, Murachi Y, Kim S, Shinozuka M (2004) Seismic risk assessment of retrofitted transportation systems. In: 13th world conference on earthquake engineering, Vancouver, BC, Canada

Chapter 2
A Computational Framework for Systemic Seismic Risk Analysis of Civil Infrastructural Systems

Paolo Franchin

Abstract This chapter presents the general framework for systemic analysis of a set of interconnected civil infrastructural systems described in this book. While the relevant following chapters provide details on specific aspects of distributed seismic hazard (Chap. 3), vulnerability of components (the companion book), functional model of each system and their interactions (Chap. 5), and socio-economic impact evaluation (Chap. 4), this chapter focuses mainly on how the overall model has been developed according to the object-oriented paradigm, and on the way uncertainty in all factors is modelled.

2.1 The SYNER-G Integrated Methodology

The goal of the general methodology developed within the SYNER-G project is to *assess the seismic vulnerability of an Infrastructure*[1] *of urban/regional extension*, accounting for inter- and intra-dependencies among infrastructural components, as well as for the uncertainties characterizing the problem.

As discussed e.g. in (Sánchez-Silva and Gómez 2013), a system can be defined as a set of interacting elements exhibiting properties such as *closure* (the system can be clearly distinguished from its environment) and *synergy*, the latter implying that the whole is more than the sum of the parts. This 'being more' in practice

[1] The term 'Infrastructure' is a short-hand used in the following for the phrase 'set of interconnected civil infrastructural systems', this use being introduced first in (PCCIP 1997).

P. Franchin (✉)
Department of Structural and Geotechnical Engineering, Sapienza University of Rome,
Via Gramsci 53, 00197 Rome, Italy
e-mail: paolo.franchin@uniroma1.it

K. Pitilakis et al. (eds.), *SYNER-G: Systemic Seismic Vulnerability and Risk Assessment of Complex Urban, Utility, Lifeline Systems and Critical Facilities*, Geotechnical, Geological and Earthquake Engineering 31, DOI 10.1007/978-94-017-8835-9_2,

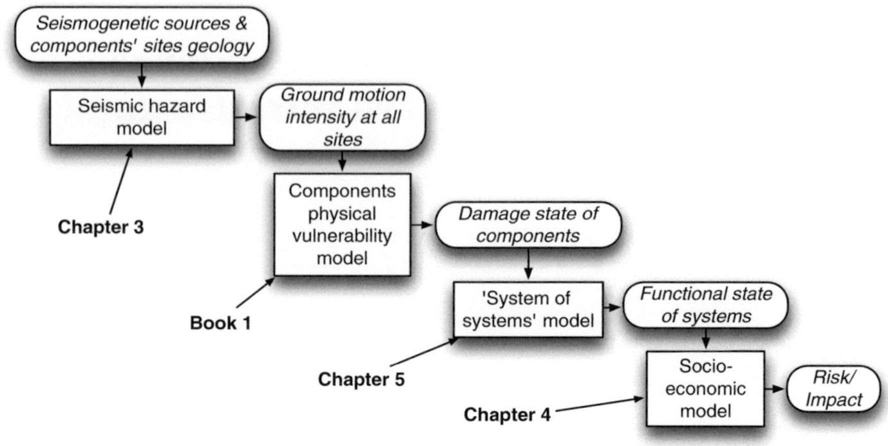

Fig. 2.1 The sequence of models with associated input and output quantities

means that the system performs a function that no part by itself can do. Opposed to reductionism, systems thinking does not only study the parts of a system, but also their relationships, leading to the concepts of e.g. 'whole' and 'hierarchy'. A hierarchy exists between parts of a system based on their relationships and often parts can be considered themselves as 'whole', or systems, in that they contain other parts and perform a function/serve a purpose. The term 'system of systems' is thus in use within systems theory to describe this fact. Furthermore, the set of all potential states of a system is called its *variety*, while the (smaller or equal) set of actual states that the system can take in practice is called its *complexity* (Ashby 1964).

Based on the above the Infrastructure can be regarded as a complex system of systems, i.e. a set of components that are themselves systems, arranged in a hierarchical fashion, and having a vary large number of possible states in practice. The description of its behaviour is a challenging task even under purely deterministic conditions. The first task was therefore that of setting up a *model* of the Infrastructure and of the hazard acting upon it, while the second one was that of *enhancing* it with the introduction of the *uncertainty* and of the analysis methods that can evaluate the system performance accounting for such uncertainty.

In its final form the SYNER-G methodology is based on a sequence of three models: (a) seismic hazard model, (b) components' physical vulnerability model, and (c) system (functional and socio-economic) model. Figure 2.1 illustrates this sequence of models and the main input and output that are exchanged. The figure also indicates where each building block is described in detail.

The sequence of models in Fig. 2.1 supports the implementation of an integrated procedure that leads from the evaluation of the seismic hazard to that of the socio-economic impact. In an attempt to constrain the size and complexity of the task, the choice was made to limit the focus of the methodological developments to the short-term period corresponding to the emergency phase. An even further simplification

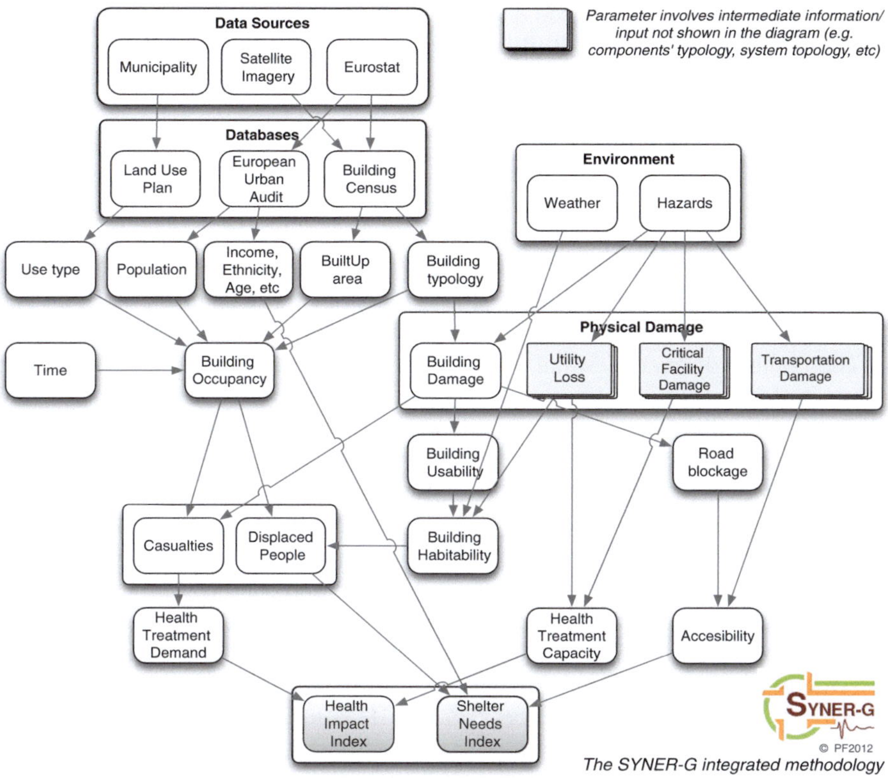

Fig. 2.2 Integrated evaluation of physical and socio-economic performance indicators

was that of developing a time-invariant model to start with. As a result the impact or risk is evaluated in terms of social measures of interest to emergency managers, rather than of economic loss, and aftershocks as well as cumulative damage are not yet considered.

Figure 2.2 shows in qualitative terms the *integrated procedure* with reference, for illustration purposes, to two of the selected social impact models described later in Chap. 4. In particular the figure shows the chain that links seismic hazard to the demands on the shelter and health-care systems in terms of **Displaced Population** and **Casualties**, down to the assessment of social indexes like the **Health Impact** and the **Shelter Needs**.

The **Environment** acts upon the Infrastructure through the **Hazards**. These induce in the components of the Infrastructural systems a certain level of **physical damage**. In the figure, this is represented in terms of damage to **buildings**, to lifelines (**Utility loss**), to **critical facilities** and to the **transportation system**. The stack symbol in the figure employed for quantities such as damage to critical facilities and utility/transportation networks indicates that several other models, components and quantities enter their evaluation. Taking for instance buildings as an example, the level of induced damage depends not only on the hazard but also on the fragility, a function of **Building Typology**. To compute both casualties and displaced population the occupancy level of the buildings (**Building occupancy**) is used as a first input. Building occupancy also depends on the typology, and moreover on the total **built-up area**, the **Population**, the building usage (**Use type**) and the **Time** of the day. The population at risk of being displaced is computed from the building occupancy and habitability at the time of the event (**Building habitability**). Building habitability in turn depends upon the state of **Building usability**, i.e. whether the buildings are still served by fundamental utilities, and also on the **Weather** conditions. Casualties are obtained as the number of deaths and injured by combining building occupancy, building damage and building typology.

The number of casualties and displaced persons are inputs into a multi-criteria utility model to determine health impacts and shelter needs. A *Shelter Needs Index* (**SNI**) is determined by simulating a households' decision-making process and considers the *Resistance to Evacuate* (**RE**) as a multi-criteria function of the individual's vulnerabilities and coping capacities (e.g., age, housing type, housing tenure and household type), as well as external spatial and temporal factors in the community. Furthermore, not all persons who leave their homes will seek public shelter, and some may find alternative shelter accommodations (rent motel rooms or apartments), stay with family and friends, or leave the affected area. Thus, the SNI also accounts for a *Shelter Seeking factors* (**SSF**) by combining major factors contributing to demand for public shelters.

The post-disaster *Health Impacts Index* (**HII**) combines the estimated casualty numbers with three additional inputs in a multi-criteria utility model: *Health Treatment Capacity* (**HTC**), hospital **accessibility**, and a health impact vulnerability model which accounts for *Health Vulnerability Factors* (**HVF**) pre-disposing the exposed population to aggravated health impacts following the earthquake disaster.

Figure 2.2 shows also how the required input information is usually contained in three distinct databases maintained by different sources. In Europe a harmonized source for physical data on the buildings and for socio-economic data on urban areas, in the form of the **Building Census** and **European Urban Audit**, respectively, is **EUROSTAT**. The information on usage is usually provided in the form of a **Land Use Plan**, maintained from a local source (the **Municipality**).

In summary, and as already shown in Fig. 2.1, the conceptual sketch can be practically implemented by developing:

1. A model for the spatially distributed seismic hazard
2. A physical model of the Infrastructure
3. Socio-economic models

Development of the hazard model has the goal of providing a tool for: (a) sampling events in terms of location (epicentre) and possibly extension, magnitude and faulting style according to the seismicity of the study region; (b) predicting maps of seismic intensities at the sites of the vulnerable components in the Infrastructure. These maps, conditional on magnitude, epicentre, etc. should describe the variability and spatial correlation of intensities at different sites. Further, when more vulnerable components exist at the same location and are sensitive to different intensities (e.g. acceleration and displacement), the model should predict intensities that are consistent at the same site.

Development of the physical model involves first the definition of an extensive taxonomy of systems and components (see the companion volume) and requires: (a) for each system, a description of the functioning of the system under both *undisturbed* and *disturbed* conditions (i.e. in the damaged state following an earthquake); (b) a model for the physical and functional (seismic) damageability of each component within each system; (c) identification of all dependencies between the systems; (d) definition of adequate performance indicators for components and systems, and the Infrastructure as a whole.

Development of the socio-economic model starts with an interface to outputs from the physical model in each of the four domains of SYNER-G (i.e., buildings, transportation systems, utility systems and critical facilities). Thus, four main performance indicators – Building Usability, Transportation Accessibility, Utility Functionality and Health Treatment Capacity – are used to determine both direct and indirect impacts on society. Direct social losses are computed in terms of casualties and displaced populations. Indirect social losses are considered in two models – Shelter Needs and Health Impact – which employ the multi-criteria decision analysis (MCDA) theory for combining performance indicators from the physical and social vulnerability models.

In order to tackle the complexity of the described problem, the object-oriented paradigm (OOP) has been adopted. In abstract terms, within such a paradigm, the problem is described as a set of *objects*, characterized in terms of *attributes* and *methods*, interacting with each other. Objects are *instances* (concrete realizations) of *classes* (abstract models, or *templates* for all objects with the same set of properties and methods). The developed model of the problem is described in the next Sect. 2.2.

2.2 Object-Oriented Model

An important choice in the development of the methodology was the adoption of the *object-oriented modelling* paradigm. The paradigm has been already used in previous studies on infrastructural systems (Chang and Chamberlin 2004; Schläpfer et al. 2008). This approach to modelling, simply known as the *object model*, emerged in the computer science community starting from the 1960s and attained widespread acceptance only since the 1990s. Its fundamental principles and some basic concepts are introduced next (Sect. 2.2.1) for the sake of self-containment of this chapter. Then, Sect. 2.2.2 introduces the classes that form the developed SYNER-G model. Readers familiar with the OOP may skip Sect. 2.2.1, while others may find it a bit abstract. This is due to the very general character of the OOP, which is not directed at modelling any specific physical or non-physical system. The section, however, introduces the minimum notions needed to understand terms used in the remainder of the chapter to describe the model and may be looked at as a stimulus to pursue a more in-depth discussion (Booch et al. 2007).

2.2.1 Basic Concepts in Object-Oriented Modelling

The seven principles of **abstraction**, **encapsulation**, **modularity**, **hierarchy**, **typing**, **concurrency**, and **persistence** are at the basis of the object-oriented modelling paradigm (Booch et al. 2007). While the last three principles play a minor role, the *first four* play a *major* role and really define the object model of a given domain:

1. **Abstraction**: An abstraction denotes the essential characteristics of an object that distinguish it from all other kinds of objects and thus provide crisply defined conceptual boundaries, relative to the perspective of the viewer. Abstraction is at the very core of object-oriented model: without it the idea of class of objects could not exist. Deciding on the right set of abstractions for a given domain is the central problem in object-oriented design. Amongst various types of abstraction, **entity abstractions** are very appealing since they directly parallel the vocabulary of a given problem domain (e.g. the classes Infrastructure and its subclasses, or the class Hazard and its subclasses).
2. **Encapsulation**: Encapsulation is the process of compartmentalizing the elements of an abstraction that constitute its structure and behaviour. *"No part of a complex system should depend on the internal details of any other part"*. Whereas abstraction "helps people to think about what they are doing," encapsulation "allows model changes to be reliably made with limited effort".
3. **Modularity**: Modularization consists of dividing a program[2] into modules which can be compiled separately, but which have connections with other modules. It is fair to say that modularity and encapsulation go hand in hand. Deciding on the right set of modules for a given problem is almost as hard a problem as deciding on the right set of abstractions. Modules serve as the physical containers in which we declare the classes and objects of our logical design. The overall goal of the decomposition into modules is the reduction of software cost by allowing modules to be designed and revised independently. In the seismic vulnerability problem at hand, almost naturally, the three modules coincide with the seismic hazard, the physical vulnerability and the systemic behaviour sub-models.
4. **Hierarchy**: Hierarchy is a ranking or ordering of abstractions. The two most important hierarchies in a complex system are its class structure (the "is a" hierarchy) and its object structure (the "part of" hierarchy). Abstraction is a good thing, but in all except the most trivial applications, one may find many more different abstractions than can comprehend at one time. Encapsulation helps to manage this complexity by hiding the inside view of our abstractions. Modularity

[2]The term 'program' is used here after Booch et al. (2007). The modelling paradigm was born in the computer science community and it is intimately related to software design. This chapter, however, does not describe a software but, rather, the conceptual design of the platform-independent model that can (and indeed has been) implemented in a platform-specific application. This separation between platform-independent conceptual models and platform-specific implementation(s) is typical of the OOP.

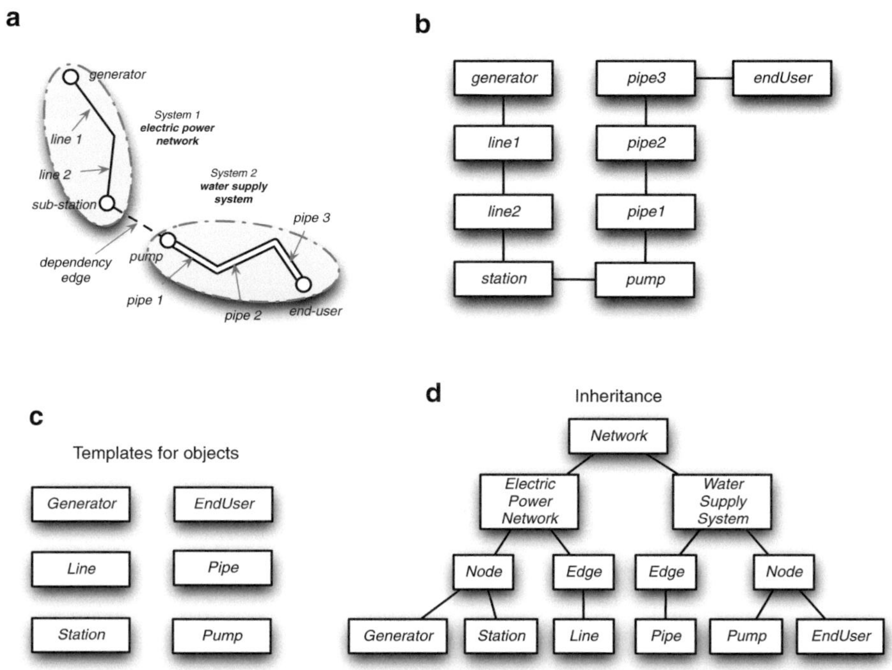

Fig. 2.3 The process of abstraction applied to a simple set of interconnected infrastructural systems (**a**) Physical system of systems. (**b**) Digital model: collection of objects. (**c**) Abstraction, first level: classes. (**d**) Abstraction, second level: hierarchy of classes

helps also, by giving us a way to cluster logically related abstractions. Still, this is not enough. A set of abstractions often forms a hierarchy, and by identifying these hierarchies in our design, we greatly simplify our understanding of the problem (e.g. Infrastructure, above critical facilities/networks/inhabited areas, each above its component subclasses, etc.).

Figure 2.3 gives a graphical representation of the process of abstraction and hierarchy identification in the design of the classes for a given object model.

In the following the object model developed within SYNER-G is described by means of *class diagrams*. These are just one of multiple representations available in the standard modelling language adopted in object model development, the *Unified Modelling Language* (UML). It should be apparent how it is impossible to capture all the subtle details of a complex system, and of the software that models it, in just one large diagram. There are both the structure and the function of the objects involved, the taxonomic structure of the class objects, the inheritance mechanisms used, the individual behaviours of objects, and the dynamic behaviour of the system as a whole. Thus, the UML has numerous types of diagrams, each providing a certain view of your system. A class diagram is used to show the existence of classes and their relationships. Its two essential elements are *classes* and *their basic relationships*.

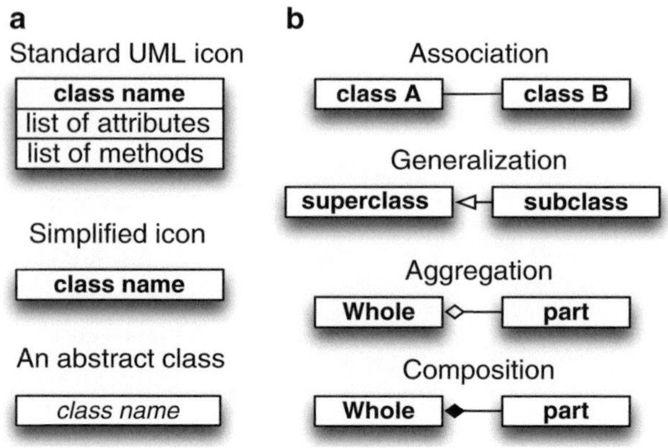

Fig. 2.4 Notation used in the class diagrams to follow (**a**) Class icons (**b**) Class relationships

As shown in Fig. 2.4, the *class* icon (used to represent a class in a class diagram) usually consists of three compartments, with the first occupied by the class *name*, the second by the *properties* (or *attributes*), and the third by the *methods* (or *functions/operations*). In the following, a simplified notation will be used whereby a single rectangle with the class name will be used as class icon, i.e. methods and attributes are not listed explicitly in the class diagram. Chapter 5 will provide details on the main methods and attributes of each class.

An *abstract class* is one for which no instances may be created. Because such classes are so important to engineering good class *inheritance trees*, there is a special way to designate an abstract class, by italicizing the class name.

Classes rarely stand alone; instead, they collaborate with other classes in a variety of ways. The essential connections among classes include *association*, *generalization*, *aggregation*, and *composition*.

The *association* icon connects two classes and *denotes a semantic connection*. The remaining three relationships are drawn as refinements of the association icon. Indeed, during development, this is exactly how relationships tend to evolve. We first assert the existence of a semantic connection between two classes and then, as we make tactical decisions about the exact nature of their relationship, often refine them into generalization, aggregation, or composition relationships.

The *generalization* icon denotes a generalization/specialization relationship (the "is a" relationship) and appears as an association with a closed arrowhead. The arrowhead points to the *superclass*, and the opposite end of the association designates the *subclass*. The subclass inherits the structure and behaviour of its superclass. Also, a class may have one (single inheritance) or more (multiple inheritance) superclasses.

Aggregation, as manifested in the "part of" relationship, is a constrained form of the more general association relationship. The aggregation icon *denotes a whole/part hierarchy* and also implies the ability to navigate from the aggregate to

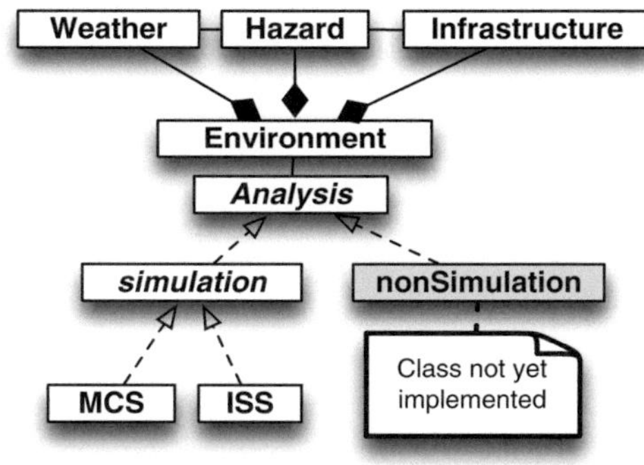

Fig. 2.5 Highest level class diagram for the Infrastructural vulnerability assessment problem (the *grey* hatch, as indicated, denotes classes that have been included at the conceptual level but have not yet been implemented)

its parts. It appears as an association with an unfilled diamond at the end denoting the aggregate (the whole). The class at the other end denotes the class whose instances are part of the aggregate object. This whole/part hierarchy *does not mean physical containment*: a professional society has a number of members, but by no means does the society own its members. While the choice of aggregation is usually an analysis or architectural design decision, the choice of *composition* (physical containment) is usually a detailed, tactical issue. Distinguishing physical containment is important because it has semantics that play a role in the construction and destruction of an aggregate's parts. The composition icon denoting a containment relationship appears as an association with a filled diamond at the end denoting the aggregate.

2.2.2 The Classes in the SYNER-G Model

A high-level representation of the object-oriented model is shown in Fig. 2.5. The first class in the scheme is the *Analysis* one. This class is an abstract one with two generalizations: *simulation* and *nonSimulation*. As explained later (see Sect. 2.3.2), these are the two large groups of methods that can be used to perform a probabilistic analysis of the system, and even though only simulation methods have been implemented so far, the object-oriented model has been set up in a general manner. The figure shows that also the *simulationMethod* class is abstract, though it contains both abstract and concrete methods, and that two concrete classes are provided so far: the plain Monte Carlo simulation method (class *MCS*) and the importance sampling simulation method (*ISS*), enhanced with the K-means clustering procedure (Sect. 2.3.2).

The object of the analysis is the *Environment,* which is *composed* of three classes, the *Infrastructure*, the *Hazard* and the *Weather*.

The Environment is the portion of physical space, inclusive of the Earth crust and the atmosphere, which needs to be considered in evaluating the impact of the hazard on the Infrastructure. Within this general scheme, ideally, one could

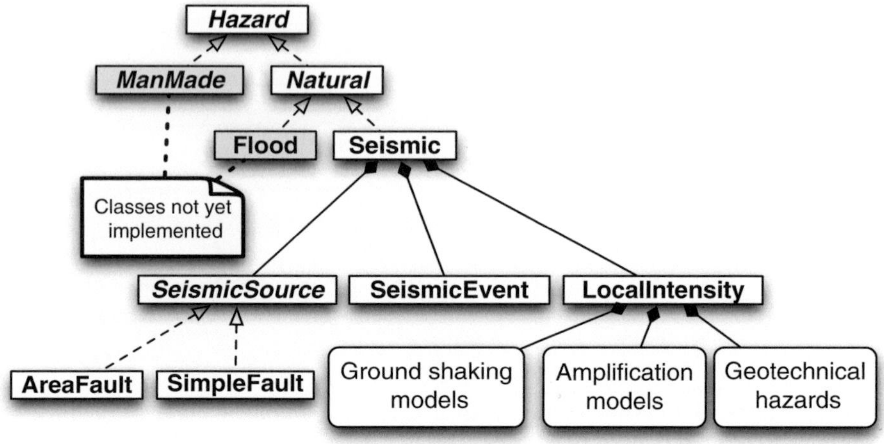

Fig. 2.6 Class diagram for the Hazard class (the *grey* hatch, as indicated, denotes classes that have been included at the conceptual level but have not yet been implemented)

evaluate the impact of a chain of events such as an earthquake occurring on a fault, inducing physical damage in the Infrastructure, triggering e.g. fires and the dispersion of pollutants, as affected by *weather* conditions, in the atmosphere. Within the SYNER-G project the scope was only the evaluation of the direct physical damage due to an earthquake and of its direct and indirect consequences, but an attempt has been made to set up a model that is general and leaves room for later extensions to multi-hazard contexts.

Figure 2.6 shows the *Hazard* class in more detail. This class is the *composition* of two *abstract* classes: *man-made* and *natural* hazards. The class Natural contains environmental hazards such as the seismic one, volcano eruption, floods, etc. The Seismic hazard, in turn, is modelled as the composition of three classes: one class for seismo-genetic sources, one for events and the third one for the local intensity at each site.

Objects from the SeismicSource class are, as the name says, sources that can generate earthquakes. So far two subclasses have been designed to account for fault models of different complexity. In any given problem a set of objects from this class will describe the regional seismicity. The class SeismicEvent is the class from which earthquakes in terms of localization and magnitude are instantiated. The passage from macro-seismic parameters to intensity values at each site of interest is performed by objects instantiated from the third class LocalIntensity. The class is further detailed in a sub-hierarchy, as shown in Fig. 2.7.

Local intensity of ground motion at a site at distance *R* from the epicentre/fault surface, during an event of magnitude *M*, can be predicted both in terms of a ground shaking parameter by means of ground motion prediction equations (GMPE), and in terms of a geotechnical hazard, that is a displacement measure such as permanent

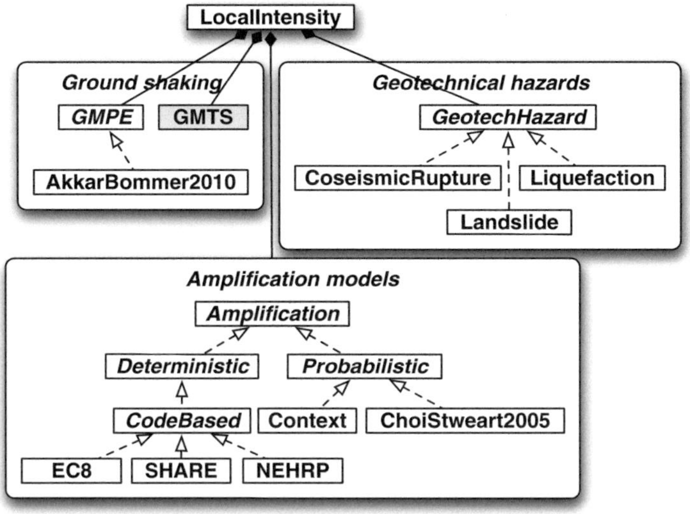

Fig. 2.7 Class diagram for the LocalIntensity class (the *grey* hatch denotes classes that have been included at the conceptual level but have not yet been implemented)

ground deformation.[3] The exact measure of the local intensity, commonly denoted as Intensity Measure (IM), depends on the components at the site of interest, and in particular on the required input to their fragility models. Furthermore, local intensity strongly depends on the site response, which may be just a function of the soil profile in the upper layers (in which case one-dimensional approaches to site response are appropriate), or of more extended portions of soil around the site (basin and/or topographic effects, that shall be treated with more refined amplification models). As a result the LocalIntensity class is the composition of the GMPE, Amplification and GeotechHazard classes.

The GMPE class is an abstract one, with concrete realizations providing alternative GMPEs. Currently only one model has been implemented, based on the GMPE by Akkar and Bommer (2010), which is fit for use in the European, Mediterranean and Middle-East regions. In the already mentioned strive for generality, it was foreseen that local shaking could be described by more than just one scalar or vector IM, i.e. in terms of an entire ground motion time series. Models that take as an input the same parameters as a GMPE, but produce such time-series matching mean and variance of natural motions have appeared recently and could be easily implemented and integrated into the SYNER-G OO-model. In this respect, the SYNER-G model

[3]Figure 2.7 shows also a class not yet developed, included to allow for "generalized ground motion prediction models", i.e. models that based on the same input of GMPE provide the full one- or multi-component time-series of motion at the site. The class is called GMTS, which stands for Ground Motion Time Series.

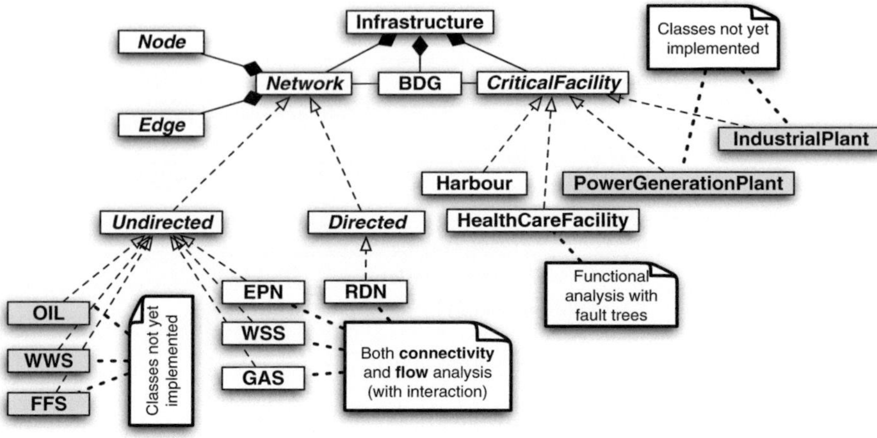

Fig. 2.8 Class diagram for the Infrastructure portion of the model (the *grey* hatch denotes classes included at the conceptual level but not yet implemented)

can be seen as versatile container framework where existing models and new ones can be easily integrated for the purpose of systemic vulnerability analysis.

The Amplification class is abstract. Amplification can be treated, based on the amount of available information, either in simple deterministic manner (which includes even basic code-type scalar amplification factors), or in mode refined probabilistic manner. In all cases the model includes so far only the simple one-dimensional amplification case.

Finally, the GeotechHazard class is also abstract, with three concrete subclasses corresponding to the physical counterparts of co-seismic rupture, landslide and liquefaction, all of which produce permanent ground deformations. Details on the theoretical aspects behind this module (Hazard) of the model can be found in Chap. 3.

Figure 2.8 shows the Infrastructure class and subclasses. The Infrastructure is made up of a number of systems that can be subdivided into three groups from a geometric point of view: point-like, line-like and area-like systems. Correspondingly, the *Infrastructure* class is the composition of three classes: the *Critical facility* class (point-like), the *Network* class (line-like) and the *BDG* class (area-like). The Network and Critical facility classes are abstract ones, and are the generalizations of all types of networks and of critical facilities.

The indicated association relationships between Network, BDG and CriticalFacility extend to their subclasses. This means that an object from one of these classes can call the methods from another class. This allows an object describing the set of buildings in a neighbourhood (an object from the class BDG) to "ask" an electric power network object (from the corresponding class) whether power is still fed to the neighbourhood after the event, or, for example, it allows a pumping station object within the water supply system object (from Network/WSS), to make a similar query to the distribution station object within the electric power network object.

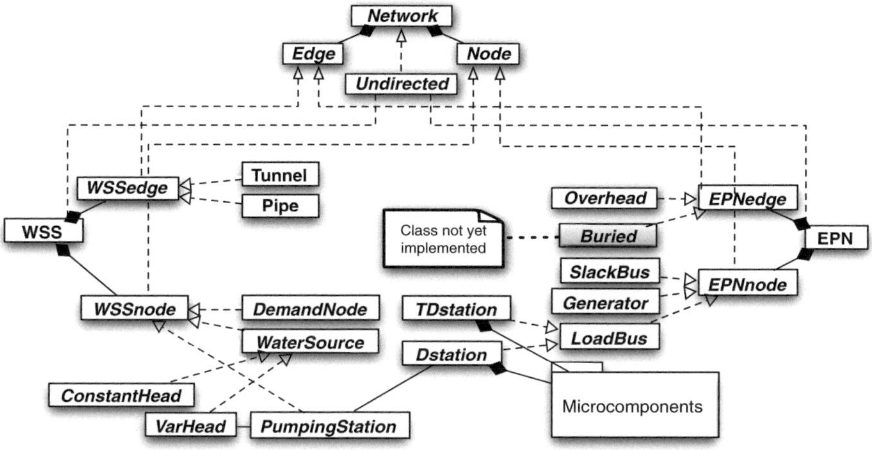

Fig. 2.9 Class diagram for the WSS and EPN classes (the *grey* hatch denotes classes included at the conceptual level but not yet implemented)

As shown in the figure, the Network class is the generalization of all the networks in the Taxonomy: the road and railway networks (transportation networks), the electric power networks, the oil and gas distribution networks, the water and wastewater networks and the fire-fighting network (utility/lifeline networks). At the very basic level of analysis, which is the connectivity one, all networks are described as graphs and a host of graph-theoretic results can be exploited for their analysis. For this reason a higher level of abstraction has been used in Network and its subclasses. In particular, the class is abstract and composed of two abstract classes for *Nodes* and *Edges*, the basic components of all graphs. Further, *Network* is defined as the generalization/abstraction of two intermediate abstractions: they correspond to the two groups of *Directed* (in the considered domain this is for transportation networks) and *Undirected* (lifelines/utilities) graphs. This abstraction allows, together with the principle of inheritance, to design the lower level subclasses with a much lower effort, since the appropriate basic network structure is already set up for all networks at a higher level. Two concrete classes are described in the following in more detail for illustration purposes: the water supply system and the electric power network. Their class diagrams are shown together in Fig. 2.9.

Both the WSS and EPN classes are composition of the corresponding node and link classes. The latter, in general, are abstract classes, since there can be several types of links and nodes within a network, differentiated in terms of properties and functions. For instance, for the water supply system, links can either be pipes or tunnels, while nodes can be end-user demand nodes, water sources or pumping stations. The *Water source* class is an abstract one, since there can be both variable head (like tanks) and constant head (like large reservoirs) water sources.

As far as the electric-power network is concerned, links can be either overhead or buried lines, and there are three node types: end-user demand nodes (load bus), electric power sources (generator) and a balance node, called the slack bus, which is usually chosen as one of the power sources and is used, as explained later, as a mathematical expedient to solve the nonlinear alternate current (AC) equations (it provides the balance of flows during iterations in the solution). Load buses can be either distribution (D) or transformation/distribution sub-stations (TD). These nodes are assemblies of several microcomponents, each one with its electrical function.

The two networks are shown in the same figure to highlight the association between the pumping station class in the WSS and the distribution and transformation/distribution classes in the EPN. This is an example of the way interdependencies of the *physical* type are modelled within the methodology (see also next section and Chap. 5). The association means that a pumping station object can query the reference EPN station to know its state and the actual power fed at any time, so that operational level of the pumps can be established. At the same time, a variable head water source (a tank, a well-field) can be out-of-service if the corresponding pumping station is not fed with power, and the connection between these objects is shown with another association in the figure.

The specification of the full set of class diagrams down to the basic component level for all systems in the SYNER-G Taxonomy is illustrated in Chap. 5.

2.2.3 Modelling Interdependencies Between Systems

Interaction within and between systems can been classified as Physical, Cyber, Geographic, Logical, Societal, Policy-related. Not all of them have been modelled within the methodology.

Geographic interactions (physical proximity) are modelled in the seismic case by correctly incorporating within the seismic hazard model the statistical dependence structure between intensities at the same or close sites, as illustrated in Chap. 3.

Societal interactions are accounted for, e.g. in passing from a potential number of shelter-seeking population to the actual figure, incorporating factors such as anxiety, neighbourhood effects, income, etc. as shown in and explained in Chap. 4. In this section the focus is on Physical (functional) interactions, which are those related to the physical modelling of the Infrastructure.

Table 2.1 reports the interdependencies between some of the systems in the Taxonomy: the i-th row presents the influences of the i-th system on the other systems, while the j-th column collects the influences from other systems on the j-th system. The letter codes stand for: Physical (P), Demand (D) and Geographical (G) interactions. For instance the D/G label in the BDG-RDN entry of the table means that transportation demand is generated in buildings (D) and that in a urban setting, structural damage to buildings produces debris that can cause road blockages (G). Chapter 5 provides a complete description of all the interactions in Table 2.1.

Table 2.1 Interdependencies for sample systems from the Taxonomy

	BDG	EPN	WSS	GAS	OIL	RDN	HBR	HCS
BDG	–	D	D	D		D/G		D
EPN	P	–	P	P	P		P	P
WSS	P		–					
GAS	P	P		–				
OIL		P			–			
RDN	P					–	P	P
HBR						D	–	
HCS						D		–

The evaluation of the above interactions requires establishing a sequence of actions and messages between the objects making up the model. This sequence establishes an order in the evaluation of states of the objects, something that is described within UML with a so-called *state diagram*. Figure 2.10 presents such a diagram. In any given system-of-systems evaluation an initialization phase is performed first, with the BDG object setting the region discretization into cells and passing their centroids to the other systems, which in turn compile a list of tributary cells for each of their demand nodes and assign this demand node as a reference node to the cells. Demand for goods and services is then evaluated and an analysis of all systems in the pre-earthquake undisturbed conditions is carried out. Then, for all considered events, generation of shake field (local intensities at all relevant locations, i.e. the systems' components sites and cell centroids) is followed by evaluation of: (1) the EPN; (2) all other utilities, with direct damage and possible power losses from the EPN; (3) the BDG, with direct damage and utility loss; (4) the RDN, with demand from the BDG and closures due to direct damage to its elements as well as from road blockages; (5) the HCS with demand from the BDG system, service level from all utilities and accessibility from the RDN.

This fixed evaluation sequence is an important and consequential modelling choice. In this respect the model is static in the sense that there is no *global time* flowing, with events taking place concurrently and influencing/impacting each other in random fashions. Future research will have to address this issue, but the level of complexity involved is an order of magnitude larger. For instance, a consistent attempt to develop a truly dynamic model of the system of systems would involve formulating also all flow equations in non-stationary conditions. On the other hand, the introduction of the time dimension is mandatory if the recovery process after the event has to be considered.

2.2.4 *Modelling Transportation and Utilities: The Network Class and Subclasses*

While Chap. 5 provides details on each system in the Taxonomy, this section illustrates a selection of attributes and methods (reported in Tables 2.2 and 2.3)

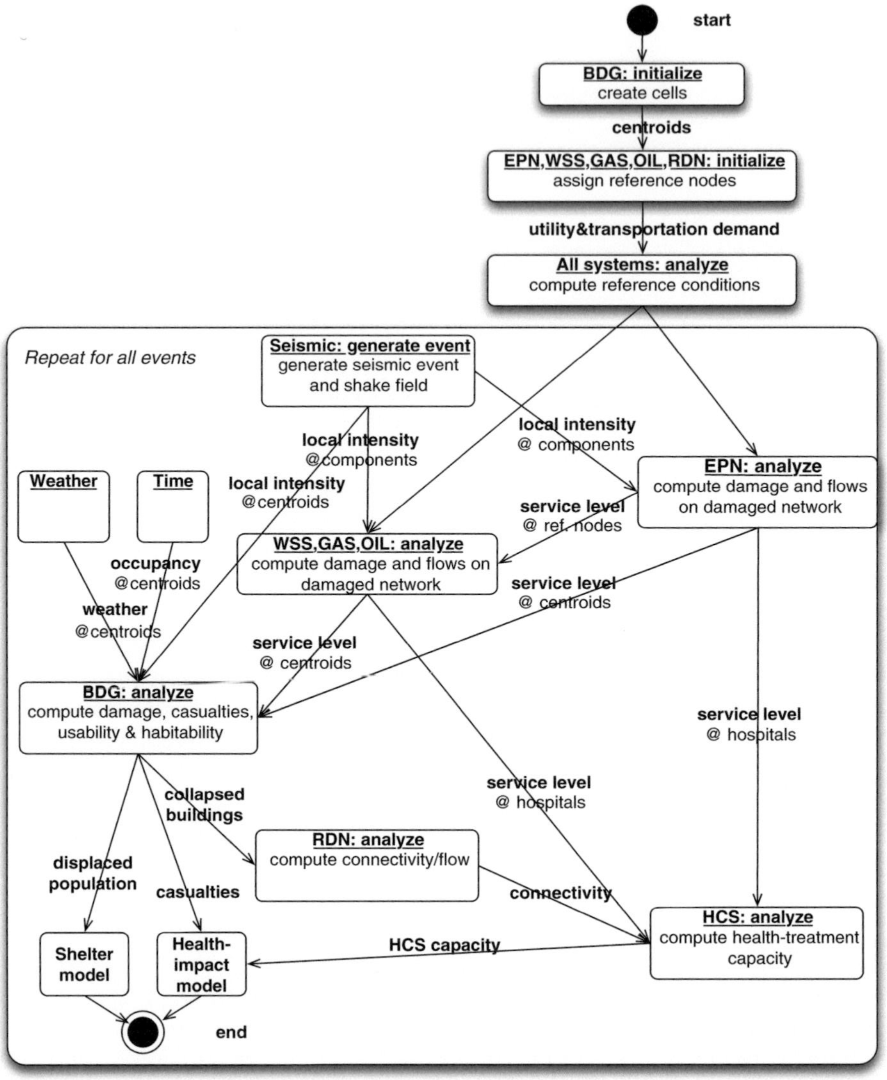

Fig. 2.10 State diagram to model the interdependencies: sequence in the evaluation of states of the objects and messages (quantities) transmitted between objects

of the *Network*, *Node* and *Edge* superclasses, which are inherited and exploited by all child classes.

Nodes and edges in all the specific systems, such as WSS, EPN, RDN etc. are concrete counterparts of the *Node* and *Edge* abstract classes from which they inherit a number of properties and methods. Most properties are those stored also in the form of vectors at the network level and are reported in the previous tables.

Table 2.2 Attributes/properties of the Network abstract class

Group	Attribute(s)	Description
Global properties	nEdges, nNodes	Number of edges and of nodes in the network
	edges, adjacencyMatrix, incidenceMatrix, incidenceList	Connectivity matrix and alternative representations (they are all equivalent and only one needs to be specified)
	deadEnds, articPTS, bridges	List of network dead ends, articulation points (nodes whose removal increases the number of connected sub-networks) and bridges (edges whose removal increases the number of connected sub-networks)
	vulnSites	List of sites of network vulnerable components
	ULweight	Influence w_i of the considered network (only for utilities) on total utility loss (UL) in buildings (must sum to one)
Edge properties stored at Network level[a]	edgeLength, edgeCentroid	–
	edgeType	WSS: pipe, tunnel, channel, etc.; EPN: overhead line, buried line, etc.; RDN: trench, bridge, embankment, tunnel, etc.
	edgeIsVulnerable, edgeIMtype	Boolean, if true an IM of the specified type must be predicted for the purpose of damage evaluation
	edgeSiteClass, edgeVs30	Geotechnical characterization for the amplification model
	edgeDepth2GW, edgeLiqSuscClass, edgeLandslideSuscClass, edgeYieldAcc	Input to geotechnical hazards model: depth to ground water, liquefaction and landsliding susceptibilities, critical acceleration
Edge properties stored at Network level[a]	nodePosition, nodeAltitude	–
	nodeType	Sink, source, junction, etc.
	nodeIsVulnerable, nodeIMtype	–
	nodeSiteClass, nodeVs30	–

All objects from subclasses inherit them

[a]These properties have counterparts in the Edge/Node classes. Here are vectors of length nEdges/nNodes collecting values that are computationally effective to store once rather than retrieve from the corresponding objects each time they are used

Table 2.3 Methods of the Network abstract class

Method	Description
anySDFS	Performs the Depth First Search algorithm, to know which vertices can be reached by a path starting from any source
connectedNodes	Lists of nodes belonging to different (eventually present) connected components in the network; a connected component of an undirected graph is a subgraph in which any two vertices are connected to each other by paths, and which is connected to no additional vertices in the supergraph
edges2Adjacency, edges2IncidenceList, edges2IncidenceMatrix	Perform transformation between equivalent network form representations
findDeadEnds, findArtcPTS, findBridges	–
isConnected	Determines whether a path exists between any two nodes in an undirected graph
minPath	Finds the minimum path between a pair of nodes
subnetwork	Returns the list of edges connecting a subset of network nodes
retrieveLandSuscEdges, retrieveLiqSusEdges, retrieveSiteClassEdges, retrieveSiteClassNodes, . . .	Retrieves the corresponding property from a map (shape file) and assigns it to edges/nodes when it is not directly specified
discretizeEdges	Subdivides all edges with length larger than a specified threshold into smaller segments, so as to allow a higher spatial resolution in the evaluation of damage
updateConnectivity	Updates the connectivity matrix for the damaged network
computeDemand	
computeFlow	

Names of abstract methods, whose implementation is different in each network class, are italicized; some of them are specific for undirected networks and are actually included in the Undirected subclass

Inspection of the previous tables should make apparent how most of properties and methods of the Network class are unspecific. They relate to the mathematical graph model underlying the physical network. Indeed, most methods implement graph-theoretic results and algorithms that have nothing to do with the physical specificities of the particular network. Considering that each network can be described in terms of *form* and *flow* (Sánchez-Silva and Gómez 2013) the Network abstract class and Directed/Undirected subclasses implement the form-related portion of the network description. This use of the hierarchy principle and the inheritance design pattern in designing the object model enhances greatly the maintainability of the model and eases the introduction of new systems. Every additional network needs only to be described in terms of *damageability* and *flow*.

Take for instance water supply systems (WSS). Physical *damageability* and *flow* behaviour are described within the object model with a number of additional classes (see Fig. 2.9), of which the main ones are the WSS, WSSedge and WSSnode classes. These are subclasses of Undirected, Edge and Node, respectively. While WSS is a concrete class from which a water supply system object will be instantiated, WSSedge and WSSnode are once again abstract classes that generalize different concrete components, described by the classes Pipe, Tunnel, PumpingStation and DemandNode. The third node type, the water source, can be either constant head (e.g. a lake) or variable head (like a tank). Tables 2.4, 2.5, and 2.6 report a selection of the attributes of the WSS, WSSedge and WSSnode classes, respectively.

The WSS class has three main methods, shown in Table 2.7. Classes WSSedge and WSSnode instead, have no methods, beside the constructor (the standard method that all concrete classes have, used to instantiate objects of the class).

As far as the damageability model is concerned, buried pipelines are usually considered the most and often the only vulnerable components in a seismic reliability analysis of water supply systems.

Separate pipeline fragility relations exist for permanent ground deformation and ground shaking effects. For instance, reference is made here to fragility functions for damage induced by ground shaking. Damage can result in leaks, with partial loss of water flow as a function of leakage area, or in outright breaks. The number of pipe breaks or leaks per unit length is denominated interchangeably as *repair rate (RR)*, *damage rate*, *damage ratio* or *failure rate*. Some of the available fragility models employ a linear model of the type $RR = a \cdot IM$, whereas others employ a power model of type $RR = b \cdot IM^c$. Two pipeline fragility relations among those included within the SYNER-G toolbox are that proposed in HAZUS (FEMA 2003) and that proposed by the American Lifelines Alliance (ALA 2001). Both take PGV in input in cm/s and return repair rate RR in km^{-1}. Other fragility models are formulated in terms of multiple IMs, in particular, typically, PGV and PGA.

Based on such relations the *isBreakAndLeaksNumber* method of the Pipe class, establishes whether the pipe is broken or not, and for non-broken pipes estimates the number of leaks based on the assumption that the leaks number along a pipe is Poisson distributed with rate RR. The Pipe method *computeLeakageArea* computes then the total outflow area, e.g. assuming an area for each leak along the pipe equal to 3 % of the whole pipe section.

Once the damaged configuration is known, flow can be evaluated. It is important to observe how this time-consuming task (usually flow-equations are nonlinear) is carried out only after checking that there is still a possibility of convergence to a solution. This is done through methods of the higher-level Network class that allow a fast evaluation based on the damaged topology of the presence of connected components with sources and demands. Attempts to solve the whole set of equations on a disconnected network are thus avoided.

Table 2.4 Attributes/properties of the WSS class

Group	Attribute(s)	Description
Global properties	sourceHead	Water head at source nodes
	endUserDemand, hydricEquipment	Required water flow at demand nodes, either assigned or evaluated by aggregating over tributary cells, employing population and hydricEquipment for the region (expressed in [l/inhab./day])
	refEPNnode	Pointers to EPN node(s) feeding power to pumping stations (for interdependence modelling)
Pointers	pipe, demand, source, pump	List of pointers to objects of class Pipe, DemandNode, constantHeadWaterSource or variableHeadWaterSource (concrete subclasses of the abstract class Source), Pump
Edge properties stored at WSS level[a]	edgeMaterial, edgeDiameter, edgeRoughness, edgeDepth	Length, centroid, etc. are attributes inherited from the Network class. Here the network-specific properties are listed (roughness, diameter, laying depth, etc.)
Node properties stored at WSS level[a]	nodeMinimalHead	minimal head required at nodes for delivery of the assigned demand water flow; this property is a function of the average building elevation in the region of interest
	nodeDepth	–
State variables recording WSS state	states	$n_E \times 1$ collection of properties that describe the current state for each of the n_E events. Fields: AHR (scalar, Average Head Ratio), SSI (scalar, System Serviceability Index), DCI (Damage Consequence Index) and UBI (Upgrade Benefit Index), $n_{Edges} \times 1$ vectors, and associated statistics (mean, standard deviation, experimental cumulative distribution function, etc.)

[a]These properties have counterparts in the WSSedge/WSSnode classes. Here are vectors of length nEdges/nNodes collecting values that are computationally effective to store once rather than retrieve from the corresponding objects each time they are used

Table 2.5 Attributes/properties of the WSSedge class

Group	Attribute(s)	Description
Properties stored also at WSS level[a]	material, diameter, roughness, depth	Length, centroid, etc. are attributes inherited from the Edge class. Here the network-specific properties are listed (material, roughness, diameter, laying depth, etc.)
State variables recording edge state	states	$n_E \times 1$ collection of properties that describe the current state for each of the n_E events. Fields: flow (l/s), primaryIM, localIMs, leaksNumber (integer), leakageArea (m^2), broken (Boolean), DCI, UBI

[a]These properties have counterparts in the WSS class

Table 2.6 Attributes/properties of the WSSnode class

Group	Attribute(s)	Description
Properties stored also at WSS level[a]	minimalHead, tributaryCells	Only for demand nodes
	waterHead	Only for source nodes
	refSource, refEPNnode	Only for pump nodes: source node served and EPN node feeding power to the pump.
State variables recording node state	states	$n_E \times 1$ collection of properties that describe the current state for each of the n_E events. Fields: demandFlow (l/s), waterHead (m), outflow (l/s, outgoing water flow as a result of pipe damage, equals half the sum of flows over all edges converging in the node, it aggregates in the node the flows on the damaged edges and sums to demandFlow for the purpose of flow equations solution), HR (head ratio of damaged to undamaged conditions, expresses service level)

[a]These properties have counterparts in the WSS class

Table 2.7 Main methods of the WSS class

Method	Description
computeDemand	Aggregates demand from tributary cells in demand nodes
evaluateDamage	Evaluates damage at each edge/node and returns the damaged configuration
computeFlow	The core of the functional evaluation of the system: implements flow equations

Flow in a WSS can be described analytically by a set of $N+L$ nonlinear equations in $N+L$ unknowns, written in matrix form as:

$$\begin{cases} \mathbf{A}_N^T\mathbf{q} - \mathbf{Q}(\mathbf{h}_N) = \mathbf{0} \\ \mathbf{R}\,|\mathbf{q}|\,\mathbf{q} + (\mathbf{A}_N\mathbf{h}_N + \mathbf{A}_S\mathbf{h}_S) = \mathbf{0} \end{cases} \tag{2.1}$$

where *N*, *L* and *S* are the number of internal (non-source) nodes, the number of links and the number of water sources, respectively. The first *N* equations are balance equations and express flow balance at the internal nodes (sum of incoming and outgoing flows equal to zero or the end-user demands $\mathbf{Q}(\mathbf{h}_N)$ in end-user nodes), while the second *L* equations express resistance in the links. The $L \times N$ and $L \times S$ matrices $\mathbf{A}_N$ and $\mathbf{A}_S$ are sub-matrices of the $L \times (N+S)$ matrix $\mathbf{A}$ which contains 0, 1 and -1 terms as a function of the network connectivity. The $N \times 1$ and $S \times 1$ vectors $\mathbf{h}_N$ and $\mathbf{h}_S$ are the corresponding partitions of the $(N+S) \times 1$ vector $\mathbf{h}$ collecting the *N* unknown heads in the internal nodes and the *S* known heads in the water-source nodes. The $L \times 1$ vector $\mathbf{q}$ collects the unknown flows in the *L* links and $\mathbf{R}$ is the $L \times L$ diagonal matrix of resistances, with terms $r_i = u_i\, L_i$, where $u_i = \beta\, D^{-5}$ (according to Darcy's law) and L_i is the i-th link length.

It is customary in the analysis of WSS for the purpose of design to treat the end-user demands $\mathbf{Q}$ as fixed boundary conditions (the system must be proportioned in order to satisfy them). The solution of the system with $\mathbf{Q}$ independent of $\mathbf{h}_N$ is called "demand-driven". In the above set of equations, on the contrary, the end-user demands are written as $\mathbf{Q}(\mathbf{h}_N)$, i.e. as functions of the unknown heads in the internal nodes. The solution of the system in this form is called "head-driven", and is employed here since in the perturbed seismic conditions satisfaction of prescribed demands is not guaranteed.

The set of nonlinear equations holds in so-called stationary conditions, i.e. it assumes constant end-user demands. This is a simplification, which is valid as long as the boundary conditions vary smoothly with time, in which case one speaks of quasi-stationary conditions. In seismic conditions this is not the case but the abrupt variation due to ruptures and leakages is soon replaced by a new stationary state.

Solution of the above set of equations by a numerical algorithm allows verification of the service level in each end-user node.

2.2.5 Seismic Hazard Model

Seismic hazard characterization for a spatially distributed system requires appropriate description of the joint occurrence of ground motion intensities and geotechnical hazards at multiple sites. This requires accounting for the statistical dependence between different intensity measures at different sites. If intensities can be modelled as joint lognormal, a spatial cross-correlation model provides a complete description of the above dependence structure. While Sect. 2.2.2 lists the classes devised to implement the seismic hazard model (classes SeismicSource, SeismicEvent and LocalIntensity, and subclasses), the latter is described in detail in Chap. 3, to which the reader is referred. Herein it is just recalled that the general approach adopted, termed conditional simulation, splits the simulation in two steps, the first accounting for spatial correlation and the second for cross-IM correlation.

2.3 Probabilistic Analysis

Multiple uncertainties affect the seismic vulnerability assessment problem and need to be properly included in the described model. They can be classified in three groups.

The first group collects uncertainty related to seismic hazard, such as that on occurrence in time and space of events, on their intensity, or on the induced level of shaking and secondary hazards at all sites of interest. The second group contains uncertainty on physical damageability of components of all systems. The third group collects all *consequence-related* uncertainty, i.e. uncertainty on the functional and socio-economic consequences of the physical damage induced by the earthquake. In all three groups, finally, uncertainty stems from both inherent variability in the associated natural phenomena, and incomplete knowledge leading to imperfect modelling.

The developed model accounts properly for the first two groups, as described in the next section. On the other hand, characterization of consequence-related uncertainty for all but the simplest single-site systems like individual buildings (e.g. the PEER approach to loss estimation, with conditional probability models of monetary loss given damage to structural and non structural elements) is still in its infancy and has been only tentatively included at the conceptual level. Its treatment within the context of spatially distributed systems and lifelines would involve modelling, for instance, uncertain network topologies, or strength of interdependence between different systems (like e.g. in Dueñas-Osorio et al. 2007a, b).

The goal of the probabilistic analysis is to evaluate statistics (mean, variance, etc.), probabilities or mean annual rates of events E defined in terms of the performance of the system.

One of the end results of such an analysis could be, for example, the mean annual rate of a performance measure, such as e.g. the connectivity loss CL or the system serviceability index SSI of a utility network (see Chap. 5 for a comprehensive list of performance indicators employed), exceeding or falling below a given threshold, respectively. An informative derivative result would then be threshold values associated with assigned return period, e.g. the CL value with a 500-years return period, and the associated causative scenarios, in terms of shake fields and damage maps.

The starting point of such an analysis, within the context of the adopted time-invariant simplifying assumption, is a joint probability model (distribution) of all uncertainty, denoted by $f(\mathbf{x})$, where $\mathbf{x}$ is the vector that collects all random variables in the problem. The probability of an event E can then be expressed as (the "reliability integral") in terms of the joint distribution:

$$p_E = \int_{\Omega_E} f(\mathbf{x}) d\mathbf{x} \tag{2.2}$$

where Ω_E is the portion of the sample space (the space where $\mathbf{x}$ is defined) collecting all $\mathbf{x}$ values leading to the event E. Thus, a probabilistic analysis requires setting up the uncertainty model $f(\mathbf{x})$ and choosing an appropriate method to evaluate Eq. (2.2). These two tasks are described in the next two sections.

2.3.1 Uncertainty Model

A rather basic system of systems is employed in this section to illustrate the uncertainty model. Figure 2.11 shows an Infrastructure made up of two systems only: an EPN with a generator (vulnerable component number 1) and distribution sub-station (vulnerable component 2), connected by an overhead line of negligible vulnerability; a WSS with a pumping station (vulnerable component 3) working with power from the adjacent distribution station in the EPN, and feeding water to an end-user or demand node with a pipeline (vulnerable component 4). The figure also shows a seismic source S on which an event of magnitude M, at location L, is generated. Finally, six points of a regular grid are also shown: as explained in detail in Chap. 3, the developed model produces probabilistically consistent samples of distributed seismic intensities through an approach termed co-simulation, in which a so called *primary* intensity is sampled over the entire region of interest at points of a regular grid, then all needed intensity measures at the components' sites are obtained by first interpolating the primary IM and then conditionally sampling secondary IMs.

Figure 2.12 illustrates the random variables employed to describe the uncertainty associated with the model of the system in Fig. 2.11. Variables are arranged as nodes in a graph, with arrows indicating statistical dependence between them. Arrows flowing from top to bottom follow the chain of cause and effect leading from the rupturing fault/source, to event location and magnitude, the local intensities, the components' state of physical damage, the functional consequences at system-level and finally the value of the performance indicators at the highest, Infrastructure, level (e.g. displaced population). Moreover, additional variability comes from the epistemic uncertainty on the employed models, which is of two types. The first one is associated with the parameters of the probabilistic models (the grey θ variables in the graph). The second type refers to the very form of the models, since in general alternative models are available for each component (e.g. different fragility models for pipes). This latter type can be dealt with by repeating the whole analysis for different models, effectively setting up a logic tree of model alternatives, as it is done for instance in Engineering Seismology for hazard assessment, weighting different GMPEs, hypotheses on the boundaries of seismo-genetic sources, etc.

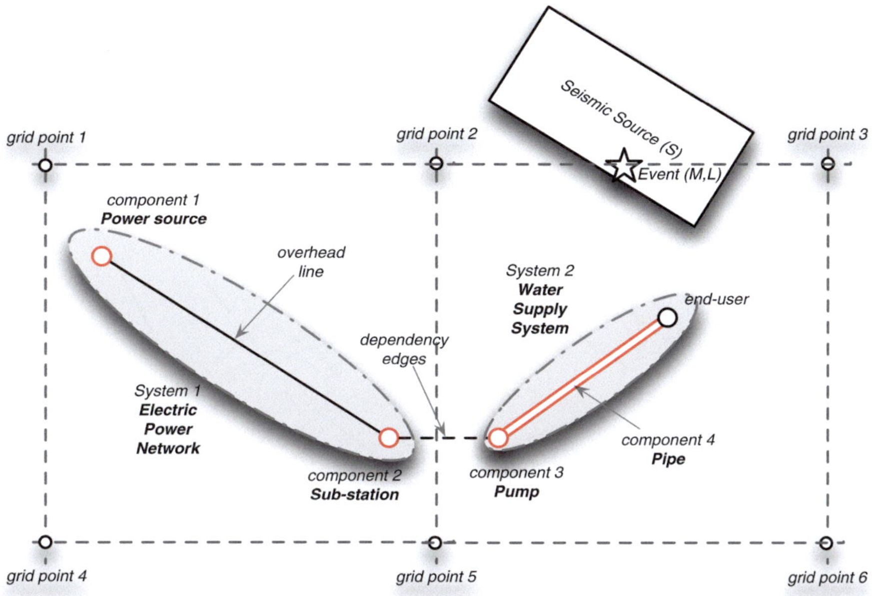

Fig. 2.11 Sample physical system for Fig. 2.12

The random variables in the graph are characterized as follows:

- *M* is the event magnitude, and is described by a weighted sum of the activity models, e.g. truncated Gutenberg-Richter or characteristic, etc., for each active source in the region. Epistemic uncertainty on its parameters regards, for instance, in the case of the Gutenberg-Richter activity model the lower and upper bound magnitude, as well as the rate α and slope β;
- *S* describes the seismo-genetic sources. It is a discrete variable characterized by a probability mass function conditional on magnitude, since not all sources can generate events for a given magnitude;
- *L*, whose distribution depends on the source, describes the event location (epicentre, hypocentre), and is usually modelled as uniform over the source area for point-source models. Epistemic uncertainty regards the definition of the source boundaries;
- Variables η and $\boldsymbol{\varepsilon}$ describe the inter-event and intra-event model error terms of the ground motion prediction equation employed to relate macro-seismic parameters with local (primary) intensity on the grid points.[4] They are both Gaussian. Vector $\boldsymbol{\varepsilon}$ is represented as dependent on a standard normal vector $\mathbf{u}$, because this is

[4]Crowley and Bommer (2006) were probably the first to point out the need to use GMPEs with separate characterization of inter- and intra-event variability in the analysis of spatially distributed systems.

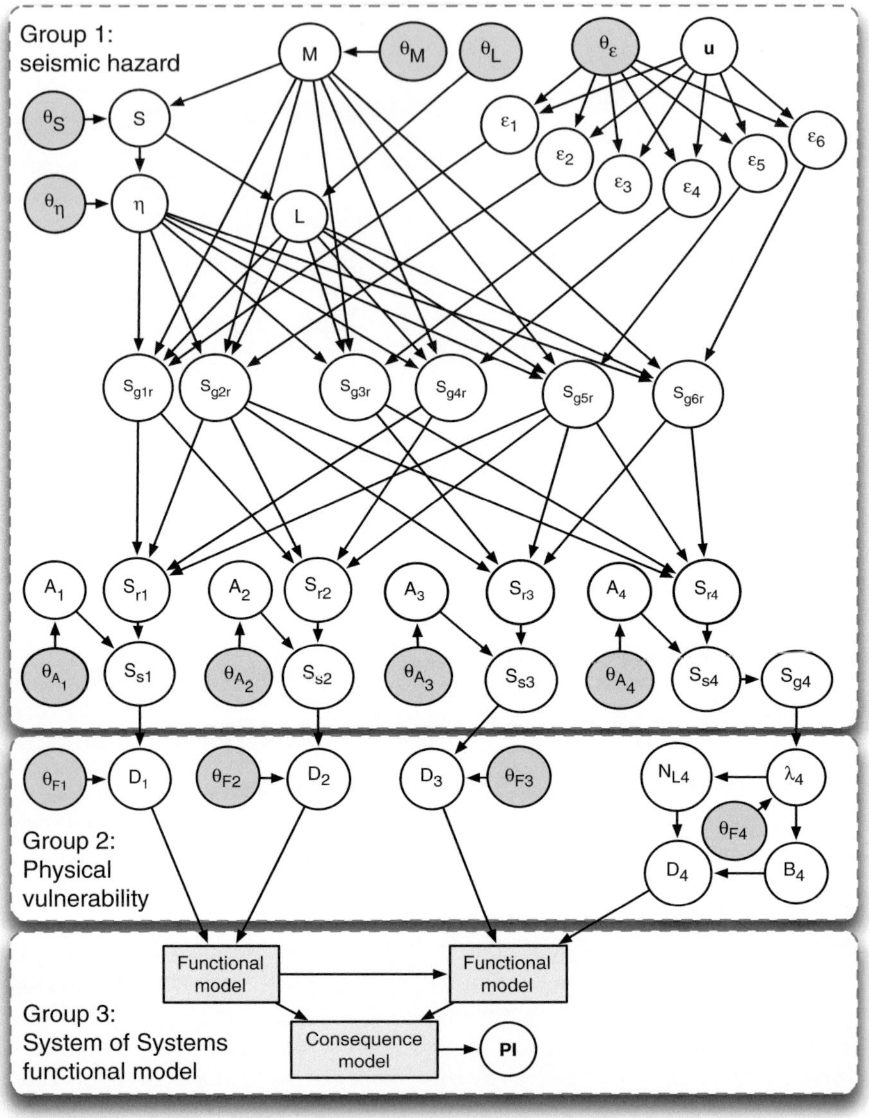

Fig. 2.12 The network of random variables modelling the uncertainty in the regional seismic vulnerability assessment problem

usually the way correlated normal vectors are sampled, and intra-event error terms are spatially correlated as described in Chap. 3. Epistemic uncertainty regards the variance of both η and $\boldsymbol{\varepsilon}$, as well as the correlation structure of the intra-event terms. More generally, several alternative GMPEs can be applicable for a given region;

- S_{gi} are the values of the primary IM at the *grid* points, and their distribution, governed by the chosen GMPE, is in general lognormal, given magnitude and distance, since η and $\boldsymbol{\varepsilon}$ are Gaussian.
- S_{ri} are the primary IM values interpolated at the components' location. These values are on rock/stiff-soil since this is how the primary IM is predicted on the grid.
- A_i are *amplification* functions for each site.
- $\mathbf{S}_{si}$ is the vector of IMs needed at each site, at the surface or rock outcrop (i.e. it depends also on A_i). As explained in Chap. 3, assuming joint normality of the logarithms of S_{ri} and $\mathbf{S}_{si}$, the latter can be easily predicted conditional on the values of the first.
- At some locations a geotechnical hazard model is required to describe IMs required by some components (e.g. pipes), such as permanent ground deformation (variable S_{g4} in Fig. 2.12, for component 4 in Fig. 2.11).
- D_i is the uncertain state of physical *damage* of the i-th component, described by either a set of fragility functions for increasing performance levels/limit-states, in case of point-like components (e.g. the first three components: the generation plant and the sub-station in the EPN, as well as the pump in the WSS), or a Poisson model for line-like ones (Der Kiureghian 2009). The damage state for components 1–3 is determined from a discrete probability distribution, function of the sampled IM values (and ATC-58 like procedure, ATC 2007 see Chap. 5). For the last, fourth component, the pipe, a repair or damage rate per unit length is determined as a function of e.g. permanent ground deformation and/or peak ground velocity, and used as the rate of a Poisson process of damage along the pipe.
- B_4 is a Boolean variable (rupture/unruptured)
- N_{L4} is the number of damages along the pipe segment. If unruptured ($B_4 = 0$) the pipe is assigned a leakage area equal to the sum of the areas at each of the N_{L4} leaks, and the corresponding additional demand is lumped at the end nodes of the sub-segment (produced by the discretization algorithm and to which an initial zero demand is assigned), otherwise the pipe is removed from the damaged network.

Finally, given the state of physical damage to their components, systems are evaluated according to their functional models and they all concur to determine the performance of the Infrastructure. Uncertainty of group 3 should be characterized at this last stage, e.g. by adopting a random topology of the networks for areas where incomplete information is available (a not so rare case for older water supply systems in some European cities).

The graph in Fig. 2.12, by highlighting the *statistical dependence* between variables, allows an efficient writing of the joint density $f(\mathbf{x})$. In general, the joint density $f(\mathbf{x})$ can be decomposed as the product of a generic set of conditional distributions in the form:

$$\begin{aligned} f(\mathbf{x}) &= f(x_n|\,x_1,\ldots,x_{n-1})f(x_1,\ldots,x_{n-1}) \\ &= f(x_n|\,x_1,\ldots,x_{n-1})f(x_{n-1}|\,x_1,\ldots,x_{n-2})f(x_1,\ldots,x_{n-2}) \\ &= f(x_1)\prod\nolimits_{i=2}^{n} f(x_i|\,x_1,\ldots,x_{i-1}) \end{aligned} \tag{2.3}$$

The network of statistical dependencies represented in Fig. 2.12 allows a drastic reduction in the terms of Eq. (2.3), which can be rewritten as:

$$\begin{aligned} f(\mathbf{x}) &= f_{SYS|PhVM}(\mathbf{x}_{SYS}|\,\mathbf{x}_{PhVM})\, f_{PhVM|SH}(\mathbf{x}_{PhVM}|\,\mathbf{x}_{SH})\, f_{SH}(\mathbf{x}_{SH}) \\ &= f(\mathbf{PI}|\,\mathbf{D})\prod_{i=1}^{4} f(D_i|\,S_i) f(S_i|\,S_{ir}, A_i) f(A_i) f(S_{ir}|\,\mathbf{S}_g) \\ &\quad \cdot \left[\prod_{j=1}^{4} f(S_{gj}|\,M, L, \varepsilon_j, \eta) f(\varepsilon_j|\,\mathbf{u}) f(L|\,S) f(\eta|\,S) f(S|\,M) f(M)\right] \end{aligned} \tag{2.4}$$

where $f_{\mathrm{SYS|PhVM}}$ is the conditional probability distribution expressing the dependence of the system of systems on the physical vulnerability of its components, i.e. the distribution of the performance indicators vector **PI** given the state of physical damage vector **D** of the systems' components:

$$f_{SYS|PhVM}(\mathbf{x}_{SYS}|\,\mathbf{x}_{PhVM}) = f(\mathbf{PI}|\,\mathbf{D}) \tag{2.5}$$

The distributions $f_{\mathrm{PhVM|SH}}$, expressing the dependence of the state of physical damage on the seismic hazard, and f_{SH} can be written, neglecting epistemic uncertainty, as:

$$f_{PhVM|SH}(\mathbf{x}_{PhVM}|\,\mathbf{x}_{SH}) = \prod_{i=1}^{4} f(D_i|\,S_i) f(S_i|\,S_{ir}, A_i) f(A_i) f(S_{ir}|\,\mathbf{S}_g) \tag{2.6}$$

$$f_{SH}(\mathbf{x}_{SH}) = \prod_{j=1}^{4} f(S_{gj}|\,M, L, \varepsilon_j, \eta) f(\varepsilon_j|\,\mathbf{u}) f(L|\,S) f(\eta|\,S) f(S|\,M) f(M) \tag{2.7}$$

or accounting for epistemic uncertainty as:

$$\begin{aligned} &f_{PhVM|SH}(\mathbf{x}_{PhVM}|\,\mathbf{x}_{SH}) \\ &= \prod_{i=1}^{4} f(D_i|\,S_i, \theta_{Fi}) f(\theta_{Fi}) f(S_i|\,S_{ir}, A_i) f(A_i|\,\theta_{A_i}) f(\theta_{Ai}) f(S_{ir}|\,\mathbf{S}_g) \end{aligned} \tag{2.8}$$

$$\begin{aligned} f_{SH}(\mathbf{x}_{SH}) &= \left[\prod_{j=1}^{4} f(S_{gj}|\,M, L, \varepsilon_j, \eta) f(\varepsilon_j|\,\mathbf{u}, \theta_\varepsilon)\right] \\ &\quad \cdot f(L|\,S, \theta_L) f(\theta_L) f(\eta|\,S, \theta_\eta) f(\theta_\eta) f(S|\,M, \theta_M) f(\theta_M) f(M) \end{aligned} \tag{2.9}$$

Finally, it can also be observed how the diagram in Fig. 2.12 describes a directed acyclic graph (DAG) and represents a Bayesian hierarchical model or a Bayesian network (Nielsen 2007). Several authors have started using these networks in the reliability analysis of spatially distributed systems. In particular, Straub et al. (2008) have devised the model for spatially distributed seismic hazard employed at the top of Fig. 2.12 (variables M, L, η and $\boldsymbol{\varepsilon}$ and their influence on local intensity, even though the model in Fig. 2.12 is modified to account for dependence of course on magnitude and to accommodate the co-simulation approach). As long as evidence is not input and Bayesian updating is not performed, however, the graph is used just as a description of the (forward) sampling scheme within the typical simulation run, as explained in the next section.

2.3.2 Simulation Methods

Probabilistic evaluation of system performance can be carried out with simulation methods and non-simulation methods. The former have been adopted in conjunction with the described model.

Simulation is a robust way to explore the behaviour of systems of any complexity. It is based on the observation of system response to input $\mathbf{x}$. Simulation of a set of inputs from $f(\mathbf{x})$ and evaluation of corresponding outputs allows determining through statistical post-processing the distribution of the output.

Simulation methods start from Eq. (2.2) by introducing the so-called *indicator function* $I_E(\mathbf{x})$, which equals one if $\mathbf{x}$ leads to the event E, and zero otherwise, by which p_E becomes the expected value of I_E:

$$p_E = \int_{\Omega_E} f(\mathbf{x})d\mathbf{x} = \int I_E(\mathbf{x})f(\mathbf{x})d\mathbf{x} = \mathrm{E}\left[I_E(\mathbf{x})\right] \tag{2.10}$$

Monte Carlo (MC) simulation (Rubinstein 1981) is the crudest possible way of approximating p_E, in that it amounts to estimating the expectation of I_E as an arithmetic average $\widehat{p}_E$ over a sufficiently large number N of $\mathbf{x}$ samples:

$$p_E = \mathrm{E}\left[I_E\left(\mathbf{x}\right)\right] \cong \frac{1}{N}\sum\nolimits_{i=1}^{N} I_E\left(\mathbf{x}_i\right) = \frac{N_E}{N} = \widehat{p}_E \tag{2.11}$$

The problem is thus reduced to that of sampling realizations $\mathbf{x}_i$ of $\mathbf{x}$ from the distribution $f(\mathbf{x})$, and evaluating for each realization the PI based on which event E is defined, in order to assign a value to the indicator function I_E.

Sampling of realizations from the joint distribution $f(\mathbf{x})$ can be performed according to different algorithms. These are not independent of the simulation method and, actually, "smart" sampling of the realizations can result in drastic improvements of efficiency, in terms of the required computational effort. For the problem at hand, and plain Monte Carlo simulation, sampling is carried out in a conditional fashion, according to the representation of the joint density given in Eq. (2.4): the first variables to be sampled from their respective marginal distributions are magnitude M and the standard normal vector $\mathbf{u}$, then, conditional on the sampled magnitude, source S, location L within the source and inter-event error η are sampled from the appropriate conditional distribution, while correlated intra-event errors $\boldsymbol{\varepsilon}$ follow from $\mathbf{u}$. Finally, through rock values at the grid and amplification by the variables A_i, local intensity S_{si} at each site follows, together with the component damage state D_i, etc.

A basic well-known result about plain MC simulation is that the *minimum* number of samples required for a specified confidence in the estimate (in particular to have 30 % probability that $\widehat{p}_E \in [0.77, 1.33] \cdot p_E$) is given by:

$$N \geq 10 \frac{1 - p_E}{p_E} \cong \frac{10}{p_E} \tag{2.12}$$

In order to reduce the required minimum N one must act on the variance of $\widehat{p}_E$. This is why the wide range of enhanced simulation methods that have been advanced in the last decades fall under the name of "variance reduction techniques". One such technique is *Importance sampling* (IS). This is a form of simulation based on the idea that when values of $\mathbf{x}$ that fall into Ω_E are rare and difficult to sample, they can be conveniently sampled according to a more favourable distribution, somehow shifted towards Ω_E. Of course the different way $\mathbf{x}$ values are sampled must be accounted for in estimating p_E according to:

$$\begin{aligned} p_E &= \int I_E(\mathbf{x}) f(\mathbf{x}) d\mathbf{x} = \int I_E(\mathbf{x}) \frac{f(\mathbf{x})}{h(\mathbf{x})} h(\mathbf{x}) d\mathbf{x} \\ &= \mathrm{E}_h \left[I_E(\mathbf{x}) \frac{f(\mathbf{x})}{h(\mathbf{x})} \right] = \mathrm{E}_h \left[I_E(\mathbf{x})\, \alpha(\mathbf{x}) \right] \cong \frac{1}{N} \sum_{i=1}^{N} I_E(\mathbf{x}_i)\, \alpha(\mathbf{x}_i) \end{aligned} \tag{2.13}$$

where now p_E is expressed as the expectation of the quantity $I_E(\mathbf{x})\, \alpha(\mathbf{x})$ with respect to the distribution $h(\mathbf{x})$, called sampling density. The IS ratio $\alpha(\mathbf{x})$ corrects the estimate to account for the different probability content of the neighbourhood of $\mathbf{x}$ assigned by the original and the sampling density. The difficulty with the IS method is to devise a good sampling density $h(\mathbf{x})$, since it inevitably requires some knowledge of the domain E. One way to do this is to start with a sampling density and change it during the simulation while samples closer to E occur. This is called *adaptive importance sampling*.

Jayaram and Baker (2010) proposed a simulation-based framework for developing a small catalogue of earthquake ground motion intensity maps for risk assessment of spatially distributed systems. Importance Sampling is first used to preferentially sample 'important' ground motion intensity maps, with a reduction of two orders of magnitude in the required number of maps with respect to plain MC simulation. Then the so-called statistical technique of K-Means Clustering

(McQueen 1967) is used to identify and combine, in an unsupervised manner,[5] redundant maps in order to obtain a small catalogue. The effects of sampling and clustering are accounted for through a weighting on each remaining map, so that the resulting catalogue is still a probabilistically correct representation.

The importance sampling density h on magnitude M,[6] inter-event error η and intra-event errors $\boldsymbol{\varepsilon}$, is built as the product of three sampling densities on each variable.

Both plain MC simulation and a highly effective brand of IS, specifically developed for spatially distributed systems analysis (Jayaram and Baker 2010), have been included in the model and are implemented in the SYNER-G toolbox.

2.4 Conclusions

The assessment of the impact of an earthquake is a complex task that calls for comprehensive modelling of the analyzed system. The latter is the set of all physical and non-physical systems that make up our society and the Infrastructure supporting it. The associated modelling effort is daunting, but also challenging.

The approach presented in this book, and the model framework illustrated in this chapter, represent a step in the direction of comprehensiveness. The main goal was that of setting up a framework where multiple interacting systems could be analyzed concurrently and consistently, describing also their inherent and modelling-related uncertainty, in order to evaluate their global state of damage and reach into socio-economic consequences.

One main choice made at the outset was that of constraining the problem in time and the focus was put exclusively on the short-term period after the event. This is reflected in the choice of impact metrics that are those typically related to emergency. A further (strong) simplifying assumption made was that a time-invariant approach is adopted, whereby aftershocks and the associated cumulative damage are not yet considered. On the other hand, a rather extensive list of systems was considered and analyzed (see Fig. 2.8, and Chap. 5). This list being large but not exhaustive, the need for a model that had the flexibility to easily accommodate future extensions was apparent.

The developed model is object-oriented. This choice, not unique to this research, has proven instrumental to tackle the complexity and size of the considered system of systems. Also, some of the principles of the object-oriented design paradigm, such as inheritance and composition, have allowed a high degree of abstraction and

[5]Previous attempts to drastically reduce the number of scenarios to represent the regional seismicity, e.g. Shiraki et al. (2007), were characterized by a more or less high degree of *subjectivity*.

[6]Kiremidjian et al. (2007) are the first to employ IS to selectively sample in the larger magnitude range.

hierarchical decomposition in the model, leading to a more maintainable and easily extensible framework.

The main strengths and limitations of the model can be summarized as follows:

1. The model encompasses a large number of systems, larger than any integrated model to date. Lifelines, including water supply systems, electric power networks and gas distribution systems, road networks, health-care facilities, harbors, and finally buildings. The integration of all these systems within the same model and simulation software allows consistent evaluation of demands across all systems starting from the same basic data on buildings, population and activities. It also allows re-evaluation of demands in the different post-event conditions. This is still at a basic level in the current stage of development but will eventually allow for the extension of the considered time span and the simulation of the recovery process, the changed business and activity pattern, to arrive at the evaluation of economic loss and resilience.
2. The specific Building sub-model (class BDG) is a very effective and scalable way of treating the large number of buildings to be considered in a study of regional or urban extension. It efficiently solves the problem of merging different data sources. Related also to point 4, it allows for different levels of granularity in the input data to be considered, through the use of dependency edges between building geo-cells and the reference nodes in all the other systems. Meshing of each component system can be different and appropriate to the level of information about the considered system, with the model taking care of connecting the systems together whenever the analyst does not provide the relevant information. This is the mechanism that allows for demands from tributary cells to be aggregated to demand nodes in each network, and to determine the residual service level of each utility, or degree of residual connectivity/accessibility of each geo-cell. In this respect, the model creates traffic analysis zones for road network analysis automatically.
3. The model for the evaluation of social metrics (explained in Chap. 4 and only hinted at herein, see Fig. 2.2) represents a novelty. Population displacement occurs due to several factors beside pure direct physical damage to buildings. While a collapsed or severely damaged building is obviously non usable, and thus not habitable, partially or even fully usable buildings can be non-habitable due to lack of basic services. The model allows, through the described interaction mechanism, to describe this aspect.
4. Interaction between systems was one of the aspects deemed central in the model development. It also, quite expectedly turned out to be one of the most difficult to tackle. While a general mechanism has been set up, and a number of interactions have been listed, in practice the approach covers three types of interactions only: the interaction for demand and utility loss evaluation between buildings and network-like systems; the interaction between the electric power network and dependent components in all other systems (e.g. pumps in the WSS); the interaction of geographical type between buildings and roads in urban environments, with debris from collapsed buildings that can induce additional

obstruction/damage to roads in the road blockage model. Hence interactions are certainly an important but still under-developed aspect of the model, that will need further work to widen the scope and the type of interactions considered. One aspect, for instance, is that of the *strength* of interaction, introduced in (Dueñas-Osorio et al. 2007a, b). In the presented model, the consequence of a failure in the dominant system deterministically follows in the dependent system. This could be weakened, especially when dealing with situations where the input data are incomplete.

5. Physical modelling of networks is carried out on a detailed level. Other comparable efforts generally employ simple connectivity models for all networks. While these may be appropriate in some cases, notably for understanding weaknesses of each network in terms of form, i.e. topology and hence can give very valuable indications on retrofit strategies, the modelling of flow is important to predict the real service level, which may be insufficient even in the presence of a surviving direct connection between a sink and a source. This has an influence on the performance of dependent systems.
6. Uncertainty is modelled with a network of random variables. Currently this sub-model is entangled with that of the physical system, with random variables being included as further attributes within the classes of the corresponding objects (e.g. magnitude M is an attribute of the event object). This aspect needs to be improved in order to have the uncertainty network stand alone as a further non-physical network in the model, connected to the physical layers through new dependency edges. A stand-alone network will allow easier introduction of probabilistic analysis methods that will have to deal only with objects on this level. At the same time, the current network, even if embedded in the physical portion of the model, is a rather comprehensive representation of uncertainty in the problem, with a refined and effective seismic hazard model (Chap. 3) and vulnerability model (Chap. 5), including epistemic/modelling uncertainty in a hierarchical fashion. Such a network, used only in a forward simulation, from the marginal variables to the performance metrics (top-down in Fig. 2.12) sets the ground for introduction of evidence and Bayesian inference, thus paving the way for a real-time use of the model as a decision support system.
7. Finally, the model encodes a fixed, pre-determined system evaluation sequence. The chosen sequence is very reasonable and grounded on considerations of intrinsic characteristic time of each considered system (simply stated, indirect damage in the electric power network propagates faster than in other systems, building collapse occurs and blocks roads before post-event travel is undertaken, etc.). Nonetheless, it is pre-determined and may not apply to all situations. On the other hand, it allows for fixed, though disturbed (with respect to pre-event ones) boundary conditions and the solution of stationary rather than non-stationary flow equations in each system.

References

Akkar S, Bommer JJ (2010) Empirical equations for the prediction of PGA, PGV and spectral accelerations in Europe, the Mediterranean Region, and the Middle East. Seismol Res Lett 81(2):195–206

ALA (American Lifelines Alliance) (2001) Seismic fragility formulations for water systems. Part 1 – Guideline. ASCE-FEMA, Reston, 104 pp

Ashby WR (1964) An introduction to cybernetics. Methuen, London

ATC (2007) Development of next generation performance-based seismic design procedures for new and existing buildings, Applied Technology Council, report no. ATC-58, Redwood City, CA

Booch G, Maksimchuk RA, Engle MW, Young BJ, Conallen J, Houston KA (2007) Object-oriented analysis and design with applications, 3rd edn. Addison Wesley, Reading

Chang SE, Chamberlin C (2004) Assessing the role of lifeline systems in community disaster resilience, MCEER research progress and accomplishments: 2003–2004, MCEER-04-SP01. MCEER, Buffalo, pp 87–94

Crowley H, Bommer JJ (2006) Modelling seismic hazard in earthquake loss models with spatially distributed exposure. Bull Earthq Eng 4(3):249–273

Der Kiureghian A (2009) Seismic risk assessment and management of infrastructure systems: review and new perspectives. In: Proceedings of the 10th international conference on structural safety and reliability, Osaka, Japan, 13–17 Sept

Dueñas-Osorio L, Craig JI, Goodno BJ (2007a) Seismic response of critical interdependent networks. Earthq Eng Struct Dyn 36:285–306

Dueñas-Osorio L, Craig JI, Goodno BJ, Bostrom A (2007b) Interdependent response of net-worked systems. ASCE J Infrastruct Syst 13:185–194

FEMA (2003) HAZUSMH MR4 multi-hazard loss estimation methodology – earthquake model – technical manual FEMA and NIBS, Washington, DC

Jayaram N, Baker JW (2010) Efficient sampling and data reduction techniques for probabilistic seismic lifelines assessment. Earthq Eng Struct Dyn 39(10):1109–1131

Kiremidjian AS, Stergiou E, Lee R (2007) Chapter 19: Issues in seismic risk assessment of transportation networks. In: Earthquake geotechnical engineering. Springer, Berlin, pp 939–964

McQueen JB (1967) Some methods for classification and analysis of multivariate observations. In: Proceedings of the 5th Berkeley symposium on mathematical statistics and probability, Berkeley, CA

Nielsen J (2007) Bayesian networks and decision graphs, 2nd edn. Springer, New York

PCCIP (1997) Critical foundations: protecting America's infrastructures, report of the President's Commission on critical infrastructure protection. Available from: http://www.fas.org/sgp/library/pccip.pdf. Last accessed 7 Sept 2011

Rubinstein RY (1981) Simulation and Monte Carlo method. Wiley, New York

Sánchez-Silva M, Gómez C (2013) Risk assessment and management of civil infrastructure networks: a systems approach. In: Tesfamariam S, Goda K (eds) Handbook of seismic risk analysis and management of civil infrastructure systems. Woodhead Publishing Limited, Cambridge/Philadelphia. doi:10.1533/9780857098986.4.437

Schläpfer M, Kessler T, Kröger W (2008) Reliability analysis of electric power systems using an object-oriented hybrid modeling approach. In: Proceedings of the 16th power systems computation conference, Glasgow, Scotland

Shiraki N, Shinozuka M, Moore JE II, Chang SE, Kameda H, Tanaka S (2007) System risk curves: probabilistic performance scenarios for highway networks subject to earthquake damage. J Infrastruct Syst 213(1):43–54

Straub D, Bensi M, Der Kiureghian A (2008) Spatial modelling of earthquake hazard and infrastructure performance through Bayesian network. In: Proceedings of the inaugural international conference of the Engineering Mechanics Institute ASCE, Minneapolis MN, USA, 19–21 May 2002

Chapter 3
Framework for Seismic Hazard Analysis of Spatially Distributed Systems

Graeme Weatherill, Simona Esposito, Iunio Iervolino, Paolo Franchin, and Francesco Cavalieri

Abstract The analysis of seismic risk to multiple systems of spatially distributed infrastructures presents new challenges in the characterisation of the seismic hazard input. For this purpose a general procedure entitled "Shakefield" is established, which allows for the generation of samples of ground motion fields for both single scenario events, and for stochastically generated sets of events needed for probabilistic seismic risk analysis. For a spatially distributed infrastructure of vulnerable elements, the spatial correlation of the ground motion fields for different measures of the ground motion intensity is incorporated into the simulation procedure. This is extended further to consider spatial cross-correlation between different measures of ground motion intensity. In addition to the characterisation of the seismic hazard from transient ground motion, the simulation procedure is extended to consider secondary geotechnical effects from earthquake shaking. Thus the Shakefield procedure can also characterise the effects site amplification and transient strain, and also provide estimates of permanent ground displacement due to liquefaction, slope displacement and coseismic fault rupture.

G. Weatherill (✉)
European Centre for Training & Research in Earthquake Engineering (EUCENTRE), Pavia, Italy
e-mail: graeme.weatherill@eucentre.it

S. Esposito • I. Iervolino
Dipartimento di Strutture per l'Ingegneria e l'Architettura, Università degli Studi di Napoli Federico II, Via Claudio 21, 80125 Naples, Italy

P. Franchin • F. Cavalieri
Department of Structural and Geotechnical Engineering, Sapienza University of Rome, Via Gramsci 53, 00197 Rome, Italy

K. Pitilakis et al. (eds.), *SYNER-G: Systemic Seismic Vulnerability and Risk Assessment of Complex Urban, Utility, Lifeline Systems and Critical Facilities*, Geotechnical, Geological and Earthquake Engineering 31, DOI 10.1007/978-94-017-8835-9_3,

3.1 Generating Seismic Scenarios for Spatially Distributed System Risk Analysis: An Overview of Concepts and Data Needs

3.1.1 Seismic Scenarios for Spatially Distributed System Analysis

The analysis of seismic risk to urban infrastructures within a region represents an important development in the effective mitigation of both economic and human losses due to earthquakes. To model the consequences of damage and failure to multiple interconnected systems it is important to define the seismic scenarios in a manner that is consistent with the properties of observed ground motions. The extension of seismic risk analysis to multiple systems of spatially distributed infrastructures presents new challenges in the characterisation of the seismic hazard input, particularly with respect to the spatial correlation structure of the ground motion values that form the basis for the systemic risk analysis.

An increasingly common approach for the estimation of losses for a spatially distributed portfolio of vulnerable elements is in the use of scenario events, or, if a probabilistic loss analysis is required, from a set of stochastically generated events that are representative of the earthquake recurrence models for the seismogenic sources under consideration. From these event sets it is possible to create realisations of ground motion fields, which are then input into the models of seismic risk.

3.1.2 Uncertainty and Spatial Correlation

For spatially distributed portfolios or systems, the spatial correlation structure of the ground motion fields may impact upon analyses of loss, and therefore cannot be neglected from the modelling process.[1] There is clear precedent in recent research into seismic risk, to take into account the spatial correlation between ground motions on an urban scale (Park et al. 2007; Crowley et al. 2008a,b; Goda and Hong 2008a,b; Jayaram and Baker 2010a).

When modelling seismic risk to urban infrastructure and complex interconnected systems, new challenges emerge in the process of generating the spatially correlated fields of ground motion. In the examples cited previously, the consideration of spatial correlation has been limited to fields of ground motion represented by a single ground motion intensity measure (IM), such as peak ground acceleration (PGA) or spectral acceleration ($S_a(T)$) or displacement ($S_d(T)$) at a specified period of oscillation (T). If considering multiple infrastructures, however, fragility

[1]Spatially distributed ground motion intensities are modelled as joint lognormal random fields herein; thus, correlation provides a complete description of their statistical structure.

models for different components of each infrastructure may require the characterisation of different measures of ground motion intensity, often depending on the physical properties (such as fundamental elastic period or, in the case of pipelines, the transient strain) of the element in question. Therefore in generating the ground motion fields, it may be inappropriate to characterise the spatial distribution of ground motion intensity using a single intensity measure, but instead the fields of many spatially correlated intensity measures may be required.

As the interconnection between infrastructural systems cannot be neglected, neither can the correlation between spatial fields of different intensity measures generated from the same event. Therefore, a critical challenge of this work is to extend process of generating fields of spatially correlated ground motion, to one in which the spatial cross-correlation structure is also represented.

The consideration of seismic risk to infrastructures introduces a further set of requirements in terms of seismic input. These relate to the geotechnical hazards associated with earthquakes, namely amplification and modulation of ground shaking at the Earth's surface due to the local site conditions (e.g., superficial geology, topography, deep geological structure etc.), and permanent ground displacement (PGD) arising due to liquefaction, slope displacement and co-seismic rupture around the fault surface. These hazards have a significant impact upon a many different infrastructures, particularly those with buried linear elements, which may traverse different geological formations within the interconnected system and are therefore vulnerable to permanent differential ground displacements. An assessment of risk to multiple systems of interconnected spatially distributed elements must therefore incorporate the geotechnical hazards in a manner that is also consistent with the ground shaking hazard.

3.1.3 The Shakefield Procedure

Initially, the focus of this analysis is upon generating fields of strong ground motion using earthquake scenarios. Each Shakefield represents a simulated strong ground motion field for a given rupture scenario. For a Shakefield, no recorded strong motion data are integrated into the strong motion fields that are generated. Shakefields can be adopted for single earthquake scenarios, a scenario in this context representing the source and magnitude of a single event, or can be applied within a probabilistic seismic hazard analysis, which considers many source and magnitude scenarios to effectively sample the variability in the seismogenic source. Each Shakefield is implemented in the following steps, and illustrated in Fig. 3.1:

1. Generate a source event with a given magnitude and source geometry. The geometry may take the form of a point or a finite rupture surface, the latter being preferred for consistency with the physical properties of the earthquake source. The source event may be a single scenario event (e.g. a historical earthquake or a hypothetical adverse case), or may be a sample from a probability distribution

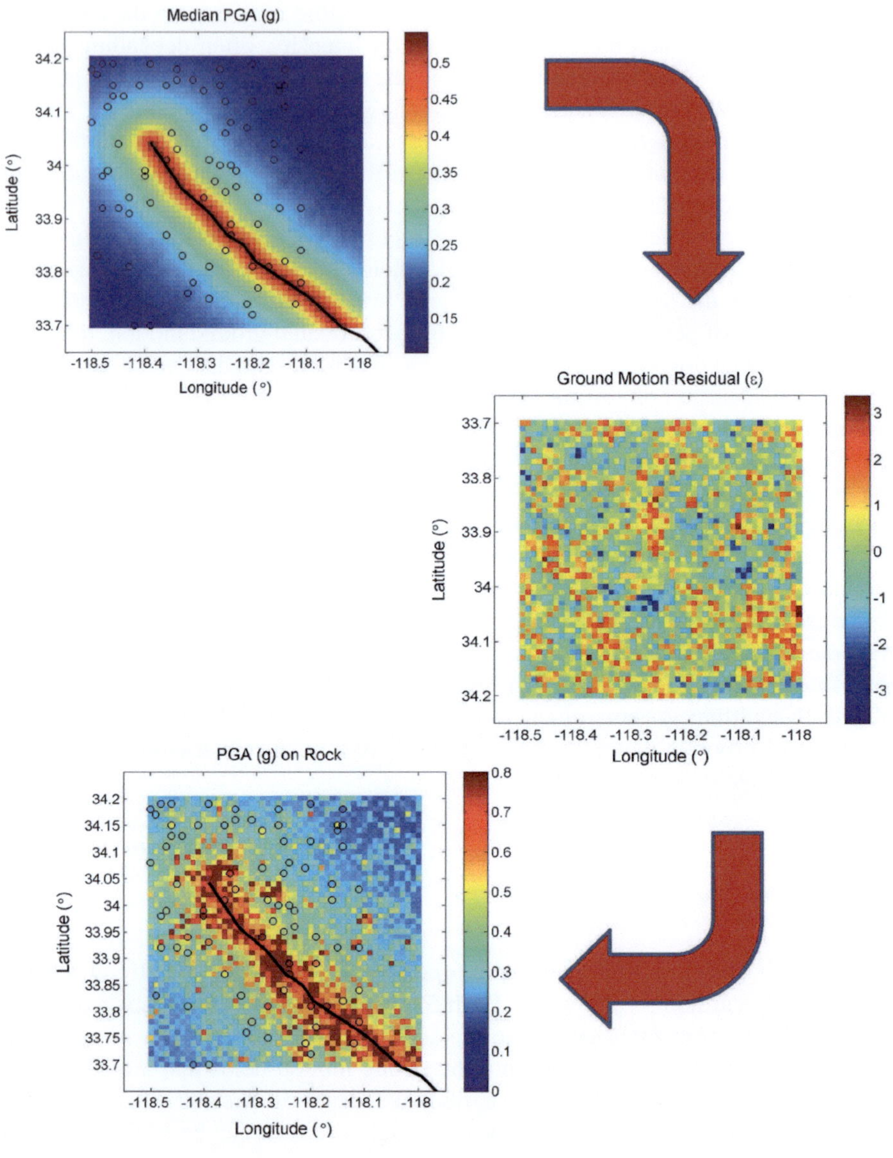

Fig. 3.1 Overview of the Shakefield process for strong motion on rock: attenuation of median ground motion (*top*), generation of field of spatially correlated ground motion residuals (*middle*) and calculation of ground motion on rock (*bottom*). Fault source indicated by *black line*, target sites indicated by *black circles*

representing the total magnitude frequency distribution of the seismogenic source and the set of potential seismogenic sources under consideration (see also Chap. 2).

2. Given a source and magnitude, and using a ground motion prediction equation (GMPE), attenuate the median ground motion field across spatial grid of locations considered in the system. Depending on the context and the number of locations under consideration, the nodes of the field may be the locations of the elements themselves, or a regularly spaced grid of points from which the ground motions at the element locations can be interpolated.
3. Generate a realisation of a standard Gaussian field representing the spatial correlation structure of a required intensity measure. For each site these variates are multiplied by the uncertainty terms of the GMPE and added to the median value on rock, thus sampling the ground motion uncertainty.
4. For each site, generate the ground motion values for the secondary intensity measures, conditional upon the simulated ground motion intensity measure of the primary IM.
5. Using an amplification factor appropriate to the soil conditions, and earthquake intensity, at each site, scale the ground motions to give the ground motion field on soil.
6. If required, proceed to define a geotechnical hazard parameter (e.g. permanent ground deformation resulting from liquefaction or slope displacement) conditional upon the intensity of ground motion at the site.

Given the scenario event, it is necessary to determine the rupture geometry. The complexity of the rupture model, and the degree of detail resolved, may depend strongly on the amount of information available to define the seismogenic source. This may, in simplest cases, define the rupture as simply a point or line source from which energy radiates in an isotropic manner. In most applications, however, the seismic source may be defined as a simple geometric rupture plane in three dimensions. This may correspond to a fault segment, or partial segment, or may be an approximation to the rupture plane of a previous event. Simple geometrical approximations to the rupture surface may be useful in hazard-based approaches where aleatory variability arising from different rupture scenarios may be constrained to the selected fault plane, without requiring a priori assumptions about the rupture dynamics.

After defining the scenario rupture, the ground motion can be attenuated from the source. Again, several methods may be employed at this point, some of which may be more appropriate than others given the limitations of the available fault rupture model. Numerical simulations of ground motion attenuation may naturally extend from those of the rupture model. These methods are reliant on detailed geological and geotechnical models of the medium through which the waves travel and of the site conditions. For site-specific analysis such detailed models may be feasible to determine. On a larger scale, including that of a city, the accuracy of the geological models may be harder to constrain at the resolution needed for numerical analysis.

For seismic sources defined according to a simple rupture geometry, the most common method of generating a Shakefield is via the use of GMPEs. A typical ground motion model is derived from observed strong motion records, with parameters fitted by regression between the intensity measure of interest and the source and site characteristics:

$$\log\left(Y_{ij}\right) = \overline{\log Y_{ij}}\left(M_i, R_{ij}, \theta_{i1,i2,\ldots,iN}\right) + \tau\eta_i + \sigma\epsilon_{ij} \tag{3.1}$$

where Y_{ij} is the ground motion intensity measure at site j during the earthquake i, and $\overline{\log Y_{ij}}$ the mean of the logarithms of the IM, at distance R_{ij} from earthquake with magnitude M_i. $\theta_{i1,i2,\ldots,iN}$ are additional parameters that may describe characteristics of the site or earthquake source. Uncertainty in the ground motion is separated into an inter-event $\tau\eta_i$ and intra-event $\sigma\epsilon_{ij}$ residual term, where η_i and ϵ_{ij} are normally distributed random variables with means of zero and standard deviations of unity, and τ and σ are the standard deviations of the inter- and intra-event terms.

3.2 Empirical Modelling of Ground Motion Spatial Correlation and Spectral Correlation

The Shakefield process is intended to provide the ground motion input for the fragility models of the vulnerable elements of each infrastructure in a complex interconnected system of infrastructures, such as a city comprising a building stock and multiple lifeline networks. As the systems themselves, and the elements within, are connected, it is critical that spatial correlation is incorporated into the simulation process. The simulation of the spatially correlated fields of ground motion, presented in detail in Sect. 3.3, requires the use of empirically derived models of correlation between intensity measures and also across spatial domains within each intensity measure. These models are typically derived from databases of observed ground motion records, or in the case of spatial correlation models from strong motion records from individual events. In this section existing models of intensity measure intra-event residuals correlation (Sect. 3.2.1) and spatial correlation (Sect. 3.2.2) are summarised, and the spatial correlation models of Esposito and Iervolino (2011, 2012) presented, which are derived from European records.

3.2.1 Correlation Between Intensity Measures Recorded at the Same Site

From the definition of an empirical ground motion intensity model given in Eq. 3.1, correlations between ground motion intensity measures pertain to the residual term

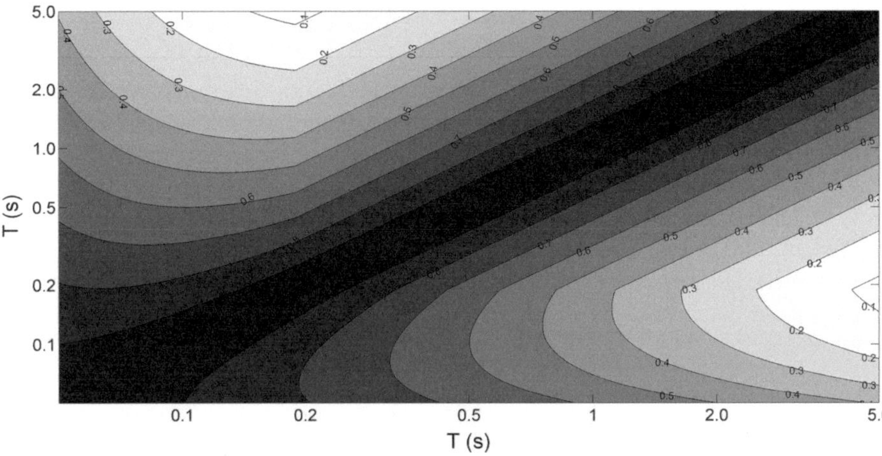

Fig. 3.2 Coefficient of correlation in the GMPE residual between spectral accelerations for two different periods of motion (T_1 and T_2) recorded at the same site, according to the model of Baker and Cornell (2006) given in Eq. 3.2

$\varepsilon_{ij}\sigma_T$, where $\sigma_T = \sqrt{\tau^2 + \sigma^2}$. Several models have been proposed in previous literature for correlation of pseudo-spectral accelerations recorded at the same site (Inoue and Cornell 1990; Baker and Cornell 2006; Baker and Jayaram 2008). The spectral acceleration correlation model of Baker and Cornell (2006) is preferred in the current analysis, shown here and in Fig. 3.2:

$$\rho_{\varepsilon_x,\varepsilon_y} = 1 - \cos\left(\frac{\pi}{2} - \left(0.359 + 0.163 I_{(T_{min}<0.189)} \ln \frac{T_{max}}{T_{min}}\right) \ln \frac{T_{max}}{T_{min}}\right) \tag{3.2}$$

where $\rho_{\varepsilon_x,\varepsilon_y}$ is the coefficient of correlation of the ground motion residual term between spectral acceleration between two periods $S_a(T_1)$ and $S_a(T_2)$ for two different horizontal components of motion recorded at the same site, $T_{max} = \max[T_1, T_2]$ and $T_{min} = \min[T_1, T_2]$ and I takes the value of 1 if $T_{min} < 0.189$ and 0 otherwise. This model is preferred here for application in Europe over the model of Baker and Jayaram (2008), in part due to the selection of the Akkar and Bommer (2010) GMPE, which uses the ordinary geometric mean of the horizontal components of ground motion, as opposed to the models from which the Baker and Jayaram (2008) model is derived, which themselves use the rotationally independent geometric mean.

The IM correlation models do not extend only to spectral accelerations. Baker (2007) presents a similar model to describe the correlation between Arias Intensity (I_a) and spectral acceleration, whilst Bradley (2011a,b) computes the correlation between spectral acceleration and other ground motions intensity measures such as duration, acceleration spectrum intensity and velocity spectrum intensity.

3.2.2 *Spatial Correlation Models in Ground Motions*

Correlation of ground motion intensity measures in a single earthquake is due to the parameters that appear in the GMPEs, such as the magnitude, and spatial correlation of the intra-event residuals ϵ_{ij}. The multivariate Gaussian distribution has been found suitable (Jayaram and Baker 2008) to model a single intensity measure across multiple sites. Many analyses of spatial correlation are limited to a few earthquakes for which there exists a dense network or strong motion records. The scarcity of densely recorded events, particularly in Europe, requires that empirical models be derived from databases of ground motion record, combining data from different earthquakes.

3.2.2.1 Empirical Models of Spatial Correlation from European Strong Motion Records

Spatial correlation models of intra-event residuals available in literature have been mostly estimated using single non-European earthquakes (e.g., Northridge 1994, Japanese earthquakes occurred from 2000 to 2003, Chi-Chi 1999), for which many records were available from dense seismic networks (e.g. Boore et al. 2003; Wang and Takada 2005; Goda and Hong 2008a; Jayaram and Baker 2009). Those models depend on inter-site separation distance (h) and provide the limit at which correlation may technically considered to be lost (i.e., distance beyond which intra-event residuals of IMs at different sites may be considered uncorrelated).

The different decay rates of spatial correlation models available in literature could depend on choices made in the estimation or on the GMPE used; moreover there could be a dependency related to the particular region or earthquake characteristics. The estimation of an appropriate spatial correlation model requires several tasks be undertaken, each with associated possible options: (i) choice of dataset to obtain the intra-event residuals: this means choosing which and how many records to consider (single or multiple earthquakes); (ii) choice of GMPE to obtain residuals e.g. from existing GMPEs or ad hoc attenuation laws (i.e., estimated on the chosen dataset); (iii) working assumptions of the estimation.

For example, Sokolov et al. (2010), starting from the strong-motion database of Taiwan Strong Motion Instrumentation Program (TSMIP) network, estimated correlation for various areas, site classes, and geological structures, asserting that a single generalized spatial model may not be adequate for all of Taiwan territory. Conversely, Goda and Atkinson (2009) also investigated the effects of earthquake types on correlation parameters using datasets from K-NET and KiK-net Japanese strong-motion networks without finding any significant dependency. In some cases (e.g. Wang and Takada 2005; Jayaram and Baker 2009), existing GMPEs are used, while in others (e.g. Goda and Hong 2008a; Goda and Atkinson 2009; Sokolov et al. 2010), an ad hoc fit on the chosen dataset is adopted. Generally, regression analysis used to develop prediction equations does not incorporate the correlation structure of residuals as a hypothesis. Hong et al. (2009) and Jayaram and Baker (2010b) evaluated the influence of considering the correlation in fitting a GMPE, finding a minor influence on regression coefficients and a more significant effect on the variance components. Goda and Atkinson (2010) investigated the influence of the estimation approach, emphasising its importance when residuals are strongly correlated.

The joint distribution of the logs of the IMs at all locations of interest that are assumed to be values of Gaussian random field (GRF), have a covariance matrix that is characterised, in the simplest of these spatial correlation models, by one parameter: the *range*. This parameter represents the inter-site distance at which the spatial correlation is practically lost.

In the European strong motion datasets, no dense observations of single earthquakes are available from which reliable estimates of spatial correlation of intensity measures can be obtained. Therefore, an ensemble of strong motion records from multiple events and regions was used to obtain a unique model (Esposito and Iervolino 2011, 2012) using the European Strong-motion Database (ESD) and the Italian Accelerometric Archive (ITACA). The GMPEs with respect to which residuals were computed were those of Akkar and Bommer (2010) for ESD, and Bindi et al. (2010b, 2011) for ITACA. Subsets of the same records used to estimate the considered GMPEs were used to perform intra-event spatial correlation models.

The analysis of correlation was performed for different IMs (i.e. PGA, PGV and spectral acceleration, $S_a(T)$) via the use of geostatistical tools. In particular, the semivariogram, $\gamma(h)$, was used to model the covariance structure of the different IMs through mathematically tractable functions fitted to empirical observations.

In building a unique semivariogram from multiple events, individual events were first treated separately, before aggregating all the possible distance pairs into a single experimental data set. An empirical trend was fitted to the data assuming an exponential model as expressed here:

$$\gamma(h) = 1 - e^{-3h/b} \tag{3.3}$$

The resulting *range (b)* was found to be 13.5 km (ESD) and 11.5 km (ITACA) for PGA and 21.5 km (ESD) and 14.5 km (ITACA) for PGV (Esposito and Iervolino 2011).

For $S_a(T)$ (5 % damped) residuals the evaluation of the spatial correlation was carried out at seven structural periods, ranging from 0.1 to 2 s, for the Italian dataset, and 0.1–2.85 s (nine periods) for the European dataset. The results are summarised in Table 3.1. Correlation length (b) tends to increase with period as also confirmed by other studies (Jayaram and Baker 2009).

Finally, in order to capture the trend of the range as a function of structural period T, for each dataset a linear-predictive model was fitted as expressed by:

$$b(T) = d_1 + d_2 T \tag{3.4}$$

Table 3.1 Estimated ranges for spectral acceleration residuals using the Akkar and Bommer (2010) and Bindi et al. (2011) GMPEs (Esposito and Iervolino 2012)

GMPE	Period (s)	b (km)	GMPE	Period [s]	b [km]
	0.1	13.7		0.1	11.4
	0.2	11.6		0.2	9.0
	0.3	15.3		0.3	13.2
Akkar and Bommer (2010)	0.5	12.5	Bindi et al. (2011)	0.5	11.9
	1.0	33.9		1.0	17.8
	1.5	27.0		1.5	25.7
	2.0	39.0		2.0	33.7
	2.5	40.5			
	2.85	48.8			

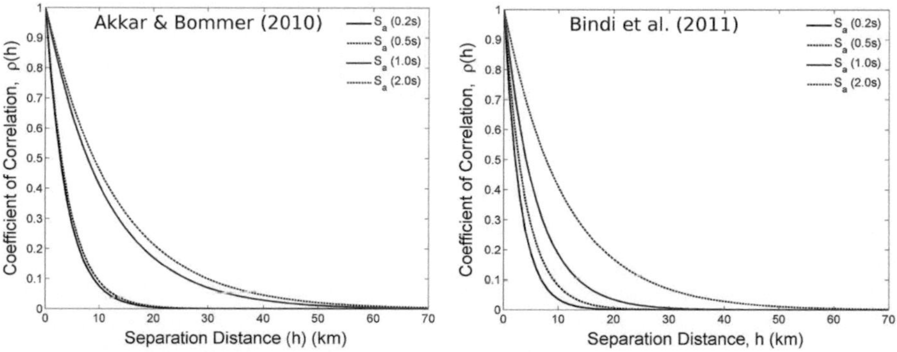

Fig. 3.3 Coefficient of spatial correlation for selected spectral periods derived using the Akkar and Bommer (2010) (*left*) and Bindi et al. (2011) (*right*) GMPEs from the correlation models of Esposito and Iervolino (2012)

where models parameters d_1 and d_2 are found to be 11.7 and 12.7, respectively, for the ESD dataset and 8.6 and 11.6, respectively, for the ITACA dataset. Comparisons of the spatial correlation models for the different periods are seen in Fig. 3.3.

3.3 Simulation of Spatially Correlated and Spatially Cross-Correlated Ground Motion Fields

The stochastic simulation of spatially correlated fields is a challenging task that has many applications in various fields of engineering and geosciences. Different methods are available to implement this approach, of which the simplest is the Matrix Decomposition method (Davis 1987). The problem is approached only in two dimensions (i.e. only considering the distribution of intra-event residuals at the surface) for convenience and efficiency. Whilst there may be merit in describing the simulation of ground motion at depths below the surface, the paucity of strong motion borehole data limits the extent to which the correlation between ground motion at different depths can be characterised. It is assumed that where subterranean lifeline elements such as gas, oil or water pipelines are considered, they are at sufficiently shallow depths as to be adequately modelled via analysis of surface ground motion. In the cases of spatially distributed subterranean systems it may be more appropriate to model vulnerability by numerical simulation of ground motion across a 3D mesh. Full constraint of the geological properties of the environment requires detailed investigation, which may only be considered follow in special circumstances.

There are many different techniques available for generating random fields with given correlation properties. It is to be expected that the manner in which multiple intensity measures are generated across a spatially distribute system will have a non-negligible impact on the resulting analyses of system performance. Two possible methods are presented and appraised here on the basis of their underlying assumptions and the challenges when applying them to real urban infrastructure analyses.

3.3.1 Conditional Simulation

The process of conditional simulation represents, in effect, a stochastic sampling of the joint probability distribution function of two correlated variables. The variables are assumed to be distributed according to a bivariate Gaussian distribution. For two given IMs (IM_1 and IM_2), each assumed to be lognormally distributed with a mean ($\mu_{\log IM_1}$ and $\mu_{\log IM_2}$) and variance ($\sigma_{\log IM_1}$ and $\sigma_{\log IM_2}$) and correlation coefficient $\rho_{\log IM_1, \log IM_2}$, the conditional distribution of IM_2 upon the value (z) generated for IM_1, given a magnitude M and source to site distance R, is described by Iervolino et al. (2010):

$$\mu_{\log IM_2 | \log IM_1, M, R} = \mu_{\log IM_2 | M, R} + \rho_{\log IM_1, \log IM_2} \sigma_{\log IM_2} \frac{z - \mu_{\log IM_1, M, R}}{\sigma_{T \log IM_1}} \tag{3.5}$$

$$\sigma_{T \log IM_2 | \log IM_1} = \sigma_{T \log IM_2} \sqrt{1 - \rho^2_{\log IM_1, \log IM_2}} \tag{3.6}$$

Note that these equations refers to the case in which the GMPE does not distinguish between inter-event and intra-event residuals, yet the method is readily applicable also in such a case.

The simulation of IM_1 and IM_2 is implemented via the following procedure:

Algorithm 3.1 Description of co-simulation via the sequential simulation process, conditional upon on a primary intensity measure

Data: Primary IM selection, Spatial correlation model, IM-correlation model, event source, magnitude, GMPE
Result: Ground Motion Fields for each IM
for $i \leftarrow 1$ **to** $N_{Earthquakes}$ **do**
  Generate Source
  Calculate Median Ground Motion at Sites for Primary IM $\mu_{\log IM_1|M,R}$
  Generate Spatially Correlated Field of Ground Motion Residuals
  Multiply residuals by standard deviation of Primary IM σ_{IM_1}
  for $IM_j \in SecondaryIMs$ **do**
    Calculate Median Ground Motion Field at Sites $\mu_{IM_j|M,R}$
    Calculate the conditional mean ($\mu_{\log IM_j|\log IM_1,M,R}$) and standard deviation ($\sigma_{\log IM_j|\log IM_1}$ of IM_j) using (3.5) and (3.6)
    Sample IM_2 from a Gaussian distribution with a mean and standard deviation of $\mu_{\log IM_j|\log IM_1,M,R}$ and $\sigma_{\log IM_j|\log IM_1}$ respectively.
  end
end

This approach has several practical advantages that make it suitable for application here. The LU decomposition need only be performed once for the primary IM at the initiation of the simulation, and only one spatially correlated field generated per event. For application to multi-system infrastructures, this is particularly advantageous as it permits for the generation of secondary IMs at sites only where required by the fragility model of the system element at a particular site.

The second advantage is that the sequential approach avoids making explicit assumptions about the nature of the cross-correlation structure between two IM fields. Instead the spatial simulation can be conditioned upon the IM for which the spatial correlation model is best constrained.

The formulation of the conditional approach requires the assumption that the ground motion field of a secondary IM is conditional only upon the primary IM at the same site, and conditionally independent of all else. If this is the case, the spatial correlation structure of the secondary IM is respected in this approach. In practice, however, this may be a strong assumption, which should be verified before applying the conditional simulation approach.

3.3.2 Extended Matrix Decomposition

This approach is an extension to two fields of the LU decomposition presented previously for simulation of a single field. A Gaussian random field can be defined as:

$$\mathbf{Y_1} = \boldsymbol{\mu}_{\mathbf{log}}\mathbf{IM_1} + \mathbf{L_1 Z_1} \tag{3.7}$$

where $\boldsymbol{\mu}_{\mathbf{log}}\mathbf{IM_1}$ is the mean of the field (assumed here to be zero-vector for the ground motion residuals), $\mathbf{Z_1}$ is a vector of independently Gaussian distributed random variables and $\mathbf{L_1}$ is the root of the positive-definite covariance matrix ($\mathbf{C_{11}}$) such that $\mathbf{L_1 L_1^T} = \mathbf{C_{11}}$ (e.g. Park et al. 2007; Crowley et al. 2008a).

Consider then a second field ($\mathbf{Y_2}$), with mean $\boldsymbol{\mu}_{\mathbf{log}}\mathbf{IM_2}$ and covariance matrix $\mathbf{C_{22}}$, for which the correlation between a common two elements in $\mathbf{Y_1}$ and $\mathbf{Y_2}$ separated by a distance (h) of 0 km is given by $\rho_{\log IM_1, \log IM_2}$ ($\rho_{1,2}$ hereafter) the following model is generated (Oliver 2003):

$$\begin{bmatrix} \mathbf{Y_1} \\ \mathbf{Y_2} \end{bmatrix} = \begin{bmatrix} \boldsymbol{\mu}_{\mathbf{log}}\mathbf{IM_1} \\ \boldsymbol{\mu}_{\mathbf{log}}\mathbf{IM_2} \end{bmatrix} + \begin{bmatrix} \mathbf{L_1} & \mathbf{0} \\ \rho_{1,2}\mathbf{L_2} & \sqrt{1-\rho_{1,2}^2}\mathbf{L_2} \end{bmatrix} \begin{bmatrix} \mathbf{Z_1} \\ \mathbf{Z_2} \end{bmatrix} \tag{3.8}$$

where Z_2 is a vector of independently Gaussian distributed random variables and $\rho_{1,2}$ the cross-covariance between fields Y_1 and Y_2. The total cross-covariance matrix $\mathbf{L}$ is therefore described by:

$$\mathbf{LL^T} = \begin{bmatrix} \mathbf{L_1 L_1^T} & \rho_{1,2}\mathbf{L_1 L_2^T} \\ \rho_{1,2}\mathbf{L_2 L_1^T} & \mathbf{L_2 L_2^T} \end{bmatrix} \tag{3.9}$$

This formulation can be expanded further to consider multiple IMs, with the full cross-covariance matrix given by:

$$\mathbf{LL^T} = \begin{bmatrix} \mathbf{L_1 L_1^T} & \rho_{1,2}\mathbf{L_1 L_2^T} & \cdots & \rho_{1,j}\mathbf{L_1 L_j^T} \\ \rho_{2,1}\mathbf{L_2 L_1^T} & \mathbf{L_2 L_2^T} & \cdots & \rho_{2,j}\mathbf{L_2}, \mathbf{L_j^T} \\ \vdots & \vdots & \ddots & \vdots \\ \rho_{i,1}\mathbf{L_i L_1^T} & \rho_{i,2}\mathbf{L_i L_2^T} & \cdots & \mathbf{L_i L_i^T} \end{bmatrix} \tag{3.10}$$

which within the simulation becomes:

$$\begin{bmatrix} \mathbf{Y_1} \\ \mathbf{Y_2} \\ \vdots \\ \mathbf{Y_i} \end{bmatrix} = \begin{bmatrix} \boldsymbol{\mu}_{\mathbf{log}}\mathbf{IM_1} \\ \boldsymbol{\mu}_{\mathbf{log}}\mathbf{IM_2} \\ \vdots \\ \boldsymbol{\mu}_{\mathbf{log}}\mathbf{IM_i} \end{bmatrix} + \mathbf{L} \begin{bmatrix} \mathbf{Z_1} \\ \mathbf{Z_2} \\ \vdots \\ \mathbf{Z_i} \end{bmatrix} \tag{3.11}$$

where $\mathbf{L}$ is determined from Cholesky factorization of the full block cross-correlation matrix $\mathbf{LL^T}$. Assuming the covariance structure of each of the fields ($\mathbf{C_{ii}}$) is real-valued and positive-definite, $\mathbf{L}$ will be real. If the correlation between the strong motion residuals is assumed to be isotropic, an assumption largely supported by the analysis of Jayaram and Baker (2009), then C_{ii} will be positive-definite. An example of three spatially cross-correlated, unconditional random fields of strong ground motion residuals is shown in Fig. 3.4.

In this method there is no requirement to assign a primary IM or secondary IM. As a means of defining the spatial cross-correlation for multiple fields, this method

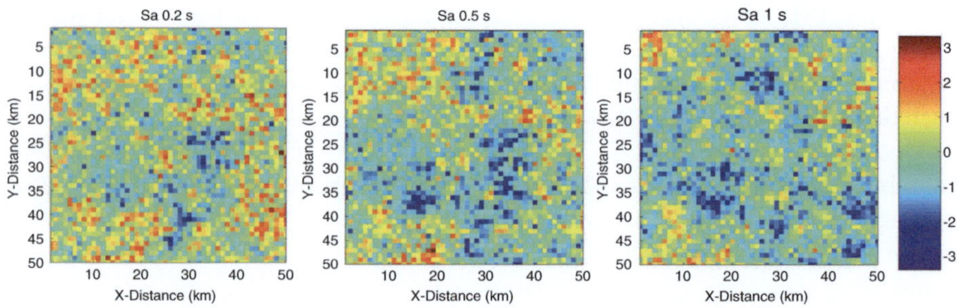

Fig. 3.4 Unconditional random co-simulation of strong ground motion residuals using the extended LU decomposition method, for spectral accelerations at 0.2, 0.5 and 1.0 s respectively

shares many advantages with that of the linear model of co-regionalisation (LMCR) approach adopted by Loth and Baker (2013). Both are able to define the full cross-covariance matrix, and can be implemented similarly. The primary difference is that the latter is derived directly from the empirical cross-correlograms, whereas the former can utilise separate models for the spatial correlation and cross-correlation.

In implementation of the co-simulation methodology there are certain costs associated with this approach. Mostly these relate to the computational inefficiencies of the methodology. The full cross-correlation matrix must be constructed and factorised, which, if considering a large number of locations and/or IMs, requires considerable computation and memory allocation.

The choice of whether to adopt a conditional simulation or a full cross-correlation methodology will inevitably depend on the circumstances of the application. Primarily, whether the assumptions required by the conditional approach can be assumed. The impact and limitations of each methodology upon seismic risk analysis for multi-system infrastructures is an area that requires considerably more analysis and experimentation.

3.4 Stochastic Modelling of Geotechnical Hazards: Site Amplification and Permanent Ground Deformation (PGD)

Fragility models for many linear lifeline systems (e.g. roads, pipelines, rail tracks, underground electricity lines, oil, gas, water and waste water) will often require the input of the permanent ground deformation (PGD), in addition to the transient shaking. It is therefore imperative that when simulating the Shakefields the influence of the local geology is taken into consideration.

The characterisation of surficial geology and its influence on both the transient and permanent ground motion across a region of the scale of a large metropolitan area brings with it significant practical limitations. The selection of models relating the surface ground displacement to the seismic hazard input must take into consideration the fact that, for much of a region, many properties of the geology are either not known, or must be conditioned upon a set of observations within the region. Therefore the reduction in variability in the model of site amplification and/or permanent displacement that accompanies a well-constrained geotechnical profile is offset by the paucity of observation points with respect to the number of sites considered in the analysis. It is within this context that a parsimony-driven approach is favoured here. A strong emphasis is placed on selecting practical methodologies for implementing site amplification and geotechnical hazard analysis across an urban region, and empirical models requiring complex characterisation of the soil profile may not necessarily be appropriate.

The need for practical implementation of a geotechnical hazard analysis across a wide spatial scale is, to a certain extent, addressed within the HAZUS (NIBS 2004) methodology. It has therefore been considered appropriate to retain specific elements of this process, such as the landsliding and liquefaction susceptibility classifications, and, in certain cases, the permanent ground displacement models where alternative formulations were not available. There is an important distinction, however, between the HAZUS methodology and the implementation here, and that is in the adoption of the stochastic framework for seismic hazard inputs. This allows us to utilise both the probability definitions (e.g. probability of slope failure given an acceleration greater than the yield acceleration of a site) and the variability of the empirical displacement models, directly within the simulation procedure. Therefore with each stochastically generated Shakefield an additional field of permanent ground displacements can be produced.

3.4.1 Site Amplification

The site amplification factor $AF(T)$, which describes the ratio of the spectral acceleration of ground motion at the ground surface $S_a(T)_{SURFACE}$ with to that of the ground motion in the engineering bedrock $S_a(T)_{ROCK}$ is defined according to (e.g. Bazzurro and Cornell 2004b):

$$AF(T) = \frac{S_a(T)_{SURFACE}}{S_a(T)_{ROCK}} \tag{3.12}$$

$$\ln S_a(T)_{SURFACE} = \ln S_a(T)_{ROCK} + \overline{\ln AF(T)} + \sigma_{T \ln AF(T)} \tag{3.13}$$

where $\overline{\ln AF(T)}$ represents the median amplification factor and $\sigma_{T \ln AF(T)}$ its corresponding total uncertainty. For practical application the bedrock is assumed to correspond to Eurocode 8 (EC8) Type A soil. Several different estimators of $AF(T)$ may be implemented. The selection of the appropriate method is dependent on the case study, but for the present purpose it is assumed that the input microzonation should provide sufficient information to define a Eurocode 8 site classification for each site at which the ground motion is required.

3.4.1.1 Direct Incorporation of GMPE Site Amplification Factors

Arguably the simplest approach is to utilise the site amplifcation models directly within the GMPE itself, simply by inputting the site class and/or V_{s30} value for each site in the ground motion field. For Akkar and Bommer (2010) or Bindi et al. (2010a) this requires the site class is input according to the NEHRP classifications (for NEHRP classes B, C and D only), whilst for other GMPEs it may be necessary to specify the V_{s30} directly (or use an assumed value on the basis of the site class). This approach is also the most consistent with the ground motion variability definition, as the uncertainty in the ground motion is defined given that the site parameter is known. However, the range of soil conditions for which the selected GMPEs are valid may be limited compared to the soil conditions observed in the study region. This is likely to be the case for regions with particularly soft or liquefiable soils. Few GMPEs can be applied to sites with V_{s30} less than 150 m s^{-1}, which would limit the validity of such GMPEs in harbour areas or for very soft alluvial beds.

An alternative approach would be to use an empirical amplification model such as that proposed by Choi and Stewart (2005):

$$\ln\left(AF_{ij}\right) = c\ln\left(\frac{V_{s30_{i,j}}}{V_{ref}}\right) + b\left(\frac{PHA_{r_{ij}}}{0.1}\right) + \tau_{\ln AF}\eta_i + \sigma_{\ln AF}\epsilon_{ij} \tag{3.14}$$

where $PHA_{r_{ij}}$ is the peak horizontal acceleration on rock at site j from earthquake i, and c, b and V_{ref} are constants, as defined in Choi and Stewart (2005). This model, derived from 1828 observed strong ground motions originating from 154 active shallow crustal earthquakes, describes the amplification as a continuum taking into account soil nonlinearity. The variability in the amplification function is separated into an inter- and intra-event term, $\tau_{\ln AF}\eta_i$ and $\sigma_{\ln AF}\epsilon_{ij}$ respectively. As with the GMPEs it may be limited to a narrow range of site conditions than is required for the case study in question.

3.4.1.2 Fixed Amplification Factors from Seismic Design Codes

In certain applications amplification of PGA, spectral acceleration or spectral displacements can be estimated using factors specified in appropriate design codes. For application within Europe, the amplification factors specified within Eurocode 8, and in more recent revisions (Pitilakis et al. 2012, 2013), are supported by the current methodology, as too are the amplification factors specified in the 1997 NEHRP Provisions (FEMA-450 2003), which form the basis for the amplification factors used by HAZUS.

As with the GMPE-based approach, this method has several advantages in terms of simplicity, and relatively minimal requirements in terms of the site classification. It is simple to implement across a region, with the advantage that the hazard itself must be specified only for the reference rock. This means there is little additional computational cost if it is necessary to interpolate between sites. Furthermore, the issue of linear or nonlinear amplification need not be addressed directly, as nonlinearity may be implicit in the code amplification factors.

By implementing code-based amplification factors, this approach is limited to site categories for which the code supplies such factors, and by the conservatism often implied by codes. An additional limitation is that design codes are unlikely to provide amplification factors for IMs other than PGA or spectral acceleration. Parameters such as Arias Intensity and Cumulative Absolute Velocity (CSV) have little use for structural design, even if they may be demonstrated to be relatively efficient predictors of damage or loss. This limits the extension of the code-based site amplification approach to less common IMs.

3.4.1.3 "Application-Specific" Amplification Factors

It may be the case that extensive microzonation studies have been undertaken for the region of application. The ability to integrate detailed microzonation information into the hazard analysis is an important means of reconciling the more general seismic hazard approach with the site-specific engineering approach to earthquake loss studies. For this purpose, it may be prudent to define amplifications factors, of the form described in Eq. 3.13, which are not necessarily generic or characteristic for a particular site class, but are specific to the sites for the application in question. They should be based on a higher level of geotechnical information than for the design code or GMPE approaches described previously, and should begin to incorporate information relevant to the structure of the subsurface and the dynamic properties of the material.

For each micro-zone within a given region, a characteristic site profile or set of profiles may be determined, possibly from borehole profiles or from non-invasive geophysical investigation. From this profile, and using a 1D numerical amplification tool, it is possible to estimate an amplification factor (AF_{ij}) particular to the micro-zone according to the formulation of Bazzurro and Cornell (2004a,b). Variability in the amplification factor may be attributed to record-to-record variation, and also to uncertainty in the site profile (e.g., Robinson et al. 2006). Bazzurro and Cornell (2004b) illustrate how this formulation can be combined with the hazard integral to allow for calculation of $S_a\,(T)_{SURFACE}$ for a given site.

This "application-specific" approach has several benefits over the more generic approaches to site amplification considered previously. The main benefit is that a greater quantity of geotechnical information is integrated into the analysis, which may better represent an engineering based approach. Whilst a greater computational effort is required to develop the amplification factors, they remain characteristic for the region in question. Implementation of the site amplification within the stochastic earthquake hazard simulation is a relatively simple and computationally efficient procedure.

3.4.2 Liquefaction

The estimation of permanent ground displacement due to liquefaction is adapted largely from the methodology introduced in HAZUS (NIBS 2004). Within this methodology, estimation of the probability of liquefaction is based on the analysis of Youd and Perkins (1978), who introduced classes of liquefaction susceptibility. These classes (Very High, High, Moderate, Low, Very Low and None) are categorised on the basis of deposit type, age and general distribution of cohesionless sediments. Each liquefaction susceptibility category has an associated conditional probability of liquefaction for a given PGA, and a proportion of map unit susceptible to liquefaction. This latter is intended to take into account the variability of soil properties within any given sedimentary class, which may act to inhibit liquefaction at a site.

This formulation may be the simplest approach, requiring only the susceptibility class (which can be inferred from surficial geology), PGA (the peak horizontal ground acceleration, in units of g, at the ground surface), magnitude and depth to the groundwater. Derivation of the probabilities of liquefaction given an input PGA, and the values for the proportions of susceptible map unit for each susceptibility category, can be found in NIBS (2004), as well as the categorisation scheme for liquefaction susceptibility.

The primary advantage of the current methodology is the simplicity and the dependence on few site-specific factors. It may be considered only a first order estimate on the amount of displacement associated with lateral spreading at a site. It may be considered, however, only a first-order estimate on the amount of displacement associated with lateral spreading at a site. Alternative models describing both the probability of liquefaction and the resulting permanent ground deformation could be supported in future.

Despite being a commonly observed liquefaction phenomenon, there are fewer established models that are used in assessment of ground settlement. It has been suggested by Tokimatsu and Seed (1987) that the extent of settlement shows little dependence on the strength of ground motion. This makes characterisation in a hazard framework substantially more challenging. In the present methodology a characteristic settlement is attributed to each susceptibility class, thus making the expected settlement a product of the liquefaction probability and the characteristic settlement.

3.4.3 Slope Displacement

As with the estimators of liquefaction induced displacement, the methodology adopted for estimation of slope displacement is also adopts the landsliding susceptibility classification defined within HAZUS (NIBS 2004). The landsliding susceptibility class is assigned on the basis of geology, slope angle and the position of groundwater with respect to the level of sliding (essentially a wet/dry distinction). For each susceptibility class, a critical acceleration is defined; hence, if PGA exceeds the critical acceleration then displacement is possible. As for the case of liquefaction, each susceptibility category is associated with a percentage of map area having a landslide-susceptible deposit. In the SYNER-G implementation, these factors are implemented as probabilities of observing landsliding at a site, given PGA at the site exceeds the critical acceleration defined for site.

The simplicity of the characterisation of landslide susceptibility is a key advantage of the current approach. The crucial parameter needed for estimation of slope displacement is critical acceleration, or "yield coefficient" (k_y), which corresponds to the threshold acceleration above which slope displacement is initiated. This may be estimated via many different ways, taking into consideration the properties of the slope. The yield coefficient may be specified by the user a priori, or assigned according to the defined susceptibility classes (NIBS 2004).

As with the case of liquefaction, alternative empirical models exist that may better constrain the uncertainty in the slope displacement, whilst retaining practicality for use with common intensity measures. For this purpose the scalar empirical slope displacement model of Saygill and Rathje (2008) is adopted to define the slope displacement as a function of PGA:

$$\begin{aligned}\ln(PGD) = {} & 5.52 - 4.43\left(\frac{k_y}{PGA}\right) - 20.39\left(\frac{k_y}{PGA}\right)^2 + 42.61\left(\frac{k_y}{PGA}\right)^3 \\ & - 28.74\left(\frac{k_y}{PGA}\right)^4 + 0.72\ln(PGA) + \sigma_{\ln PGD} \end{aligned} \tag{3.15}$$

where PGD is the sliding displacement in cm, and the total variability described by a Gaussian distribution with zero mean and standard deviation $\sigma_{\ln D}$ of 1.13. Alternative empirical models of slope displacement could be implemented within the same framework, if preferred.

3.4.4 Transient Strain

Transient strain ϵ_S is rarely treated explicitly in analyses of seismic risk to infrastructural systems, as fragility curves for strain-sensitive elements are usually given as functions of peak ground velocity (PGV). For application within the Shakefield approach, it is more convenient to characterise peak transient ground strain via an empirical model relating the strain to other parameters of ground motion. Using observed ground motions from selected Californian, New Zealand and Japan earthquakes, simple empirical models relating maximum in-plane horizontal strain to PGV (m s^{-1}) or PGA (m s^{-2}) have been determined by Paolucci and Smerzini (2008):

$$\log_{10}\left(\max\left(\epsilon_S\right)\right) = 0.95\log_{10}\left(PGV\right) - 3.07 + \sigma_{\epsilon_s} \tag{3.16}$$

$$\log_{10}\left(\max\left(\epsilon_S\right)\right) = 6.0 \times 10^{-5} PGA + \sigma_{\epsilon_s} \tag{3.17}$$

In both cases the standard deviation (σ_{ϵ_s}) is approximately 0.16.

3.4.5 Coseismic Rupture

Since the development of the original HAZUS methodology several studies have been undertaken that outline a Probabilistic Fault Displacement Hazard Assessment (PFDHA) methodology (Youngs et al. 2003; Petersen et al. 2011; Chen and Petersen 2011; Moss and Ross 2011). The methodology describes the hazard integral as follows Youngs et al. (2003):

$$\begin{aligned} v_k\left(d\right) = &\sum_n \alpha_n\left(m^0\right)\int_{m_0}^{m_u} f_n\left(m\right) \\ &\times \left[\int_{r_{min}}^{r_{max}} f_{kn}\left(r|m\right) P_{kn}\left(slip|m,r\right) P_{kn}\left(D > d|m,r,slip\right) dr\right] dm \end{aligned} \tag{3.18}$$

where $\alpha_n\left(m^0\right)$ is the rate of events in source n with $m \geq m_0$, $f_n\left(m\right)$ the probability density function of the earthquake magnitude (m) for the source n, $f_{kn}\left(r|m\right)$ the conditional probability density function for distance, r, from site k to an earthquake of magnitude m occurring on source n; $P_{kn}\left(slip|m,r\right)$ the probability of observing displacement at a site, given the magnitude and distance, and $P_{kn}\left(D > d|m,r,slip\right)$ the probability of exceeding a displacement level d, given that slip is observed at the site. The formulation is adapted slightly for distributed slip on smaller structures away from the principal fault, which need not necessarily be conditional upon the occurrence of slip at the surface from the principal fault but assumes slip at depth. The objective of the rupture object designed here is to attempt to adapt aspects of this formulation, using a stochastic approach to model the conditional probabilities outlined previously.

To implement this approach within a stochastic framework, the process probabilities are broken down into separate steps. The manner in which a stochastically generated fault rupture is distributed across a surface (described in more detail in Sect. 3.5) is already using the seismogenic sources to produce a set of rupture realisations sufficient to sample $P_{kn}(slip|m,r)$, the probability of observing slip on the principal fault. At the nodes of the mesh where slip is given to occur, the expected slip is sampled as a ratio of the average displacement on a fault (itself sampled from the magnitude scaling relation of Wells and Coppersmith (1994)), described by a gamma distribution with the cumulative distribution function $F(D/AD)$. The cumulative distribution function is then sampled using the empirical principal slip models from Youngs et al. (2003), Petersen et al. (2011), and Moss and Ross (2011), for normal faulting, strike-slip faulting or reverse faulting respectively.

Youngs et al. (2003) For normal faulting, the expected slip at a point on the fault is sampled from the CDF of the gamma distribution:

$$F(D/AD) = \frac{1}{\Gamma(a)} \int_0^{(D/AD)/b} e^{-t} t^{a-1} dt \tag{3.19}$$

where D/AD is the displacement as a proportion of average displacement, $\Gamma(.)$ is the gamma function, $a = \exp(-0.193 + 1.628\, l/L)$ and $b = \exp(0.009 - 0.257\, l/L)$, where l/L is the ratio of the length from site to the nearest end of the fault, with respect to the total length of the fault. This value is therefore constrained within the range $0 \leq l/L \leq 0.5$

Petersen et al. (2011) Three different functional forms are presented to describe the scaling of D/AD with respect to l/L for strike slip faulting

- Bilinear:

$$\log(D/AD) = \begin{cases} 8.2525\,(l/L) - 2.3010 \pm 1.2962\epsilon_{\log(D/AD)} & \text{for} l/L < 0.3008 \\ 0.1816 \pm 1.0013\sigma_{\log(D/AD)} & \text{for} l/L \geq 0.3008 \end{cases} \tag{3.20}$$

- Quadratic:

$$\log(D/AD) = 14.2824\,(l/L) - 19.8833\,(l/L)^2 - 2.6279 \pm 1.1419\sigma_{\log(D/AD)} \tag{3.21}$$

- Elliptical:

$$\log(D/AD) = 3.2699\sqrt{1 - \frac{1}{0.5^2}\left[(l/L) - 0.5\right)^2} - 3.2749 \pm 1.1419\sigma_{\log(D/AD)} \tag{3.22}$$

The choice of models may be a decision for the user, although the bilinear model is shown to have the lowest aleatory variability.

Moss and Ross (2011) For reverse faults, the displacement as a ratio of average displacement is again assumed to be gamma distributed assuming the same functional form as (3.19), with the shape parameter (a) and scale parameter (b) described by

$$a = \exp\left(-30.4\,(l/L)^3 + 19.9\,(l/L)^2 - 2.29\,(l/L) + 0.574\right) \tag{3.23}$$

$$b = \exp\left(50.3\,(l/L)^3 - 34.6\,(l/L)^2 + 6.6\,(l/L) - 1.05\right) \tag{3.24}$$

Other probability density functions are presented by Moss and Ross (2011); however, the gamma distribution is chosen here for consistency with the Youngs et al. (2003) model for normal faulting.

In addition to the empirical models of slip on the principal rupture, both Youngs et al. (2003) and Petersen et al. (2011) describe empirical models displacement for sites distributed around the rupture. A similar process is followed to that of the on-fault rupture, albeit the empirical models for $P_{kn}\,(slip|m,r)$ and $P_{kn}\,(D > d\,|m,r,slip)$ are sampled directly from the corresponding models for normal faults and strike-slip faults found in Youngs et al. (2003) (Eq. 7 and Appendix A of the original paper) and Petersen et al. (2011) (Eqs. 19 and 20 of the original paper) respectively.

This stochastic implementation of a probabilistic fault displacement hazard process contains many critical developments when compared to previous approaches. The most significant is the inclusion of aleatory uncertainty into the process, something that has been difficult to constrain until this point. The inclusion of distributed (off-fault) rupture is also important in validation studies, where many displacements may be observed without necessarily requiring that the primary rupture propagate to the surface (a circumstance that is more common in the extensional environments of central Italy and Greece). Furthermore, by utilising the same "simplified" rupture propagation process for the purpose of defining both the finite rupture surface for the ground motion attenuation and for the slip displacement, this methodology introduces a consistency in the transient shaking and permanent displacement generation process that is not possible via traditional means of probabilistic shaking hazard and fault displacement hazard analysis alone.

There remain limitations to this approach, however, and these mostly arise from the relative paucity of observational data, particularly in extensional faulting environments. It can be seen in the analysis of Youngs et al. (2003) that empirical models of slip probabilities and displacements are derived from a relatively small database of normal faults, to which European extensional ruptures contribute little.

3.5 Implementation of the Probabilistic Seismic Hazard Methodology

3.5.1 *The Stochastic Methodology*

The Shakefield procedure outlined in Sect. 3.1 represents a common methodology that can be applied to generate ground motion fields from potential scenario events. The use of Monte Carlo based random simulations of seismicity is also extensible to probabilistic seismic hazard analysis (e.g. Musson 2000; Weatherill and Burton 2010). For the purposes of the SYNER-G project, a general methodology was developed that allows for either to be considered. The following section elaborates upon the general hazard methodology that is used to generate seismic hazard input for multi-system analysis of seismic risk, implemented within the Object-Oriented Framework for Infrastructure Modelling and Simulation (OOFIMS) software described in this book.

One of the most important considerations in the development of a general methodology is the adaptability to different degrees of data information and quality, depending on the case study in question. A multi-tiered approach, particularly with respect to the characterisation of the site and the geotechnical hazard, is necessary in order to allow for application to larger regions. The characterisation of the seismic source and the ground motion attenuation is consistent with that required for regional scale probabilistic seismic hazard analysis. When taking into consideration spatial correlation and cross-correlation between different intensity measures, further adaptability is also necessary.

The general process for a stochastic implementation of a seismic hazard and geotechnical hazard calculator is outlined in Fig. 3.5 (for the general methodology) and Fig. 3.6 for the geotechnical extension. For each iteration the occurrence and corresponding magnitude of the event in each source is sampled from the magnitude frequency distribution of the source. In most of the sources considered in the European case studies this may correspond to the classic double-truncated exponential distribution, but other forms of the distribution can be equally well supported, and of course the same process can be applied to scenario earthquake ruptures.

3.5.1.1 Generation of the Earthquake Rupture

The use of seismogenic fault sources presents a challenge in understanding how to stochastically distribute ruptures across the fault surface in a manner that is consistent with both the properties of the rupture and of the magnitude frequency distribution. For this purpose an iterative procedure is invoked, in which a rupture is initiated from a randomly sampled point in the mesh of the surface (see Fig. 3.7). An automata-style approach is then used in which the rupture expands to the

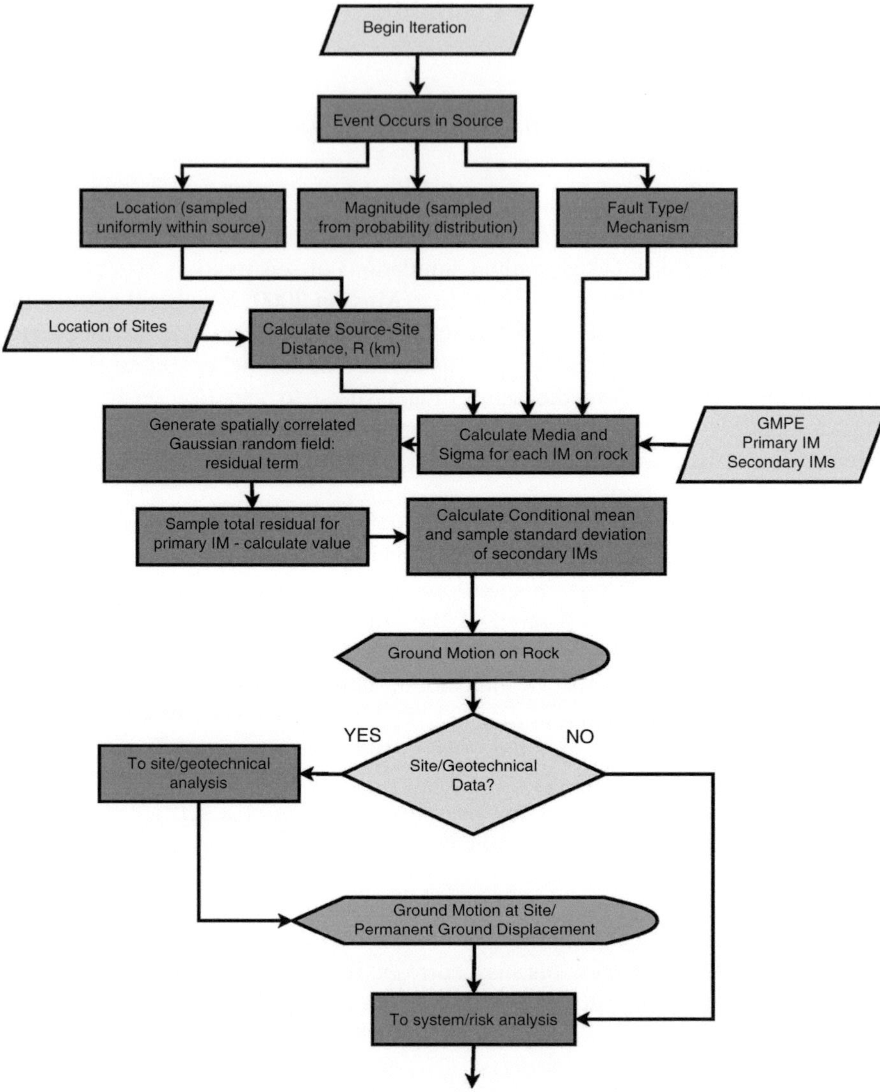

Fig. 3.5 Overview of stochastic seismic hazard simulation procedure. Input (*trapezium*), processes (*rectangle*), decisions (*triangle*) and outputs (*display*)

neighbouring points of an "active" cell on each subsequent iteration, until the total area of the rupture matches the expected area of the rupture for that magnitude. The expected area is sampled from a magnitude scaling relation, such as Wells and Coppersmith (1994) in this case. Rupture expansion is isotropic until it reaches the confining boundaries of the fault surface, at which point it will continue to propagate in the unconfined direction. In principal, certain behaviours of the kinematic rupture

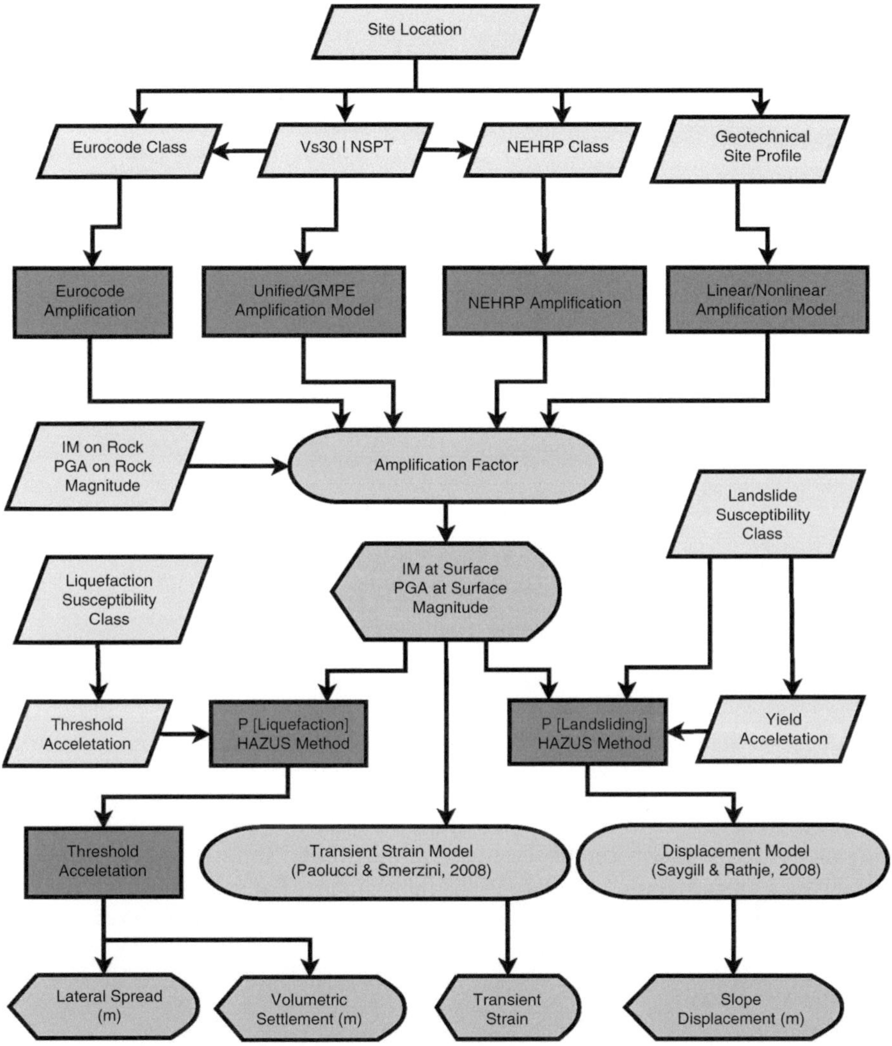

Fig. 3.6 Overview of the process to define the site amplification and geotechnical hazard. Input (*trapezium*), processes (*rectangle*), models (*spheroid*) and outputs (*display*)

process can be mimicked (e.g. unilateral propagation, temporary asperities, down-dip or along-strike dependent hypocentre probability distributions etc.) without necessarily incurring a significant computational cost.

The rupture generated on each simulation can be utilised in two sections of the calculation. The first is for defining the finite surface within the seismogenic fault, from which the finite-distance metrics of the ground motion prediction equation can be determined. The second case is in sampling, in a physically consistent manner,

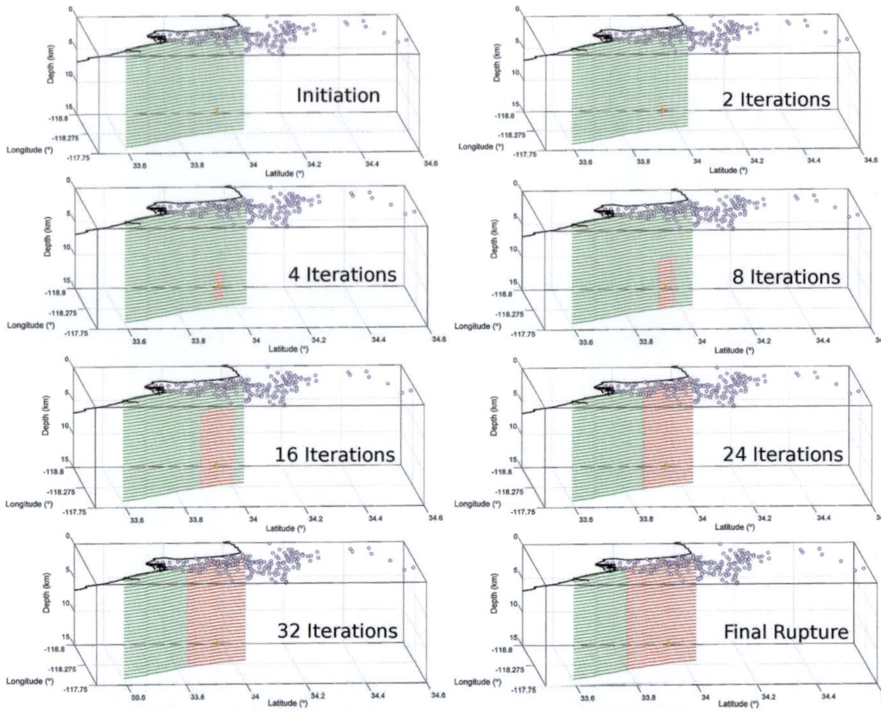

Fig. 3.7 Simple simulation process for the stochastic generation of ruptures (*red*) on a fault surface (*green*). Vertical scale exaggerated

the probability of observing displacement at the ground surface due to on-rupture slip, given a magnitude and distance to the fault rupture. For analysis of seismic risk to lifeline structures such as pipes, rails and tunnels, which require PGD as an input to the fragility functions, this information is relevant.

3.5.2 Issues Emerging in Practice

The general framework presented here for generating stochastic fields of ground motion and permanent ground displacement may still represent an idealised process, only fully applicable in areas where a considerable quantity of input data can be constrained. Arguably the most significant limiting factor in practice, particularly with respect to the geotechnical calculators, is the inability to describe the site conditions and/or source geometries with sufficient accuracy to apply the empirical models of ground motion or displacement across a regional scale, in a manner that is consistent with their derivation. It is therefore inevitable that simplifications will emerge.

Rather than exhaustively list all the possible areas in which uncertainty may be overestimated, underestimated or not even characterised, we shall focus on a few critical areas:

3.5.2.1 Earthquake Source Geometry

Whilst many of the major active fault systems in Europe are well known to geologists, the precise extent of the fault geometries are rarely well constrained. In many parts of the continent, regions of active seismicity still cannot be represented by the fault source typology described previously. Instead, distributed seismicity sources, such as uniform zones of seismicity, are used. In these regions, many state-of-the-art hazard calculators will generate synthetic finite pseudo-ruptures, with properties that are consistent with the seismotectonics of the region in question. Where such sources are invoked, it is certainly the case that rupture displacement calculators, such as the one described here, cannot be utilised.

3.5.2.2 Ground Motion Uncertainty and Correlation

The requirement to consider both the spatial correlation in ground motion and the cross-correlation between different IMs, is both a novel development and a critical requirement of multi-system analyses such as those undertaken within this project. The conditional simulation methodology is an efficient process for generating multiple ground motion fields potentially preserving some of the cross-correlation between IMs, depending on the applicability of the hypotheses of the method. This is not merely a trivial benefit, as the systemic vulnerability algorithms to which these fields are fed do, themselves require considerable computational resources. Both the choice of ground motion model and correlation model will influence the resulting loss analyses. The selection of GMPEs is a common process in seismic hazard analysis, and it could be argued that the consistency of correlation models between European records (Esposito and Iervolino 2011, 2012) and other models in active shallow zones (Jayaram and Baker 2009) would suggest that there is not a strong regionalisation of correlation models within active shallow tectonic regimes.

For the selection of GMPEs for use in the Shakefield process, it is preferable to consider those GMPEs for which PGA, PGV and $S_a\,(T)$ are defined, and form the reference ground motion model for the derivation of the spatial correlation models from European records (Sect. 3.2). For the applications within the SYNER-G project Akkar and Bommer (2010) and Bindi et al. (2010a) GMPEs are preferred.

3.5.2.3 Probabilities of Landsliding/Liquefaction

As described previously, the models for characterising the respective probabilities of liquefaction and landsliding were adopted from those of HAZUS (NIBS 2004). Without further investigation, and validation studies such as the ones undertaken for L'Aquila, Thessaloniki and Vienna within the SYNER-G project, it remains to be seen to what extent the current methodology for defining liquefaction and landsliding probabilities can be adopted in European terrains, without the need for modification.

3.5.2.4 Geotechnical Site Characterisation Across an Urban Region

One of the principal factors limiting the application of site specific analysis methodologies to urban and regional scales is the paucity of geotechnical data over wider spatial scales, and the challenge of microzoning geotechnical properties given a limited set of samples. This is, of course, dependent on the case study in question. Nevertheless, it can be seen, even in detailed case studies, that in order to include geotechnical effects on an urban scale there is a need to simplify the characterisation of the sites in terms of simpler, yet representative, soil classes and geotechnical failure susceptibility categories. The simplification naturally drives the methodology to seek out those models that may be most broadly applicable in a practical sense. This may include the use of V_{S30}, or similar approaches such as Pitilakis et al. (2013) in favour of a more localised site amplification function, and equally common the use of a generalised design code amplification function based on site category where it is not possible to adequately constrain V_{S30} or equivalent parameter.

3.6 Future Developments in Seismic Hazard Methodologies for Analysis of Seismic Risk to Multi-system Infrastructures

The analysis of seismic risk to multi-system infrastructures presents many new challenges in the characterisation of the ground motion input. Given the interconnected nature of the different spatial systems, and the heterogeneity of the vulnerable elements within each, it is clearly necessary to implement a procedure for characterising the ground motion fields across many spectral periods, all whilst preserving the correlation structure of the motion.

It is of course recognised that, in terms of correlation, improved results could be achieved for a scenario rupture using numerical simulations of ground motion with a 3D wave propagation model. Whilst these sorts of simulations may be desirable in many situations, their application remains limited when considering the a wide range of stochastic events, each with considerable uncertainty on the physical characteristics of the rupture. This is in large part due to the computational cost of the process, which severely restricts the range of rupture scenarios that can be incorporated into analyses such as these, and the specificity of the physical models of the geological morphology of a region, which limits the applicability over many areas in Europe. However, the use of ground motion simulations for urban scale analysis of risk to urban infrastructure retains great potential for use in future applications. Furthermore, many of the stochastic processes developed within the present methodology can be readily integrated into a simulation-based methodology.

The extension of the seismic hazard modelling process to provide realisations of permanent ground deformation has highlighted the need to develop more empirical models that can be applied over urban and regional scales. There have been many developments in the field of probabilistic liquefaction hazard analysis (e.g., Kramer and Mayfield 2007; Goda et al. 2011), probabilistic seismic sliding displacement hazard (e.g., Rathje and Saygill 2011) and co-seismic rupture displacement hazard (e.g., Youngs et al. 2003; Chen and Petersen 2011). In many of these examples, however, the applications are largely limited to a site-specific context. Despite the limitations in the geotechnical characterisation, empirical methodologies are emerging for characterising soil properties over spatial scales, conditional upon a limited set of observations (Baker and Faber 2008; Chen et al. 2012). These models, and subsequent developments, may assist in increasing the applicability of more site-specific geotechnical hazard analysis procedures to broader scales.

The multi-system infrastructure analyses undertaken within the SYNER-G project illustrate why it is of great benefit, to both the research community and to the end-users of the products, to begin to consolidate existing data (both pre- and post-event). It is difficult to establish the portability of these methodologies to different regions of the globe, without being able to ascertain how microzonation and geotechnical information is represented across different urban regions. The establishment of a unified taxonomy for representing geotechnical data would be a significant step in this direction, both as a means of consolidating a wide and disparate assortment of geotechnical parameters and as a means of establishing a truly hierarchical methodology for undertaking geotechnical seismic hazard assessment in different regions of the globe. This may also be true for many other elements of the systems considered in the infrastructure analysis.

The application of the Shakefield and geotechnical procedure in the case studies described in later chapters will demonstrate the challenges in implementing this approach in a real-world context. It will be shown that the incorporation of the hazard uncertainties and the correlations have a significant impact on the variability of the expected losses, both from single events and from full probabilistic approaches. The need for model evaluation and validation studies is paramount in establishing where improvements to the methodology will be needed, and where the uncertainties may have the greatest impact. Likewise, it is anticipated that future extensions of these applications to different cities both in Europe and across the globe will create many new challenges in implementation.

References

Akkar S, Bommer JJ (2010) Empirical equations for the prediction of PGA, PGV, and spectral accelerations in Europe, the Mediterranean Region, and the Middle East. Seismol Res Lett 81(2):195–206

Baker JW (2007) Correlation of ground motion intensity parameters used for predicting structural and geotechnical response. In: Kanda J, Takada T, Furuta H (eds) Applications of statistics and probability in civil engineering. Taylor & Francis, London/New York

Baker JW, Cornell CA (2006) Correlation of response spectral values for multicomponents of ground motion. Bull Seismol Soc Am 96(1):215–227
Baker JW, Faber MH (2008) Liquefaction risk assessment using geostatistics to account for soil spatial variability. J Geotech Geoenviron Eng 134(1):14–23
Baker JW, Jayaram N (2008) Correlation of spectral acceleration values from NGA ground motion models. Earthq Spectra 24(1):299–317
Bazzurro P, Cornell CA (2004a) Ground-motion amplificationin nonlinear soil sites with uncertain properties. Bull Seismol Soc Am 94(6):2090–2109
Bazzurro P, Cornell CA (2004b) Nonlinear soil-site effects in probabilistic seismic hazard analysis. Bull Seismol Soc Am 94(6):2110–2123
Bindi D, Luzi L, Massa M, Pacor F (2010a) Horizontal and vertical ground motion prediction equations derived from the Italian Accelerometric Archive (ITACA). Bull Earthq Eng 8:1209–1230
Bindi D, Luzi L, Rovelli A (2010b) Ground motion prediction equations GMPEs derived from ITACA. Technical report, Deliverable No. 14, Project S4: Italian Strong Motion Data Base
Bindi D, Pacor F, Luzi L, Puglia R, Massa M, Ameri G, Paolucci R (2011) Ground motion prediction equations derived from the Italian strong ground motion database. Bull Earthq Eng 9:1899–1920
Boore DM, Gibbs JF, Joyner WB, Tinsley JC, Ponti DJ (2003) Estimated ground motion from the 1994 Northridge, California, earthquake at the site of the Iinterstate I-10 and La Cienega Boulevard bridge colapse, West Los Angeles, California. Bull Seismol Soc Am 93(6):2737–2751
Bradley BA (2011a) Correlation of significant duration with amplitude and cumulative intensity measures and its use in ground motion selection. J Earthq Eng 15(6):809–832
Bradley BA (2011b) Empirical correlation of PGA, spectral accelerations and spectrum intensities from active shallow crustal earthquakes. Earthq Eng Struct Dyn 40(15):1707–1721
Chen R, Petersen MD (2011) Probabilistic fault displacement hazards for the Southern San Andreas fault using scenarios and empirical slips. Earthq Spectra 27(2):293–313
Chen Q, Seifried A, Andrade JE, Baker JW (2012) Characterization of random fields and their impact on the mechanics of geosystems at multiple scales. Int J Numer Anal Methods Geomech 36:140–165
Choi Y, Stewart JP (2005) Nonlinear site amplification function of 30m shear wave velocity. Earthq Spectra 21(1):1–30
Crowley H, Bommer JJ, Stafford PJ (2008a) Recent developments in the treatment of ground-motion variability in earthquake loss models. J Earthq Eng 12(S2):71–80
Crowley H, Stafford PJ, Bommer JJ (2008b) Can earthquake loss models be validated using field observations? J Earthq Eng 127:1078–1104
Davis M (1987) Production of conditional simulation via the LU decomposition of the covariance matrix. Math Geol 192:91–98
Esposito S, Iervolino I (2011) PGA and PGV spatial correlation models based on European multievent datasets. Bull Seismol Soc Am 101(5):2532–2541
Esposito S, Iervolino I (2012) Spatial correlation of spectral acceleration in European data. Bull Seismol Soc Am 102(6):2781–2788
FEMA-450 (2003) NEHRP recommended provisions for seismic regulations for new buildings and other structures. Technical report, Federal Emergency Management Agency (FEMA), Washingon, DC
Goda K, Atkinson GM (2009) Probabilistic characterisation of spatial correlated response spectra for earthquakes in Japan. Bull Seismol Soc Am 99(5):3003–3020
Goda K, Atkinson GM (2010) Intraevent spatial correlation of ground motion parameters using SK-net data. Bull Seismol Soc Am 100(6):3055–3067
Goda K, Hong HP (2008a) Spatial correlation of peak ground motions and response spectra. Bull Seismol Soc Am 98(1):354–365
Goda K, Hong HP (2008b) Estimation of seismic loss for spatially distributed buildings. Earthq Spectra 24(4):889–910

Goda K, Atkinson GM, Hunter JA, Crow H, Motazedian D (2011) Probabilistic liquefaction hazard analysis for four Canadian cities. Bull Seismol Soc Am 101(1):190–201

Hong HP, Zhang Y, Goda K (2009) Effect of spatial correlation on estimated ground motion prediction equations. Bull Seismol Soc Am 99(2A):928–934

Iervolino I, Giorgio M, Galasso C, Manfredi G (2010) Conditional hazard maps for secondary intensity measures. Bull Seismol Soc Am 100(6):3312–3319

Inoue T, Cornell CA (1990) Seismic hazard analysis of multi-degree-of-freedom structures. Technical report, RMS, Stanford

Jayaram N, Baker JW (2008) Statistical tests of the joint distribution of spectral acceleration values. Bull Seismol Soc Am 98(5):2231–2243

Jayaram N, Baker JW (2009) Correlation model of spatially distributed ground motion intensities. Earthq Eng Struct Dyn 38:1687–1708

Jayaram N, Baker JW (2010a) Efficient sampling and data reduction techniques for probabilistic seismic lifeline risk assessment. Earthq Eng Struct Dyn 39(10):1109–1131

Jayaram N, Baker JW (2010b) Cequations spatial correlation in mixed-effects regression, and impact on ground motion models. Bull Seismol Soc Am 100(6):3295–3303

Kramer SL, Mayfield RT (2007) Return period of soil liquefaction. J Geotech Geoenviron Eng 133(7):802–813

Loth C, Baker JW (2013) A spatial cross-correlation model of spectral accelerations at multiple periods. Earthq Eng Struct Dyn 42:397–417

Moss RES, Ross ZE (2011) Probabilistic fault displacement hazard analysis for reverse faults. Bull Seismol Soc Am 101(4):1542–1553

Musson R (2000) The use of Monte Carlo simulations for seismic hazard assessment in the UK. Annali di Geofisica 43(1):1–9

NIBS N (2004) HAZUS-MH: techincal manual. Technical report, Federal Emergency Management Agency (FEMA), Washington, DC

Oliver DS (2003) Gaussian cosimulation: modelling of the cross-covariance. Math Geol 356:681–698

Paolucci R, Smerzini C (2008) Earthquake-induced transient ground strains from dense seismic networks. Earthq Spectra 24(2):453–470

Park J, Bazzurro P, Baker JW (2007) Modeling spatial correlation of ground motion intensity measures for regional seismic hazard and portfolio loss estimation. In: Kanda J, Takada T, Furuta H (eds) Applications of statistics and probability in civil engineering. Taylor & Francis, London

Petersen MD, Dawson TE, Chen R, Cao T, Wills CJ, Schwartz DP, Frankel AD (2011) Fault displacement hazard for strike-slip faults. Bull Seismol Soc Am 101(2):805–825

Pitilakis K, Riga E, Anastasiadis A (2012) Design spectra and amplification factors for Eurocode 8. Bull Earthq Eng 10:1377–1400

Pitilakis K, Riga E, Anastasiadis A (2013) New code site classification, amplification factors and normalised response spectra based on a worldwide ground-motion database. Bull Earthq Eng 11(4):925–966

Rathje EM, Saygill G (2011) Estimating fully probabilistic seismic sliding displacements of slopes from a pseudoprobabilistic approach. J Geotech Geoenviron Eng 137(3):208–217

Robinson D, Dhu T, Schneider J (2006) SUA: a compter program to compute site-response and epistemc ununcertainty for probabilistic seismic hazard analysis. Comput Geosci 32:109–123

Saygill G, Rathje EM (2008) Empirical predictive models for earthquake-induced sliding displacements of slopes. J Geotech Geoenviron Eng 134(6):790–803

Sokolov V, Wenzel F, Jean WY, Wen KL (2010) Uncertainty and spatial correlation of earthquake ground motion in Taiwan. Terr Atmos Ocean Sci 21(6):905–921

Tokimatsu K, Seed HB (1987) Evaluation of settlement in sands due to earthquake shaking. J Geotech Geoenviron Eng 113(8):861–878

Wang M, Takada T (2005) Macrospatial correlation model of seismic ground motions. Earthq Spectra 21:1137–1156

Weatherill G, Burton PW (2010) An alternative approach to probabilistic seismic hazard analysis in the Aegean region using Monte Carlo simulation. Tectonophysics 492:253–278

Wells DL, Coppersmith KJ (1994) New empirical relationships among magnitude, rupture length, rupture width, rupture area, and surface displacement. Bull Seismol Soc Am 84(4):974–1002

Youd TL, Perkins DM (1978) Mapping of liquefaction induced ground failure potential. J Geotech Eng Div ASCE 1044:433–466

Youngs RR, Arabasz WJ, Ernest Anderson R, Ramelli AR, Aki JP, Slemmons DB, McCalpin JP, Doser DI, Fridrich CJ, Swan FH III, Rogers AM, Yount JC, Anderson LW, Smith KD, Bruhn RL, Knuepfer PLK, Smith RB, dePolo CM, O'Leary DW, Coppersmith KJ, Pezzopane SK, Schwartz DP, Whitney JW, Olig SS, Toro GR (2003) A methodology for probabilistic fault displacement hazard analysis (PFDHA). Earthq Spectra 19(1):191–219

Chapter 4
Framework for Systemic Socio-economic Vulnerability and Loss Assessment

Bijan Khazai, James E. Daniell, Şebnem Düzgün, Tina Kunz-Plapp, and Friedemann Wenzel

Abstract A unified approach for modeling shelter needs and health impacts caused by earthquake damage which integrates social vulnerability into the physical systems modeling approaches has been developed. The shelter needs and health impact models discussed here bring together the state-of-the-art casualty and displaced population estimation models into a comprehensive modeling approach based on multi-criteria decision support, which provides decision makers with a dynamic platform to capture post-disaster emergency shelter demand and health impact decisions. The focus in the shelter needs model is to obtain shelter demand as a consequence of building usability, building habitability and social vulnerability of the affected population rather than building damage alone. The shelter model simulates households' decision-making and considers physical, socio-economic, climatic, spatial and temporal factors in addition to modeled building damage states. The health impact model combines a new semi-empirical methodology for casualty estimation with models of health impact vulnerability, and transportation accessibility to obtain a holistic assessment of health impacts in the emergency period after earthquakes.

B. Khazai (✉)
Geophysical Institute, Karlsruhe Institute of Technology, Hertzstr. 16, 76187 Karlsruhe, Germany
e-mail: khazai@kit.edu

J.E. Daniell • T. Kunz-Plapp • F. Wenzel
Karlsruhe Institute of Technology, Hertzstr. 16, 76187 Karlsruhe, Germany
e-mail: j.e.daniell@gmail.com; tina.kunz-plapp@kit.edu; Friedemann.Wenzel@kit.edu

Ş. Düzgün
Middle East Technical University, ODTU Maden Müh. Böl., 06531 Çankaya, Ankara, Turkey
e-mail: duzgun@metu.edu.tr

K. Pitilakis et al. (eds.), *SYNER-G: Systemic Seismic Vulnerability and Risk Assessment of Complex Urban, Utility, Lifeline Systems and Critical Facilities*, Geotechnical, Geological and Earthquake Engineering 31, DOI 10.1007/978-94-017-8835-9_4,

4.1 Integrated Framework for Modeling Socio-economic Impacts

4.1.1 Introduction

Recent decades have seen an exponential growth in the physical impacts and losses from earthquakes throughout the world. The Great Wenchuan Earthquake in May 2008, the Haitian Earthquake of 2010, and the 2011 Great East Japan Earthquake provide poignant reminders of the susceptibility of communities to devastating loss of lives, livelihoods, and property from earthquake events. These disasters, plus many other smaller ones, illustrate how earthquakes adversely impact people and the communities in which they live, and the impacts of such events occur across geographic boundaries and at multiple scales affecting governments, institutions, economic sectors, livelihoods, and people.

There is a consensus within the scientific community that disasters associated with earthquakes are not wholly the product of the physical impacts of natural hazard events. Rather, these disasters are the outcome of the interaction between the earth's biophysical systems, the engineered environment, and the social conditions inherent at particular places (Hewitt and Burton 1971; Mileti 1999; Wisner et al. 2004). It is increasingly becoming clear that some people and groups are impacted differentially by damaging events, react differently in an event's aftermath, adjust to its circumstances in dissimilar ways, and recover in a differential manner. These circumstances have stimulated great interest in understanding how to manage the associated seismic risk, adverse impacts, and loss.

The current state-of-the-art in earthquake loss estimation (ELE) software provides several parameters of direct socio-economic consequences which are needed to support effective decision making. These include parameters such as casualties, displaced persons, and business failures by industry, for example. However, poor linkages between damage to physical systems and resultant social consequences remain a significant limitation with existing hazard loss estimation models (Bostrom et al. 2008). A new direction with earthquake loss estimation software, which has been developed by researchers of the Mid America Earthquake Center, is the inclusion of social vulnerability (Elnashai et al. 2009). Although definitions and applications of social vulnerability vary within the literature (Cutter 1996), the concept is often described as the potential for harm and the ability of an individual or community to protect itself from damaging events (Cutter et al. 2003; Wood et al. 2010). Given equal exposure to seismic threats, two groups may vary in their social vulnerability due to internal societal characteristics where differences according to wealth, gender, race, class, history, and socio-political organization influence the patterns of loss, mortality, and the ability of communities to reconstruct following a disaster (Burton and Cutter 2008). Inclusion of social vulnerability into quantitative loss estimation frameworks allows for a more holistic measurement of risk. However, more work is needed to develop a systemic approach that quantitatively

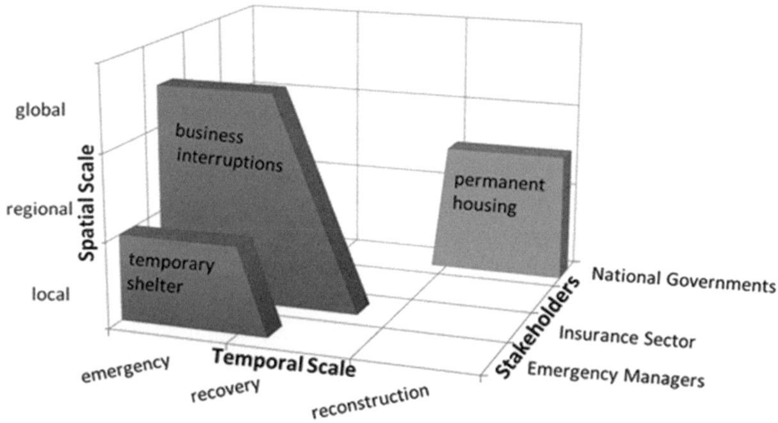

Fig. 4.1 The three dimensions in an infrastructure vulnerability and impact studies

brings together these disparate research areas in social vulnerability research into a comprehensive modeling tool, which provides decision makers with a dynamic platform to capture post-disaster decisions, interactions and changes over time.

At the core of the SYNER-G project are the development of state-of-the-art modeling capabilities and a suite of software tools that can be utilized for the assessment and communication of earthquake risk through the integration of both physical and social vulnerability. Thus, one of the main aims in SYNER-G was to compute social losses (displaced population, shelter needs and health impacts) as an integrated function of hazard intensity, systemic vulnerability of physical systems and the social vulnerability of the population at risk. This way of conceptualizing integrated risk emphasizes the importance of understanding the interrelations between physical and social systems. In other words, how direct physical losses can potentially aggravate existing vulnerabilities in society and how vulnerabilities in society can ultimately lead to greater impacts from physical damage and losses.

4.1.2 *Integrated Evaluation of Physical and Socio-economic Models*

The impact of an earthquake on the infrastructure evolves in space with the time elapsed from the event. Different stakeholders have different interests and play distinct roles in the various phases of the disaster. Correspondingly they look at impact assessments according to their own particular needs and mandates. These three dimensions (time, space, stakeholders) are represented in Fig. 4.1, which allows the vulnerability and impacts on infrastructure systems to be operationalized.

In particular, along the time-dimension three periods of a disaster – emergency, recovery and reconstruction – can be identified. The first period constitutes the immediate aftermath of the event and its short-term consequences where the damaged infrastructure operates in a state of emergency. In this phase emergency managers must deal with the demand generated by damaged infrastructure in terms of temporary shelter needs or hospitalization and treatment of victims. In the midterm recovery period, while the infrastructure progressively returns to a new state of normal functionality, the disruptions to businesses might be of interest to stakeholders in the insurance sector. In the long-term reconstruction period, national governments and multi-lateral organizations have to grapple with the costs of permanently rebuilding or upgrading/retrofitting damaged infrastructure, and mitigate the risk from the next event. From the perspective of systemic studies there are two distinct phases which are commonly addressed:

- Emergency phase: short-term (a few days/weeks) at the urban/regional scale (e.g. Franchin et al. 2006; Nuti and Vanzi 1998)
- Economic recovery phase: medium to long-term, at the regional/national scale (e.g. Karaca 2005)

Furthermore, the position on the "time axis" of the analyst/observer with respect to the time-frame changes the goal of the systemic study:

- Outside/before the time-frame: the goal of the system analyst is to forecast the impact in order to set-up preparedness, planning and mitigation measures. It is important to underline how the information basis in this case can be considered as constant.
- Within the time-frame: the goal of the system analyst is that of providing the managers with a real-time decision support system, which updates the Infrastructure state based on the continuously incoming flow of information.
- After the time-frame: the goal of the system analyst is to validate the models against occurred events.

The general methodology developed within the SYNER-G Project focuses on the **short-term only, with emergency managers as the reference stakeholders**, and with the goal of forecasting, before the event occurs, the expected impact in terms of dead, injured and displaced population, for the purpose of planning and implementing risk mitigation measures. The goal of the methodology developed within the SYNER-G project is to assess the seismic vulnerability of an infrastructure of urban/regional extension, accounting for all relevant and meaningful inter- and intra-dependencies among infrastructural components, as well as of uncertainties. As described in Chap. 2 (see Fig. 2.1), the model for the infrastructure consists of both physical models of the systems making up the infrastructure (e.g. building and building aggregates, transportation network, lifeline systems) as well as socio-economic models that take as an input the output of the physical models and provide the socio-economic consequences of the event. Thus, the integrated approach proposed in SYNER-G provides a framework to link the degree of damage and performance of inter-related physical systems to vulnerabilities and coping

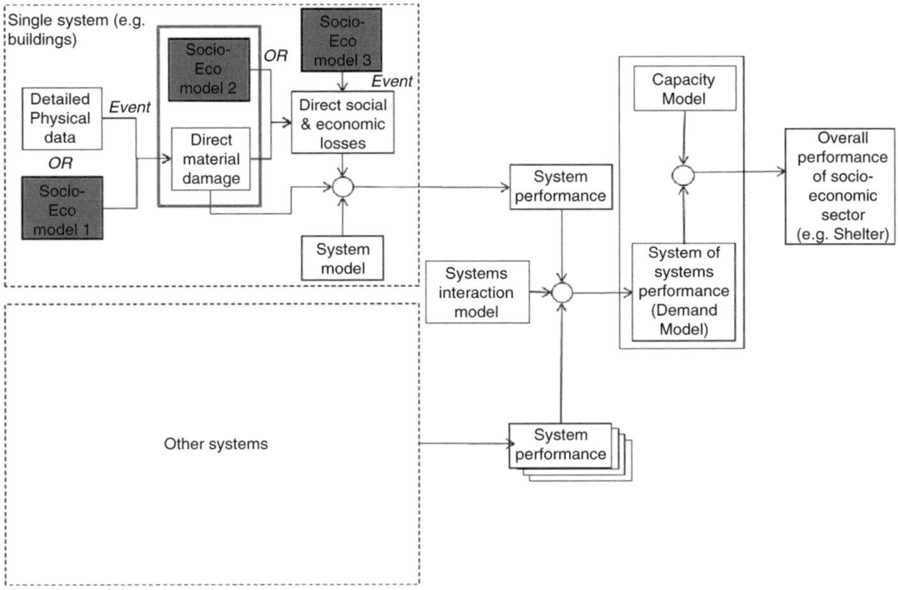

Fig. 4.2 Possible interaction of socio-economic models with physical vulnerability/loss estimation models

capacities in society to assess: (1) Impacts on displaced populations and their shelter needs, and (2) Health impacts on exposed populations and their health-care needs. Furthermore, disruption to transportation and lifeline networks (roads, pipelines, electricity and water supply) have important consequences on the recovery process and contribute to increased social disruptions within both shelter and health sectors.

There are various possible interactions of socio-economic models with physical vulnerability/loss estimation models. Figure 4.2 shows three possible entry points for socio-economic models. In many earthquake loss estimation models socio-economic models are brought in at the third entry point as linear "damage-consequence functions" for the estimation of direct social and economic losses from physical system parameters, such as, level of building damage. Bringing in socio-economic models at the first entry point shown in Fig. 4.2 as empirical models, requires the systematic collection of post-event social and economic post-earthquake data which is typically not feasible, given the perishable nature of such data and that it is currently not being collected in a systematic and coordinated fashion.

In SYNER-G the socio-economic models are being brought in at the second entry point depicted in Fig. 4.2. Here, new methods have been developed to compute social losses (e.g., number of displaced people and casualties) as an integrated function of hazard intensity, vulnerability of physical systems and the social vulnerability of the population at risk. In this framework design it is assumed that interactive and causal processes take place between society and the physical systems it interacts with. For example, the loss of building habitability (derived from physical building and utility damage models) will play a major role in the decision to evacuate one's domicile and seek public shelter. Here the interaction between building habitability, environmental factors (e.g., weather conditions) and social factors such as the occupants tenure status (home owner vs. renter), whether the occupant lives in a single family home or a multi-family apartment structure, the level of anxiety of aftershocks, etc., will ultimately form the integrated model that is used to determine an estimate of displaced populations after an earthquake.

4.1.3 Multi-criteria Decision Analysis Framework

The integration of multiple dimensions of vulnerability with diverse and complex linkages constitutes some of the most challenging questions in store for interdisciplinary research. Contributing to this is the fact that social vulnerability is a fundamentally relative phenomenon and not something that can be directly observed and measured (Birkmann 2006). Thus, one of the main objectives has been the adoption of an indicator system and common nomenclature which posits social vulnerability in relational terms with respect to both shelter and healthcare systems. In this regard, indicators are pieces of information that summarize the characteristics of a system or highlight what is happening in a system. An indicator is a quantitative or qualitative measure derived from observed facts that simplify and communicate the reality of a complex situation (Freudenberg 2003). Indicators reveal the relative position of the phenomena being measured and when evaluated over time, can illustrate the magnitude of change (a little or a lot) as well as direction of change (up or down; increasing or decreasing). The mathematical combination (or aggregation as it is termed) of a set of indicators is often termed an index or a composite indicator. Consequently, transparent and validated indicator systems, which characterize the different aspects sectors addressed here (i.e., shelter, health, transportation) have been defined.

The theoretical framework for integration of physical and social performance indicators is founded on the work of Cardona (2005). It provides an overview of not only the expected direct damages, but also the potential for aggravating impact of the direct damages by the social fragility and lack of resilience of the different sectors analyzed here. As shown in Fig. 4.3 a physical performance index is obtained, for each unit of analysis by interacting with the physical infrastructure models, whereas the total social impact index is obtained by multiplying the direct physical performance indices by an indirect impact factor, based on variables associated with the socio-economic conditions of each unit of analysis. In order to reduce the complexity of the total system for applied purposes, vulnerability in each system is operationalized by a set of discrete indicators, representing social vulnerability and coping capacities. The indicators and sub-indicators have been chosen according to the vulnerability factors and decision criteria identified for each system and are described in the subsequent sections.

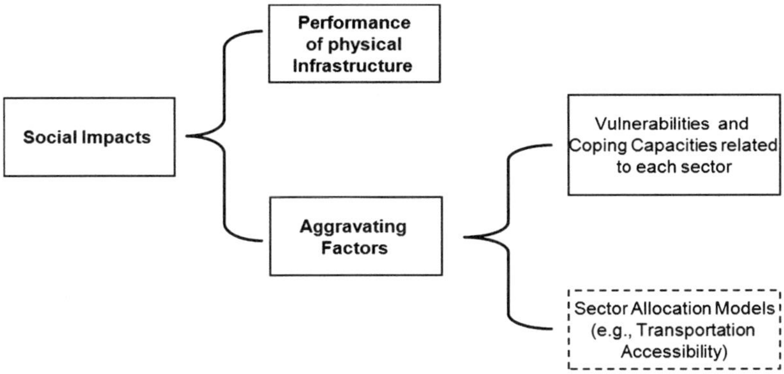

Fig. 4.3 Structure of the integrated framework for assessment of social impacts

Fig. 4.4 Methodological steps of the development of a hierarchical indicator framework

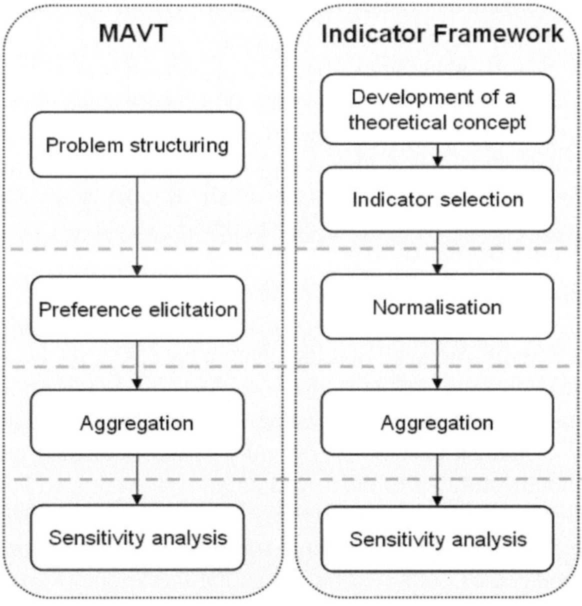

The process of developing an integrated indicator framework for assessment of overall social impacts depicted in Fig. 4.4 consists of five main steps which should be passed in an iterative manner (Nardo et al. 2005). These development steps are very similar to the main phases of the multi-criteria decision theory (MCDA) or multi-attribute-value theory (MAVT). Therefore, for the development of a hierarchical indicator framework for the shelter needs and health impacts models in SYNER-G, the methodological approaches used within a MAVT-Analysis were transferred to the vulnerability for each of the two sectors – shelter and health – analyzed. The MCDA/MAVT framework also provides the basis for developing the software system in SYNER-G as a decision-support tool for emergency planning by taking into account a broader range of expert judgment through the interactive modeling of indicator weights or complementing the existing system of indicators with additional available data (e.g., for the assessment of additional vulnerability dimensions).

4.2 Shelter Needs Model

4.2.1 Population Displacement Following Historic Earthquakes

For the planning of public shelter provisions in the aftermath of earthquakes the expected number of homeless persons and people seeking public shelter is an essential input for emergency managers. Few models exist that estimate the displaced or homeless population and the number of displaced persons seeking public shelter in an earthquake. Most Earthquake Loss Estimation (ELE) software providing input for displaced population are based on the HAZUS methodology which computes both displaced population as a linear consequence of building damage. For example 90 % of all occupants in severely damaged multi-family homes and 100 % of all occupants in extensively and completely damaged multi-family and single-family homes are assumed to be displaced according to the HAZUS model default conditions (FEMA 2003).

Looking at data from 457 historic earthquakes from 1900 to 2012 with destroyed or heavily damaged building data in the CATDAT Damaging Earthquakes Database (Daniell et al. 2011), a linear trend (on a logarithmic scale) of displacement and building damage can indeed be observed (Fig. 4.5). This data shows that the number of displaced persons is generally a little less than one order of magnitude larger than the number of destroyed or severely damaged buildings. However, the data in Fig. 4.5 also shows that in many past events the number of displaced persons is much larger than can be accounted for only through the number of occupants in severely damaged or collapsed buildings. Observations from past earthquake events found in the literature show that the number of displaced persons after an earthquake not only depend on external factors like building damage, loss of utilities, and weather conditions but also from household internal socioeconomic and individual factors such as safety concerns or fear of aftershocks (see full literature review of factors influencing displacement in Khazai et al. 2011a). The intention to leave can also be undermined through feasibility restraints, e.g. if the next shelter is too far away, if people are disabled or lack mobility. Even if households decide to leave their homes the final question is where they will find accommodation. Alternatives to public shelter are for example to stay with friend and family or in hotels. Thus, only a subset of the total population should be considered in computing demand for public shelter.

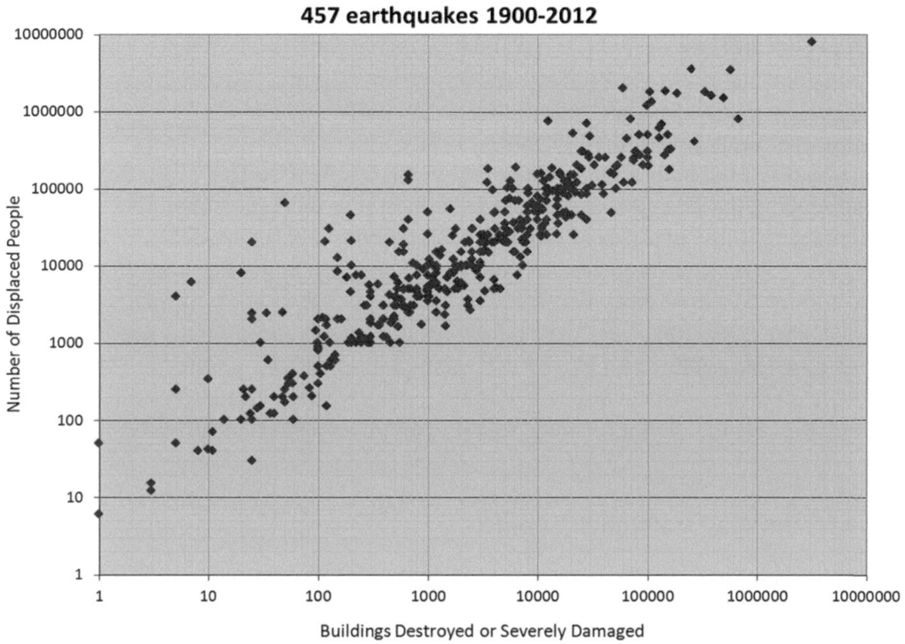

Fig. 4.5 Relationship between severely damaged and destroyed buildings and displaced persons after earthquakes (n = 457 earthquakes from 1900 to 2012)

4.2.2 Shelter Demand Models

To determine shelter needs in Earthquake Loss Estimation, most software follow the HAZUS methodology, where displaced population (determined only from building damage) is multiplied by a factor that considers age, ownership, ethnicity and income to determine demand for public shelters. These four parameters were originally developed by the American Red Cross and were based on expert opinion along with historical data from the 1994 Northridge earthquake (Harrald et al. 1992). New approaches have recently been developed which simulates households' decision-making in seeking shelter and considers socio-economic, temporal and spatial factors in addition to housing damage and lifeline loss to estimate displaced and shelter seeking populations (Chang et al. 2009; Wright and Johnston 2010; Khazai et al. 2011c). For example, the model by Chang et al. (2009) adopts an agent-based approach that utilizes census microdata on households and simulates households' decision-making about post-earthquake shelter on the basis of their dwelling condition, risk perception, mobility, and resources.

A new approach is presented for modeling emergency shelter demand in SYNER-G by integrating shelter-seeking logic models into a systemic seismic vulnerability analysis and earthquake loss estimation software tool (Khazai et al. 2012a). The selection of socio-economic vulnerability indicators and other factors in the shelter logic model are based on an in-depth literature survey of historic earthquakes and are derived and validated using statistical models. Thus a new advancement to shelter estimation methodology is being explored through three types of key inputs: (1) the "habitability" of buildings which combines inputs from the physical models (building usability, utility loss and climate factors) to provide information on the habitability of a building and can be used as a better determinant in influencing the decision to evacuate than building damage alone; (2) GIS-based shelter accessibility analysis as an input to the shelter seeking model – not discussed here; and (3) a multi-criteria decision model for implementing a shelter-seeking logic model based on complex socio-economic factors which ultimately lead to the decision to evacuate and seek public shelter. These three inputs are combined into a dynamic shelter model and software tool developed within the SYNER-G software platform to provide stakeholders an interactive framework in decision-making process for shelter planning and preparedness as well as resource allocation.

4.2.2.1 Building Habitability Model

The first step in the decision to evacuate after an earthquake is based on the structural stability of a building and functional lifeline structures, such as access to water gas and electric power services. Weather conditions can further aggravate potential displacement from damaged buildings with disrupted lifeline services. If a building is only slightly damaged and it is very cold and there are no possibilities to heat, that home will be uninhabitable. During other seasons and weather conditions the same building might be habitable. In a rare study surveying post-earthquake survivors about their shelter preferences, Chien et al. (2002) found evidence that under normal weather conditions 67 % of the interviewees after the 1999 Chi-Chi earthquake chose to stay in nearby open fields or under a tent, whereas under wet or cold weather conditions only 17 % showed a preference for staying there. Similar to the Chi-Chi earthquake, cold weather played a major role in the choices of occupants who sought shelter in both of the last two major earthquakes: The 2011 Tohoku earthquake (Khazai et al. 2011b) and the 2012 Van earthquake in Turkey (Wenzel et al. 2012).

Building habitability is determined as a combination of the functionality of buildings (building usability), utility services and impending weather conditions and constitutes the first decision step in leaving or staying at home after an earthquake. Building usability is derived from a simplified semi-empirical approach as a function of the severity of observed damage to structural and non-structural elements of buildings. The usability model was developed based on a detailed survey of 305 buildings in the densely packed suburb of Pettino obtained from the Italian Department of Civil Protection after the 2009 L'Aquila earthquake. The six usability classes considered during the survey were reduced in this model to just three: buildings which are immediately non-usable (NU), partially usable (PU) or fully usable (FU). Using the Pettino database, Usability Ratios (UR) for buildings were derived for each of the three usability classes as a function of the damage data,

Table 4.1 Empirically-derived usability ratios

	Damage state		
UR	None	Yield	Collapse
FU	0.87	0.22	0.00
PU	0.13	0.25	0.02
NU	0.00	0.53	0.98

reported according to six damage states (DS0 to DS5), which were also reduced to three damage states (none, yield, collapse). Usability ratios can be used then to estimate the number of persons in each of the three building usability classes (NFU, NPU, NNU). Using the Usability Ratios in Table 4.1, the number of persons in each of the three building usability classes can be obtained using the following expression:

$$\mathrm{N}_{FU\ or\ PU\ or\ NU} = \sum\nolimits_{i=1}^{3} N_i \cdot NO_i \cdot UR_{i.FU\ or\ PU\ or\ NU} \tag{4.1}$$

where:

- $i =$ damage level ($i = 1, \ldots, 3$)
- $N_i =$ number of buildings having damage level i,
- $NO_i =$ number of occupants (at the time of the event) in each building for each damage level i,
- $UR_i =$ usability ratio (UR) for damage level i for each usability class

To determine building habitability the usability of buildings is considered together with utility loss in a systemic seismic vulnerability analysis (Cavalieri et al. 2012). Non-usable buildings (NU) are also non-habitable. If a building is fully or partially usable, depending on the level of residual service in the utilities and the prevailing weather conditions at the time of impact, it can be habitable (H) or non-habitable (NH). For each utility, the level of residual service is satisfactory when the Utility Loss (UL), defined as one minus the ratio of satisfied to required demand, is lower than a threshold value ($UL_i < ULT_i$). The threshold values depend on Weather conditions and Building Usability and due to the subjective nature of perceptions, the Utility Loss Threshold (ULT_i) should be established on a context-specific basis by the analyst. The total Utility Loss is a weighted average of UL_i on each of the utilities, with weights w_i provided by the analyst:

$$UL = \sum\nolimits_{j=1}^{N_{UN}} UL_j \cdot w_j \tag{4.2}$$

where:

- $j =$ utility systems ($j = 1, \ldots N_{UN}$ with $N_{UN} = 2$ in this application)
- $UL_j =$ Utility Loss in system j
- $w_j =$ weight associated with the importance of loss in utility system j in making the building uninhabitable

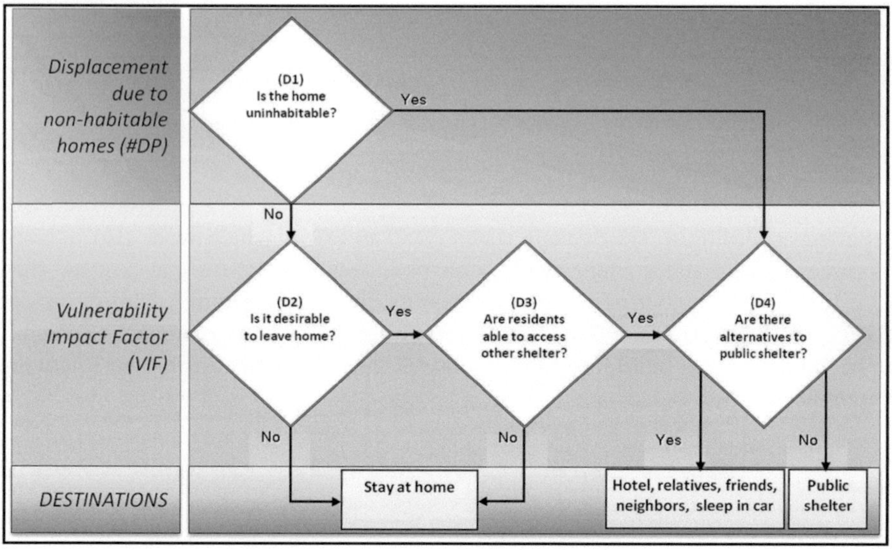

Fig. 4.6 Proposed model framework for shelter seeking logic model

The percentage fully or partially usable buildings that are non-habitable (NH_{FU} or NH_{PU}) is thus determined as the portion of buildings which have utility losses greater than the utility loss threshold value ($UL \geq UL_T$). The Uninhabitable Building Index (UBI) is computed as the ratio of occupants of buildings that are uninhabitable to the total population (N) according to the following relationship:

$$BHI = \left(N_{FU}NH_{FU} + N_{PU}NH_{PU} + N_{NU} - N_d\right)/N \tag{4.3}$$

where:

- N_{FU}, N_{PU}, N_{NU} = number of occupants in buildings that are fully, partially and non-usable
- NH_{FU} = percentage of fully usable buildings that are non-habitable, where $\mathrm{UL} \geq \mathrm{ULT}$
- NH_{PU} = percentage of partially usable buildings that are non-habitable, where $\mathrm{UL} \geq \mathrm{ULT}$
- N_d = number of dead persons estimated in a selected casualty model

4.2.2.2 Shelter-Seeking Decision Model

The basic elements of the logic model for the shelter demand model are based on the ideas of Chang et al. (2009). The shelter model combines each of the decision steps (represented as an output indicator) shown in Fig. 4.6 in a weighted multi-criteria

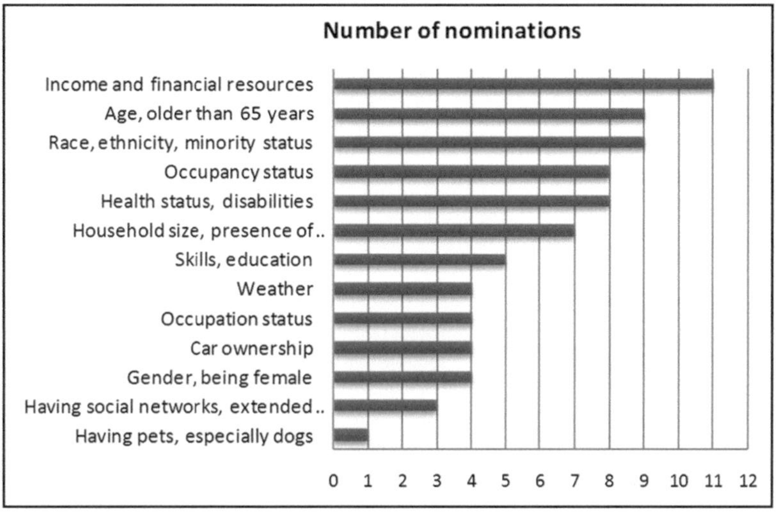

Fig. 4.7 Number of nominations found for indicators influencing post-earthquake evacuation behavior in 18 studies surveyed

decision analysis framework according to the following scheme: D1 is given by an output indicator as the proportion of population residing in uninhabitable buildings criteria; D2 and D3 are a combination of a number of internal and external factors and given by an output indicator representing the desirability to evacuate criteria; and D4 is given by an output indicator representing the desirability to seek public shelter based on the access to resources criteria.

The decision to evacuate one's home after an earthquake and to utilize public shelter is correlated with a variety of social and demographic factors (Tierney 2006). These decisions are also usually made at the household level; however, as was seen in the case of the L'Aquila earthquake the decision to evacuate can also be imposed by government authorities that make an evacuation of homes mandatory. A survey of disaster literature regarding post-earthquake sheltering demand provided an initial basis for selection of relevant socio-economic indicators related to the desirability to evacuate (Khazai et al. 2012a; Braun 2011). The main factors influencing evacuation behavior were derived from 18 key studies and are shown in Fig. 4.7.

While the literature survey provides for a comprehensive wish list of indicators, an important requirement for operationalizing the approach is that it should be possible to quantitatively populate the socio-economic indicators based on an approach that can be harmonized at the European level for the urban scale of analysis; one of the aims of this study. As such, data was compiled from the EUROSTAT Urban Audit for European cities at the sub-city districts (SCD) level and used as a next step to pre-select the most relevant indicators from the Urban Audit that were found in the literature survey. In order to narrow down the selection of the most influential indicators from the Urban Audit and to assign a set of default

Table 4.2 Results of principal component analysis of urban audit data

Subjective factors	Strongest correlated indicator	Strongest correlation value
Mortality/age	Mortality rate for <65 per year	−0.88
Education	Prop. of working age population qualified at level 3 or 4 ISCED	+0.77
Lone parent with children	Prop. of households that are lone-parent households	+0.68
Population density	Population density: total resident pop. per square km	−0.64
Migration/ethnicity	Proportion of Residents who are not EU Nationals and citizens of a country with a medium or low HDI	+0.58
Gender	Proportion of females to males in total population	+0.51
Unemployment	Unemployment rate	−0.54
Sub-standard housing	Proportion of dwellings lacking basic amenities	+0.67

weights a factor analysis was conducted with the Urban Audit data. Out of the 338 indicators described in the Urban Audit, data is available for only 44 indicators at the SCD level. The 44 indicators were analyzed for two periods: 1999–2002 (7,856 districts in 321 cities in 30 European countries); and 2003–2006 (2,972 districts in 173 cities in 24 European Countries). Principal component analysis (PCA) was used to calculate the inter-correlation between variables and a new set of transformed variables was created where the importance of each of the new variables in terms of the variability of the data is identified. It was found that close to 75 % of variation in data is represented by 8 dimensions shown in Table 4.2. Additionally, the PCA provides a possibility to model the relative influence of all data in terms of their explanatory power (i.e., how much of the statistical variation can be explained by each indicator).

The literature survey and the statistical models provide a set of candidates for operationalizing the shelter-seeking decision model presented in Fig. 4.6. The first step (D1) is determined through the building habitability analysis as discussed above. The following presents the methodology and indicator framework related to desirability to evacuate (D2 and D3) and desirability to seek public shelter (D4).

Desirability to Evacuate

The desirability to evacuate is a combination of factors related to a set of internal factors which is a reflection of perceived security and safety, as well as external factors forcing residents to leave (Table 4.3). Feeling safe at home (or the feeling that it is safer to leave) is subjective and depends on a large range of factors each with different perceived importance values and cultural contexts. As mentioned above

Table 4.3 Urban audit indicators influencing desirability to evacuate

Decision factors	Urban audit indicators for desirability to evacuate
Household tenure (owner vs. renter)	Proportion of households living in private rented housing
	Proportion of households living in owned dwellings
Housing type (single, multi-family)	Number of houses per 100 apartments
	Proportion of households living in social housing
	Proportion of dwellings lacking basic amenities
	Proportion of non-conventional dwellings
Household type (large families with children, single parents)	Avg. size of households
	Lone-parent households with children aged 18 or under
	Proportion of households living in social housing
Age (children and elderly)	Proportion of total population aged 0–4
	Proportion of total population aged 75 and over
Perceived security	Total number of recorded crimes per 1,000 population

the perception of weather conditions is compounded with the building damage and utility services disruptions. The resistance to evacuation is also influenced by sociological and economic factors, such as having strong social networks, belonging to a minority or being disabled, having enough knowledge and financial resources to protect oneself, and knowing where to obtain information. Other factors influencing the perceived security are conditions such as fear and anxiety of aftershocks or mistrust in safety evaluation of one's home (green, yellow and red tags) which are more difficult to describe and define quantitatively through indicators. Thus, the desirability to leave is a combination of a complex set of social factors and is ultimately determined by the individual's perception of the importance of each one of these factors in driving the decision to evacuate. While desirability to leave represents an internal driver to evacuation, the resistance to evacuation is also driven by external decisions imposed on the affected population, which in some cases may force them to evacuate (e.g., mandatory evacuation of an entire city centre as in 2009 L'Aquila earthquake, or a radiation advisory and evacuation radius, as in the aftermath of the 2011 Tohoku earthquake and tsunami). The equation for the desirability to evacuate (DE) is given as

$$\mathbf{DE} = \mathbf{EF} \times \sum_{i=1}^{n} w_i \cdot I_i \tag{4.4}$$

where:

- DE = Desirability to evacuate
- w_i = overall weight given to each indicator
- I_i = indicators representing the desirability to evacuate
- EF = External Factors, derived from a GIS analysis and/or different evacuation scenarios

Table 4.4 Urban audit indicators influencing desirability to seek public shelter

Decision factors	Urban audit indicators for shelter seeking index
Income	Percent of households with less than 60 % of national median annual disposable income
	Proportion of households reliant upon social security
Unemployment	Unemployment rate
Migration/ethnicity	Number of residents born abroad (not only nationals)
	Residents who are not EU Nationals and citizens of a country with a medium or low HDI
Education	Prop. of working age population qualified at level 1, 2, 3 4, 5 and 6 ISCED

Desirability to Seek Public Shelter

Not all displaced population will seek public shelter, and some may find alternative shelter accommodation (e.g., rent motel rooms or apartments), stay with family and friends, or leave the affected area. For estimations of shelter demand it is necessary to account various factors that lead to populations seeking public shelter. Desirability to seek public shelter in this study is given by an indicator model related to the "Access to Resources" which accounts for both "push" factors (such as low income, lack of mobility or having no social networks) and "pull" factors (such as being too far from the shelter sites). The "push" factors are determined in terms of socio-economic drivers, while the "pull" factor is an input from a GIS-based shelter accessibility model (Khazai et al. 2011a). The question of accessibility relates mostly to residents who are able to choose between different destinations (Table 4.4). The proximity and ease of access of shelter locations might be a key criteria for these households whose decision of leaving is not founded on aspects of vulnerability but on individual preferences. The Shelter Seeking Index (SSI) is then derived as an additive weighted sum of each of the indicators constituting the shelter seeking population and multiplied by how accessible each of the designated shelter sites are, according to:

$$\boldsymbol{SSI} = \boldsymbol{AI} \times \sum\nolimits_{j=1}^{n} \boldsymbol{w}_j \cdot \boldsymbol{I}_j \tag{4.5}$$

where:

- SSI = Shelter Seeking Index
- w_i = overall weight given to each indicator
- I_i = indicators representing shelter seeking population
- AI = Accessibility Index, derived from a GIS distance-cost analysis to shelter sites

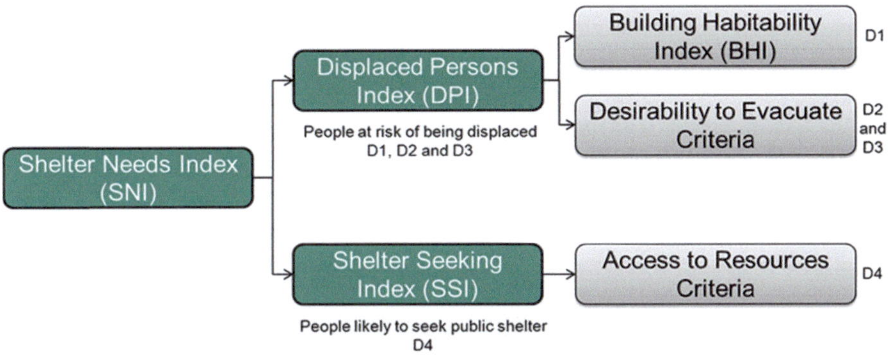

Fig. 4.8 Decision criteria for computing Shelter Needs Index (SNI)

4.2.2.3 Multi-criteria Shelter Model

The integrated shelter needs model developed here is based on a multi-criteria decision theory (MCDA) framework which allows the bringing together of parameters influencing the physical inhabitability of buildings, with social vulnerability (and coping capacity) factors of the at-risk population to determine as well as external factors to determine the desirability to evacuate and seek public shelter. As shown in Fig. 4.8, the multi-criteria framework can be described schematically as composed of the two main criteria: overall population at risk of being displaced after an earthquake (DPI) and the proportion of this population likely to seek public shelter (SSI). Subsequently, the total demand for public shelter for a particular location (i.e., city district) can be described as a product of the population at risk of being displaced (D1, D2 and D3) to the population likely to seek public shelter (D4). This can be expressed by the equation below where w_{DPI} and w_{SSI} are the weights assigned to DPI and SSI, respectively:

$$SNI = w_{DPI} \cdot DPI + w_{SSI} \cdot SSI \tag{4.6}$$

where:

SSI is derived from a weighted index related to lack of access of resources indicators in a community or neighborhood, and DPI is given as occupants in uninhabitable buildings amplified by external and internal factors related to desirability to evacuate according to the following expression:

$$DPI = BHI\ (1 + DE) \tag{4.7}$$

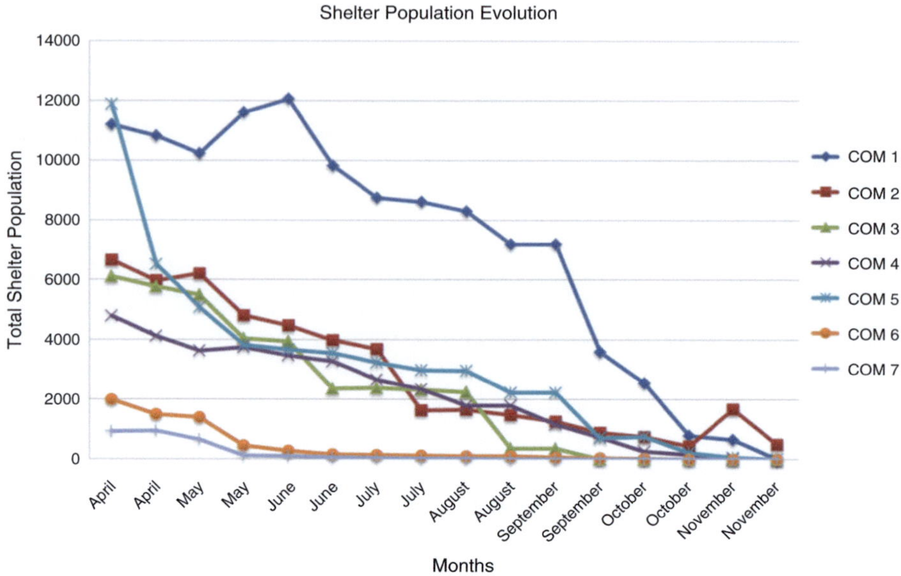

Fig. 4.9 Shelter population evolution in each COM after the 2009 L'Aquila event

4.2.3 Shelter Demand Model Implementation

To demonstrate the shelter methodology it has been applied to the 2009 L'Aquila earthquake, where detailed data on post-earthquake *Building Usability* (AEDES Survey of 1,667 buildings); *Socio-economic* data for 106 fractions (ISTAT data); and *Shelter Population* data from April to August 2009 for 107 shelter sites (Italian Civil Defence) was used to validate the model (Khazai et al. 2012c). For each Mixed Operations Centres (COM), information was collected about the number of shelter sites, the total number of shelters, and the population in the shelters. When shelter population evolution is compared between the different COMs as shown in Fig. 4.9, it can be seen that in most COMs a drop in shelter population is observed after the first month. The most drastic development is observed in COM 5, where there is a drop of almost 60 % in shelter population from April to May. Contrary is the development in COM 1 with an increase in people in shelter until end of June. This increase could partly account for the loss in COM 5 supporting the assumption that migration may have taken place between displaced populations of the different COMs.

The shelter model methodology was implemented into the Multi-Criteria Decision Analysis (MCDA) which is the basis of the software developed in SYNER-G (Khazai et al. 2012b). The tool will allow stakeholders to display the Shelter Needs ranking of different neighborhoods using various output and visualization formats. The user can assign different importance (weights) to selected indicators and the

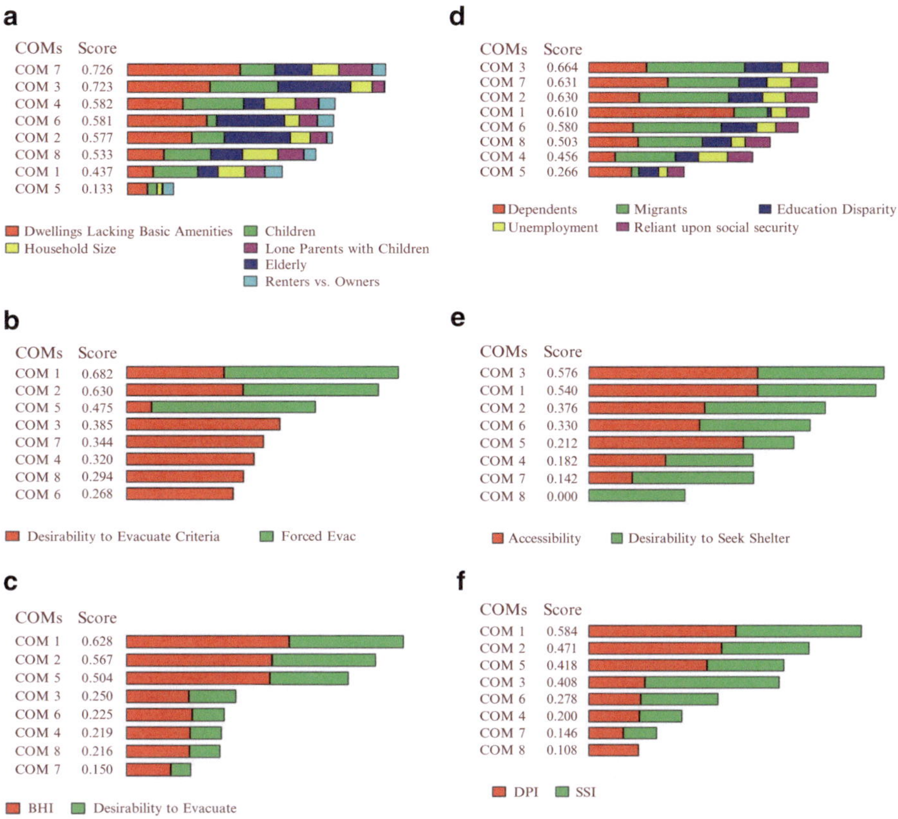

Fig. 4.10 Ranking of the displaced persons (*left*, 6a–c) based on the Building Habitability Index (*BHI*) and the desirability to evacuate criteria. Ranking of the Shelter Needs Index (*right*, 6d–f) based on the Desirability to Seek Shelter (SSI) criteria and the Displaced Persons Index. (**a**) Desirability to Evacuate. (**b**) Desirability to Evacuate (DE) given forced evacuation of city centre. (**c**) Displaced Persons Index (DPI). (**d**) Desirability to Seek Shelter. (**e**) Desirability to Seek Shelter (SSI) given Shelter Accessibility. (**f**) Shelter Needs Index (SNI)

tool can be used to discuss the weighting outcomes and interactively examine the variability of shelter demand in different areas for different weighting schemes, or for different earthquake scenarios. The rankings for shelter demand after the L'Aquila earthquake are shown in Fig. 4.10 for the eight COMs which had the overall coordinating role in their own territories for all rescue and shelter provision operations. First the Displaced Persons Index (DPI) is obtained as the number of occupants living in uninhabitable buildings (BHI) amplified by the Desirability to Evacuate Criteria. In this case, the proportion of persons in uninhabitable buildings was not modelled following the methodology but taken directly based on observed values of partially usable and non-usable buildings in each of the eight COMs from the AEDES Survey. Furthermore, in the calibration of the shelter

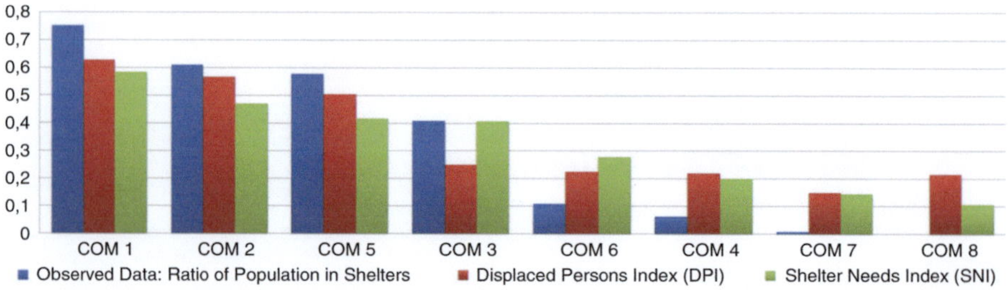

Fig. 4.11 Ratio of actual population in shelters (observed data) shown against the ranking of displaced persons and shelter needs in the eight COMs

model people living in the historical city centre were recommended to evacuate without consideration of unique building stability due to historical buildings and narrow alleys. Accordingly, the Desirability to Evacuate criteria accounts for forced evacuations in COM1, 2 and 5 (Fig. 4.10b).

To obtain the Shelter Needs Index shown in Fig. 4.10f, the Desirability to Seek Shelter Indicators (Fig. 4.10d) were obtained and amplified based on accessibility to shelter sites in the eight COMs (Fig. 4.10e). Finally, the Shelter Needs Index (SNI) is obtained as the interaction between the Displaced Persons Index and Shelter Seeking Index (SSI). Fig. 4.11 shows how the modeling approach can be used to capture the actual shelter demand conditions (given as the observed number of people in shelter camps normalized by total population in each COM). For example, based on building usability alone, COM 3 should have a lower shelter demand than COM 6 and 4. However, given the high desirability to evacuate and seek shelter based on socio-economic indicators, COM3 obtains a more realistic ranking.

4.3 Health Impact Model

4.3.1 Health and Healthcare System Impacts of Earthquakes

Health impacts after earthquakes include mortality and morbidity in term of injuries, disability, psychological effects, inadequate treatment of non-communicable and chronic diseases (e.g. problem with drug procurement), and increased transmission of infectious diseases (e.g., caused by parallel systems dysfunction such as water and sanitation). Health impacts are the result of direct consequences of the physical impact of the earthquake (collapsing buildings, falling objects, falls due to shaking, traffic accidents, secondary hazards such as tsunami or landslides), and additionally, systemic failures of healthcare systems and parallel infrastructure (water and sanitation system, emergency housing system, transport and communication system) can as well lead to increased secondary fatalities (Gunn 1995; Howard et al. 1996; Ramirez and Peek-Asa 2005; Aldrich and Benson 2008).

Studies, models, and guidelines related to health impacts in earthquake engineering have focused largely on casualty estimation methodologies (Coburn and Spence 2002; FEMA 2006), and hospital performance largely in terms of physical damage to structural systems and in some cases nonstructural systems and equipment (e.g., HAZUS model, FEMA 2003). Casualty estimation methodologies generally exclude casualties due to secondary causes, and do not account for injuries which can digress to fatalities as a result of systemic failures of healthcare systems and parallel infrastructure (e.g., transport, power, etc.). Systemic failures in healthcare delivery and lack of access to food and shelter can also lead to the exaggeration of baseline diseases and increased transmission of communicable infectious diseases. Several researchers, have proposed methods to assess interrelated systems – structural, nonstructural, lifelines, and personnel – according to performance levels indicating functionality (Chang et al. 2009; Lupoi et al. 2008; Yavari et al. 2010). Furthermore, systemic vulnerabilities in the healthcare or infrastructure systems can aggravate the overall health impact after earthquakes beyond the direct fatalities, for example, by the exaggeration of baseline diseases and increased transmission of communicable infectious diseases.

Health impacts (mortality and morbidity, population in need of medical treatment) after earthquakes are also influenced or aggravated by social factors that are best described using the term social vulnerability. This includes latent vulnerability conditions in the at-risk population and systemic failure in the healthcare delivery system. Although there is wide evidence that social vulnerability is a key component for influencing health outcome of disasters (see Sect. 4.3.2.3), it is seldom linked to common formal and quantitative seismic loss estimates of injured people which provide direct impact on emergency health care services. To link social impacts of health and health-care services to a systemic seismic vulnerability analysis, a conceptual model on health and the health care systems has been developed in SYNER-G.

4.3.2 Basic Elements of the Health Impact Model

The aim of the Health Impact Model is to extend earthquake casualty estimation methods by developing a combined engineering and social science approach for modeling earthquake health impacts. As such, the approach presents a new method for modeling health impacts caused by earthquakes by linking casualty estimation methods typically used in Earthquake Loss Estimation (ELE) to key factors of individual health and health-care systems. The methodology provides an operational framework for implementing the different factors into an analytical hierarchical process model, and deploying them using indicators following the principles of Multi-criteria Decision Analysis. The integrated approach for estimating post-earthquake health impact with a special focus on social vulnerability is composed of the following basic components which are described in the following sections.

4.3.2.1 Casualty Estimation Model

Casualty estimation methodologies in earthquake engineering (Kawasumi 1954; Whitman et al. 1975; Wiggins 1977; Christoskov and Samardjieva 1984, 2013; Coburn and Spence 2002; FEMA 2006, etc.) following earthquakes provide estimates of both injuries and fatalities, which is a key input to assist planners in determining the resources required to deal with the increased surge in the patients. Most models provide "in-door" casualty estimates from structural building collapse and some provide non-structural casualties as a ratio of structural damage. It has been observed that 80 % of fatalities attributed to earthquakes have been caused by the collapse of buildings (Coburn and Spence 2002). This has changed significantly with the past earthquakes in the last decade but the overall percentage has stayed around the same (77 %) with Marano et al. (2010) from 749 fatal events from 1968 to 2008 and Daniell et al. (2012) from 2020 fatal events from 1900 to 2012 that 71 % of deaths are due to earthquake shaking. As the construction of RC buildings is increasing, the portion of casualty victims in RC buildings is rapidly increasing. This can also be attributed to the fact that RC structures built in the poorer countries are highly vulnerable and when they collapse they are considerably more lethal and kill a higher percentage of their occupants than masonry buildings.

Starting with the first casualty model by Kawasumi (1954), earthquake casualty methodologies seek to define a relationship between a ground motion parameter and/or damaged buildings and the potential number of fatalities. In addition to models that rely on expert opinion (Whitman et al. 1975; Wiggins 1977), others have attempted to empirically derive relationships between casualties and a form of magnitude, PGA or other ground motion parameters (Christoskov and Samardjieva 1984; Coburn 1986; RGELFE 1992 and Surahman 2000). Other approaches (Alexander 1982; Ohta et al. 1983; de Bruycker et al. 1985; Coburn and Spence 1992; Zuccaro and Cacace 2010) have used the damage state of a building after the event to estimate the potential casualties via empirically derived data.

The difficulty of correlating casualties to a ground motion parameter is due to the fact that different parts of the world design buildings and structures differently, thus there can be a large discrepancy in the damage patterns seen between two countries for the same PGA and such models are generally not transferrable to another region or context. There are also many difficulties in correlating casualties against building damage states, including: damage states are notoriously difficult to classify into correct brackets; non-structural or furniture related deaths are generally not taken into account; and the fact that in-depth fatality counts are generally not known for each building post-event with only scattered studies detailing these fatalities.

Many new and old methodologies are based primarily on intensity with Berberian (1978), Ohashi and Ohta (1983) and Coburn et al. (1987) to name a few. Recent models of authors such as So (2009), Jaiswal and Wald (2010), Maqsood and Schwarz (2011) have concentrated on earthquake intensity as the hazard metric vs. population ratios or by the population in collapsed buildings using expert opinion related collapse ratios using historical data. By definition, intensity links hazard and damage in one measure and this method attempts to alleviate some of the key uncertainties in casualty estimation, which include:

- Casualty statistics (types and numbers) from past earthquakes are often inconsistent and unreliable (i.e., lack of standardization of injury data and established methodologies for reporting casualty data).
- Casualty statistics very often do not provide information about cause of death (e.g., structural, non-structural, other causes)
- Casualty statistics do not discern which building type (e.g., RC, Masonry, etc.) the casualty figures come from
- Lethality Ratios used in CEMs are often engineering factors and not based on empirical/historic data
- Uncertainties in population per building (i.e., uncertainties and incompleteness of data relating occupants to building volume)
- Uncertainties in building occupancy at the time of earthquake (day and night variability as well as seasonal variability due to inflow of tourists or students)

As a result of the many uncertainties existing casualty models have failed to convey the degree of confidence to which estimates are given. This is extremely important in any decision making process as administrators or policy makers must be aware of the margins of error to make informed decisions. The seismic community has so far failed to disseminate data and stimulate responses from national and government organizations effectively (Spence 2007).

Casualty Estimation Model in SYNER-G

As casualty ratios (CR) used in casualty loss estimation depend on the particular building typologies, building practices and living arrangements in each region, a global or pan-european casualty model is not feasible. The casualty model developed in SYNER-G provides an initial (direct) casualty estimate for occupants of buildings at the time of earthquake based on an original idea developed by Coburn and Spence (1992). However, the casualty model developed in SYNER-G has several new components and considerations compared to available casualty models. The features of the SYNER-G casualty model are described below:

- The model estimates casualties from all damage states. Most casualty models determine casualty as a function of building collapse only. While building collapse is the dominant factor, So and Spence (2013) show using historic data (Pakistan, Indonesia and Peru) that casualties can also occur in moderate and low damage states.
- The model estimates casualties using semi-empirical casualty ratios. The process of determining casualty ratios is often unclear and in many cases it is an engineering judgement based on historic evidence. The model optimizes casualty ratios for regions with comparable building construction.
- In addition to determining casualty ratios as a function of building damage, the model also considers Seismic Intensity.

Table 4.5 Building-casualty superclasses defined for the selected Italian events

Casualty potential	Superclass category	Construction typology
Very high	1-BC	Reinforced concrete
High	2-BC	Stone, brick or block masonry walls with reinforced concrete floors/roofs
Moderate	3-BC	Stone, brick or block masonry walls with timber rubble masonry, timber or steel joist floors/roofs

- The model estimates casualties by proposing a "Building-Casualty" Superclass based on the propensity of different building typologies in producing casualties.

In the SYNER-G casualty model, casualty and building damage data has been collected for two large Italian earthquakes: 214 municipalities (deaths and injured) for Irpinia 1980 and 26 municipalities (deaths) in Friuli 1976. Based on the building type designations in the survey data, 3 "Superclasses" of building typologies have been defined and shown in Table 4.5. The choice of the superclasses are based on casualties produced by different building types in Italian earthquakes and knowledge of the reduction of volume in different building types which influences the number of trapped people and hence casualties (References). It should be emphasized that building vulnerability classes (e.g., EMS classes A-F) do not represent the potential to produce casualties in buildings and only represent the vulnerability of buildings to damage. Hence a novelty in this methodology are the superclasses, which discern between different building classes of equal damage state (e.g. slight or collapse) in terms of the relative casualty level which can be produced in each of the building classes. However, as no empirical data as such is available to validate the choice of the superclasses, a sensitivity analysis is performed to determine the influence of change in superclasses.

The aim is to produce a semi-empirical approach by which Casualty Ratios would be derived from empirical data on building damage classes and Super classes of building typologies. In this approach Casualty Ratios (CR) for an Italian model are obtained by optimization with casualty and damage data for the selected Italian earthquakes which occur in comparable settings and are night-time events. The approach of estimating casualty ratios can be described in the following steps:

1. The building stock of any region is grouped in terms of its distribution in Building-Casualty super classes (Table 4.5) and 6 EMS Vulnerability Classes (A-F) at the event year.
2. A database for each historic earthquake event is constructed, where for each administrative unit (e.g., municipality) the following parameters are captured: (a) Number of dead and injured in each Municipality; (b) Number of total buildings in each Municipality, and (c) Population at time of event for each Municipality
3. Distribution of Building Damage states (none, slight, moderate, severe, collapse) in each Municipality
4. Distribution of Seismic Intensity (EMS scale) in each Municipality

Table 4.6 Casualty ratios derived from optimization algorithm

Intensity	Superclass	D0	D1	D2	D3	D4	D5
6	1-BC	0	0	0	0.0011	0.0027	0.0067
	2-BC	0	0	0	0.0005	0.0013	0.0033
	3-BC	0	0	0	0	0.0007	0.0017
7	1-BC	0	0	0.0009	0.0021	0.0053	0.0133
	2-BC	0	0	0	0.0011	0.0027	0.0067
	3-BC	0	0	0	0.0005	0.0013	0.0033
8	1-BC	0	0.0009	0.0021	0.0053	0.0133	0.0333
	2-BC	0	0	0.0011	0.0027	0.0067	0.0167
	3-BC	0	0	0.0005	0.0013	0.0033	0.0083
9	1-BC	0	0.0048	0.0073	0.0182	0.0454	0.1136
	2-BC	0	0.0024	0.0036	0.0091	0.0227	0.0568
	3-BC	0	0.002	0.003	0.0076	0.0189	0.0473

5. Allocation of Building Damage states for each Building-Casualty superclass (1-BC, 2-BC and 3-BC) and Building Vulnerability EMS class (A-F).
6. Allocation of Seismic Intensity for each Building-Casualty superclass (1-BC, 2-BC and 3-BC) and Building Vulnerability EMS class (A-F).
7. Allocation of Total Building Occupancy for each Building-Casualty superclass (1-BC, 2-BC and 3-BC) and Building Vulnerability EMS class (A-F).
8. Using the above parameters the number of dead is simulated in an initial run with an assumed Casualty Ratio matrix using values published in the literature Coburn and Spence (2002), So and Spence (2013) and ATC 13.
9. Casualty ratios are optimized in subsequent iterations using an optimization algorithm so that a best-fit is achieved between simulated and surveyed casualty numbers (Table 4.6).

The optimization algorithms use constraints and restrictions which are derived from common assumptions (e.g., the casualty ratio for moderate damage state should not be greater than casualty ratio for collapsed damage state). Lower and Upper boundary functions are also defined based Coburn and Spence (2002), So and Spence (2013), ELER (Erdik et al. 2011), and ATC-13 (HAZUS). Using the optimized Casualty Ratios for the region, the number of deaths are determined for that region, using the mean inhabitants by building type, and occupancy rate by day and night. Combining the death records for the Friuli and Irpinia events according to seismic intensity, the casualty model proposed here was compared against the 2008 Coburn and Spence casualty model developed for LessLoss (Casualty Ratios only in D5 damage state) and recorded deaths (Fig. 4.12). As the graph is plotted on a logarithmic scale, the relative comparison of both models is best seen through the average error bars. It can be seen that Coburn and Spence overestimate deaths at low seismic intensities, while they at higher intensities the average error is comparable.

$$N_d = \sum_{t=1}^{3} \sum_{d=1}^{6} \sum_{i=1}^{5} N_{t,d,i}\ CR_{t,d,i}\ NO_t \tag{4.8}$$

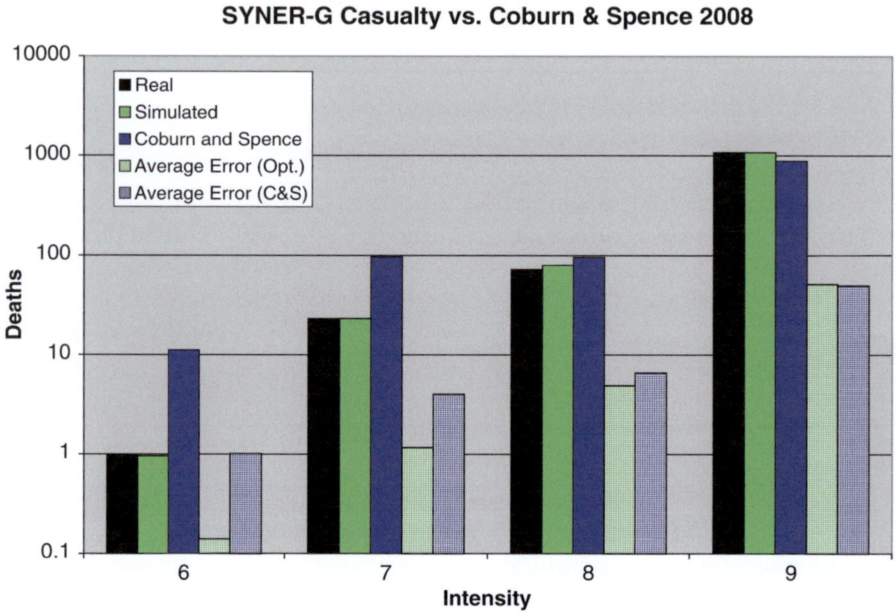

Fig. 4.12 Comparison of SYNER-G and Coburn and Spence (Less Loss) casualty models for Italy

where:

- t = building-casualty type (t = 1-BC. 2-BC. 3-BC)
- d = damage level (d = D0, D1, D2, D3, D4, D5)
- i = seismic intensity level (i = VI, VII, VIII, IX, X)
- $N_{t,d,i}$ = number of buildings of type *t* having damage level *d* at seismic intensity level *i*
- $CR_{t,i}$ = proportion of deaths by building type, damage level and seismic intensity
- NO_t = number of occupants (at the time of the event) by building type *t*

4.3.2.2 Transportation Accessibility Model

Health impacts of earthquakes like mortality and morbidity in term of injuries, disability, inadequate treatment of non-communicable and chronic diseases (e.g., problem with drug procurement), and increased transmission of infectious diseases (e.g., caused by parallel systems dysfunction such as water and sanitation, communication and transport) is directly related to transport accessibility to healthcare facilities. Transport accessibility basically refers to ability to reach health care services by using various travel costs on the considered road network. As many factors affect transport accessibility, different aspects and measures of accessibility are found in the literature (Litman 2011). However, in order to better assess accessibility to healthcare facilities it is essential to use network-based accessibility analyses. When emergency healthcare accessibility is considered, use of Geographical Information Systems (GIS) becomes essential to assess spatially accessibility zones in terms of emergency services, and help medical emergency service providers determine the extent to which a city is ready for any medical emergency. For example, physical accessibility of medical emergency services can be measured to check if urban/rural areas are highly accessible by medical emergency vehicles (ambulances) within 5 or 10 min of critical time threshold.

The analysis of healthcare accessibility has inherent complexity (Pedigo and Odoi 2010). Pedigo and Odoi (2010) refer to Guagliardo (2004) and Higgs (2009) as the researchers who defined -accessibility to healthcare as the ability to obtain potential healthcare that may be impeded by both spatial, like travel impedances, and non-spatial factors like the ability to pay. However, in the case of earthquakes, physical accessibility, which is assessed spatially, to healthcare services, has direct relation to health impacts. Pedigo and Odoi's (2010) work is a typical example of healthcare accessibility analysis related to health impact dimension of socio-economic vulnerability, where spatial accessibility to emergency care for stroke and Myocardial Infarction (MI) in the East Tennessee Appalachian Region is evaluated within a GIS to identify the spatial extent of regions which have inadequate accessibility to emergency care units for stroke and MI. McGrail (2012) provides a methodology for determining primary healthcare services accessibility by considering the volume of healthcare services provided relative to the population's size and the proximity of health services provided relative to the location of the population. The proposed methodology by McGrail (2012) is based on two-step floating catchment area (2SFCA) method, which includes a distance-decay function within a catchment area together with variable catchment sizes. Yiannakoulias et al. (2013) investigate effect of two different travel cost measures on health care service accessibility. They compare simple network travel cost metric (requiring only data on speed limit and road segment length) and a more complex network travel cost metric (requiring estimates of traffic congestion and intersection turn costs) on a gravity-based measure of accessibility to primary healthcare services. Yiannakoulias et al. (2013) find that the city center is less sensitive to change in travel cost matrix as compared to outlying regions and more complex network travel cost metric (requiring estimates of traffic congestion and intersection turn costs) provides more realistic accessibility results. Ertugay and Duzgun (2011) present health service accessibility based on a stochastic approach, where uncertainties related to the catchment boundaries are considered and probabilistic catchment boundaries are determined instead of crisp catchment boundaries in the deterministic model. Düzgün et al. (2011) provide a framework for vulnerability assessment of urban environment in neighborhood scale where in addition to building fragility and socio-economic vulnerability, the accessibility of neighborhoods by the critical services like fire brigades, ambulances etc. are modeled as one of the vulnerability component of the neighborhoods. Horner and Widener (2011) investigate the impacts network failures, which are predicted by simulations made based on various hurricane disaster scenarios, on disaster relief planning strategies, for Florida City.

They report that the modest disruptions to the transportation network yields marked changes in the number and spatial configuration of relief facilities. Bono and Eugenio (2011) suggest a method based on combining simple graph theory and GIS-based spatial analyses to define the urban accessibility landscape in the aftermath of earthquake damage. They use open source data for Port Au Prince and evaluate difference between the undamaged and damaged networks to measure potential decrease in transport performance as a result of the earthquake.

Transportation Accessibility Model in SYNER-G

In SYNER-G two accessibility analysis approaches for healthcare services in case of earthquake are developed at the urban scale. The first method is based on evaluating accessibility by using indicators of accessibility, which can be obtained from publically available socio-economic data in Europe (e.g., EUROSTAT Urban Audit Database). A comprehensive overview of factors affecting transport accessibility provided by Litman (2011), include: transportation demand and activity, mobility, transportation options, user Information, integration-terminals-parking, affordability, mobility substitutes, land use factors, transportation network connectivity, roadway design and management, prioritization, value of inaccessibility. Among the various 'Urban Audit' data groups investigated the following show a high degree of relevance to the indicators listed by Litman (2011) for assessing accessibility in urban environments: population structure (population of working age, proportion of children and elderly, demographic dependency, migrant population) which indicates transportation demand and activity; economic aspects (labor market, economic activity, income disparities and poverty), which forms the basis of various transportation option's affordability; travel patterns (vehicle ownership, mode of journey to work, percentage of commuters, length of public transportation network), which relates to transportation options and integration-terminals-parking; information society (user and infrastructure, local government and ICT sector), which is relevant to express user information. Hence by using these indicators relative accessibility levels of various urban environments can be assessed by using multi criteria decision approach.

The second approach is based on use of transportation data in GIS with prediction of road closure for various earthquake scenarios, following a three-step procedure: (1) data acquisition and integration; (2) traveling cost calculation; and (3) accessibility modeling and visualization. Data acquisition and integration phase includes compilation of data related to road network as well as supply and demand. Transportation network data also have attributes of road segments indicating their type and hence capacity such as main roads, secondary roads etc. The supply data constitutes locations of healthcare service and their capacities. The demand data involves boundaries of administrative zones and centroids of these zones so that accessibility of each administrative zone can be calculated. Usually these data are obtained in various formats, which require integration in GIS environment.

Table 4.7 Reduction in average traveling speed as function of road closure probability

Road closure probability	Transportation network cost
0 %	30 km/h (major roads)
	25 km/h (minor roads)
≤0.25	Reduce by 25 %
0.25–0.50	Reduce by 50 %
0.50–0.75	Reduce by 75 %
≥1	Reduce by 100 %

Traveling cost calculation phase involves determination of traveling costs for each road segment on the transportation network according to road closure probabilities. The road closure probabilities can be obtained based on either the structural analysis of the road segments or prediction of road closure due to building collapse, which is estimated by structural analyses of buildings or both. Finally, accessibility modeling and visualization phase is computation of health service accessibility scores and their visualization in a GIS environment.

The health-care accessibility model is implemented as a pilot application for the city of Thessaloniki based on three different accessibility modeling techniques: isochronal-based technique (travel time measure); zone-based technique (cumulative time from each health service to each administrative district centroid without considering the scale of the health service); and zone-based technique (gravity measure). The basic data used in the model are:

- Transportation network (line): The main transportation network data which is considered to be important in case of emergency (the city of Thessaloniki as case study example). The local roads, which are expected to have higher blockage probability due to building collapses, are not taken into account.
- Supply locations (point): The locations of healthcare services (Thessaloniki as a case study) and scale of healthcare service data are also taken into account. The scale of healthcare service indicates the capacity of the healthcare service and hence used as gravity measure where high capacity healthcare services attract more healthcare seekers.
- Demand locations (area): The administrative districts of the urban areas (Thessaloniki in this case) as defined by the sub-city district boundaries of the EUROSTAT Urban Audit.
- Road closure probabilities due to building collapses, soil liquefaction, bridge damages and overpass bridge collapses were calculated. The average number of damages over 10,000 runs are presented for the estimating road closure probabilities which are described in Chap. 7.
- Depending on the road closure probability different average speeds were assigned to the road segments to compute the traveling costs (Table 4.7). For the road segments that have road closure probability scores greater than zero, the scores are first normalized to have a range of 0 and 1, then classified into four categories as shown in Table 4.7 and used in calibration of the transportation network costs.

As the performance of the network is affected at the same time from all of the different types of damages, the results are aggregated into one group in order to provide a more complete view of the accessibility. In particular, the results that include the road closure probabilities due to liquefaction, building collapses, bridge and overpass damages are aggregated into a new group which is named as "overall" accessibility. When there is an overlay case during aggregation process, the maximum of the probability scores is considered in calculation of the transportation network costs. Five minutes (300 s) cost is accepted as a critical time threshold for the health service accessibility. An example accessibility analysis result is given in Fig. 4.13, the results for the combined health service accessibility are shown for both the isochrones-based and zone-based techniques.

4.3.2.3 Social Vulnerability Factors Influencing Health Issues

Beyond the parameters used in earthquake casualty estimation to assess and describe expected mortalities and injuries (earthquake intensity, building class, occupancy, time of day, building damage state), social vulnerability and health care system vulnerability can aggravate the overall health impact after a disaster. To identify the most critical factors determining post-earthquake health impacts and to deduce appropriate indicators representing these factors, a literature research focusing on case studies on impacts of earthquakes on individual health and on the health care system was carried out and supplemented by publications giving a general overview on health and health care system impacts from earthquakes and other disasters (Gunn 1995; Alexander 1996; Howard et al. 1996; Ramirez and Peek-Asa 2005; Watson et al. 2007; Aldrich and Benson 2008; Bartels and VanRooyen 2011).

The case studies used to identify the most critical factors cover 15 earthquakes[1] worldwide and rely on various research designs:

- epidemiological studies that analyze earthquake-related mortality and injury, trauma, infectious diseases studies (Chan 2008; Shi et al. 2010; Guha-Sapir and van Panhuis 2009; Phalkey et al. 2011; Karmakar et al. 2008; Matsuoka et al. 2000; Tanaka et al. 1999),
- studies that combine an epidemiological approach with socio-economic variables to identify risk factors for mortality and morbidity (Chou et al. 2004a; Liang et al. 2001; Peek-Asa et al. 2003),
- studies dealing with particular injuries such as burns (Nakamori et al. 1997) or crush syndrome (Vanholder et al. 2007; Erek et al. 2002; Oda et al. 1997),

[1] Tohoku earthquake 2011 (Japan), L'Aquila 2009 (Italy), Wenchuan 2008 (China), Noto Peninsula 2007 (Japan), Kashmir 2005 (India, Pakistan), Sumatra 2004 (Indonesia), Niigata-Chuetsu 2004, Japan, Gujarat 2001 (India), Chi-Chi 1999 (Taiwan), Izmit 1999 (Turkey), Kobe 1995 (Japan), Northridge 1994 (USA), Loma Prieta 1989 (USA), Spitak 1988 (Armenia), Irpinia 1980 (Italy).

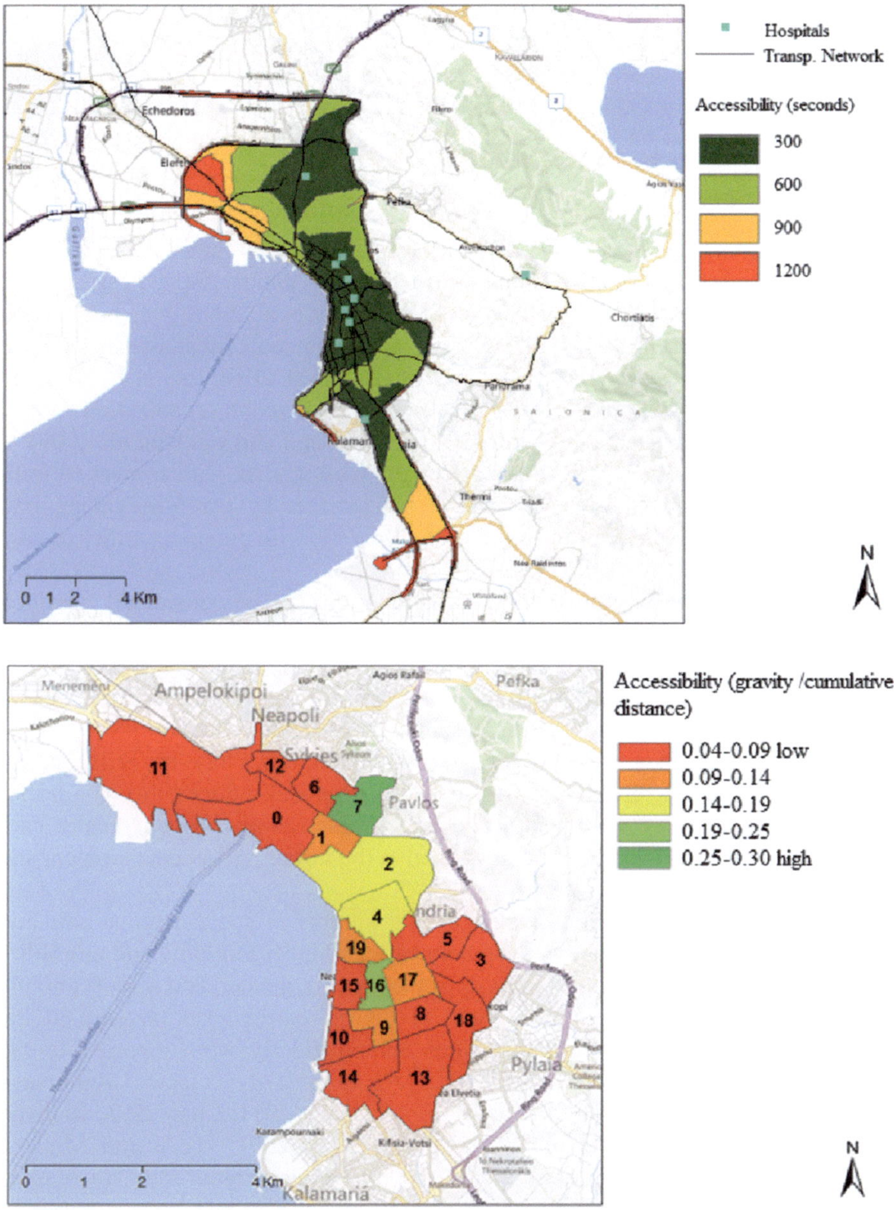

Fig. 4.13 Combined healthcare accessibility example results for Thessaloniki using the isochrones-based technique (*top*) and zone based techniques (*bottom*)

infectious diseases (Guha-Sapir and van Panhuis 2009; Karmakar et al. 2008; Vahaboglu et al. 2000; Schneider et al. 1997), impacts on mental health (Başoğlu et al. 2002; Chan et al. 2012), or on reproductive health (Liu et al. 2010; Hibino et al. 2009; Chang et al. 2002),

- studies that consider pre-existing health conditions such as chronic disease (Mori et al. 2007),
- studies that investigate long-term health impact outcome (Zhang et al. 2011; Seplaki et al. 2006; Oda et al. 2002; Dobalian et al. 2011; Armenian et al. 1997),
- studies that focus on challenges and limitations faced in medical emergency response immediately after earthquakes (Roy et al. 2002, 2005; Chan 2008; Shi et al. 2010; Phalkey et al. 2011; McCurry 2011).

Correspondingly, the data analyzed in the case studies is rather heterogeneous. It ranges from hospital records (Guha-Sapir and van Panhuis 2009; Phalkey et al. 2011; Tanaka et al. 1999), laboratory tests (Karmakar et al. 2008; Vahaboglu et al. 2000), field observations (Chan 2008) or rapid assessments (Takahashi et al. 2011; McCurry 2011), and surveys among survivors (Kun et al. 2010; Roy et al. 2002), to data of statistical offices and national and international health organizations (Chou et al. 2004b; Muramatsu and Akiyama 2011; Guha-Sapir and van Panhuis 2009).

Analyzing and comparing the results of the case studies with respect to individual, social, health and health care system factors by carefully taking into ac-count the heterogeneity of the studies and limitations in generalizability of case studies, a number of recurring variables emerged across several earthquakes that influenced the health impact of earthquakes and that can be considered classified as contributing to risk factors for negative health outcomes after earthquakes. The variables were categorized in five factors that potentially aggravate health impact and that were later on used to define indicators for the further analysis: social vulnerability, environmental health, baseline health status, health care capacity, infrastructure (Table 4.8). Three of them (social vulnerability, baseline health status, health care capacity) represent different facets of social vulnerability conditions (socially vulnerable groups, health and health care status, hospital health care capacities) and should therefore be integrated in the further model development and implementation.

The factor social vulnerability subsumes a number of socio-demographic and socio-economic variables that were associated with higher mortality and morbidity rates: age (elderly and children) (elderly and children), gender, and socioeconomic status, physical disability. Across most of the 15 earthquakes covered in the literature research, higher mortality rates among elderly were reported in the majority of the case studies that considered age as variable or risk factor (Muramatsu and Akiyama 2011; Suzuki and Kim 2012; Alexander 2011; Chan 2008; Sullivan and Hossain 2010; Chou et al. 2004a; Liang et al. 2001; Osaki and Minowa 2001; Tanaka et al. 1999; Peek-Asa et al. 2003), but also higher rates for injuries (Nakamori et al. 1997; Peek-Asa et al. 2003), pneumonia and heart diseases in the immediate phase after the earthquake (Matsuoka et al. 2000). Also for children higher mortality and injury rates were observed (Chan 2008; Sullivan and Hossain 2010; Roy et al. 2002; Chou et al. 2004b; Liang et al. 2001). Concerning gender, the literature includes a significant number of case studies that reported higher female rates in mortality (Kun et al. 2010; Chou et al. 2004b; Liang et al. 2001), injuries (Roy et al. 2002; Armenian et al. 1997; Peek-Asa et al. 2003; Nakamori et al. 1997)

Table 4.8 Indicators for the main aggravating factors for health and health care impacts after earthquakes for systemic seismic vulnerability analysis

Factor	Factors/indicators (urban audit)
Social vulnerability	
Elderly	Proportion of population above 60 years
Children	Proportion of children <5 years
	Proportion of children in schooling age
Female	Proportion of women
living alone	Proportion f single-headed households
Low education	Proportion of population with low school education
Low income, relying on social welfare	Proportion of population relying in social welfare
	Proportion of population with low income
Environmental health	
Access to water and sanitation, hygiene	Proxy: proportion of ppl. having access to water
Crowded shelter	Proxy: population density
Weather conditions	has to be included for particular case
Baseline health status	
Chronic disease, handicaps	Proportion of ppl. with chronic disease
Vaccination	Vaccination status (measles)
Health Status	Proxy: life expectancy at birth
	Proxy: child mortality
Health care capacity	
Medical personnel	No. of physicians, surgeons, orthopedist per 100,000
	No. of nurses per 100,000
Hospital capacity	No. of hospital beds per 100,000
Hospital functionality	Has to be included for particular case based on model calculations

and mental impact rates such as posttraumatic stress disorder syndrome PTSD (Tural et al. 2004). Regarding the socio-economic status, being divorced,/widowed or living alone turned out to be a negative factor for physical and mental health outcome (Kun et al. 2010; de Bruycker et al. 1985), and also low income, relying on social welfare or living in low-cost buildings were observed as variables associated with higher mortality rates (Chou et al. 2004b; Alexander 2011) or more emotional distress (Dobalian et al. 2011).

Environmental health factors cover characteristics of the natural and build environment that have an immediate impact on health in the emergency phase after earthquakes and that can aggravate the health situation. Examples from the case studies are cold and harsh weather conditions that impede immediate medical help (Chadda et al. 2007; Noji et al. 1993) or may lead to a higher rate of Acute Respiratory Infections (Matsuoka et al. 2000), warm weather conditions that favor the spreading of vector borne infections such as the Valley Fever (Coccidioidomycosis) outbreak after the Northridge earthquake (Jibson 2002; Schneider et al. 1997). Also interruption of access to clean water and sanitation has been reported as factor across the case studies that had an influence on increasing gastro-intestinal infection rates (Kun et al. 2010; Liu et al. 2010; Karmakar et al. 2008; Guha-Sapir and van Panhuis 2009; Matsuoka et al. 2000). Finally, changed living conditions in crowed shelters has been observed as factor associated with increased death rates, pneumonia and heart disease (Matsuoka et al. 2000).

The factor pre-earthquake *baseline health status* as influencing factor refers to constrained individual health conditions (people with chronic diseases, functional limitations, bedridden people) which may lead to increased mortality and an increased rate of negatively aggravated health effects of an earthquake (Muramatsu and Akiyama 2011; Suzuki and Kim 2012; Osaki and Minowa 2001; Dobalian et al. 2011; Chou et al. 2004a), but also to indicators such as a general high rate of major infections (Wen et al. 2009) or a low vaccination status against Tetanus (Chan 2008) as factors that could aggravate the health impact of earthquakes. Results subsumed to this factor include also an increased risk due to hospitalization in the month before the earthquake (Chou et al. 2004a) and pre-quake emotional distress (Dobalian et al. 2011).

The factor *health care capacity* summarizes those characteristics that limit and hinder the health care sector to respond immediately and adequately to the medical needs of the injured: collapse of hospitals, partial non- functionality or disruption of normal functionalities due to the physical impact of the earthquake (McCurry 2011; Suzuki and Kim 2012; Rossetto et al. 2011; Kuwagata et al. 1997; Noji et al. 1993; Roy et al. 2002; Chan 2008), and lack of medical doctors, nurses and hospital beds compared to the high influx of patients in the first days after the earthquake (Shi et al. 2010; Roy et al. 2002; Liang et al. 2001; Tanaka et al. 1998, 1999).

The factor *infrastructure* finally refers mainly to interruptions of roads, bridges, and railway tracks that are necessary to access the affected area and that, on the one hand, hamper the transportation of injured in due time to functioning hospitals for medical treatment (Takahashi et al. 2012; Shi et al. 2010; Wen et al. 2009; Vanholder et al. 2007; Chadda 2007; Roy et al. 2002; Liang et al. 2001; Noji et al. 1993), and that, on the other hand, can lead to an interruption in supply of necessary medical equipment and drugs (Suzuki and Kim 2012).

4.3.3 Health Impact Model Implementation and Conclusions

The integrated health impacts model developed here is based on a multi-criteria decision theory (MCDA) framework which allows the bringing together of parameters influencing the direct social losses represented through estimates of casualties and injuries, with factors related to overall health impact of the population at risk in an earthquake. As shown in Fig. 4.14, the multi-criteria framework can be described schematically as composed of the two main criteria: overall population at risk of mortality in an earthquake (represented by casualties) and an *Impact Factor* (IF). Subsequently, the overall health impact for a particular location (i.e., city district, county or country) can be described as the population at risk of mortality, amplified by the set of conditions that can aggravate the health impacts following a disaster which are derived as a weighted index of a set of indicators in four main

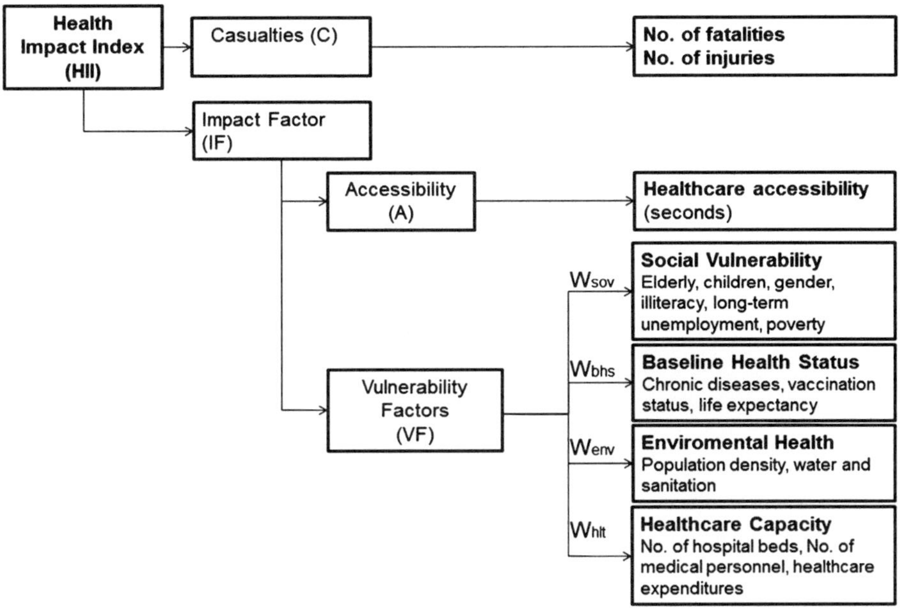

Fig. 4.14 Multi-criteria framework for health impact analysis, application on national level

categories of: social vulnerability, baseline health status, environmental parameters and healthcare accessibility. This can be expressed by:

$$HII = C \times (1 + IF) \tag{4.9}$$

$$IF = AI \times \sum\nolimits_{j=1}^{n} w_j VF_j \tag{4.10}$$

where:

- HII = Health Impact Index.
- C = Indicator representing the mortality ratio from an event. When using a casualty estimation model this is taken as the ratio of the fatalities to occupants at a given location.
- IF = Impact Factor.
- AI = Accessibility Index, derived from accessibility model.
- w_j = overall weight given to each indicator.
- VF_j = represented as the weighted sum of indicators representing the categories social vulnerability, baseline health status and environmental health and healthcare capacity.

According to the overall aim to implement a combined engineering and social science approach the next steps would have been to combine the developed Casualty

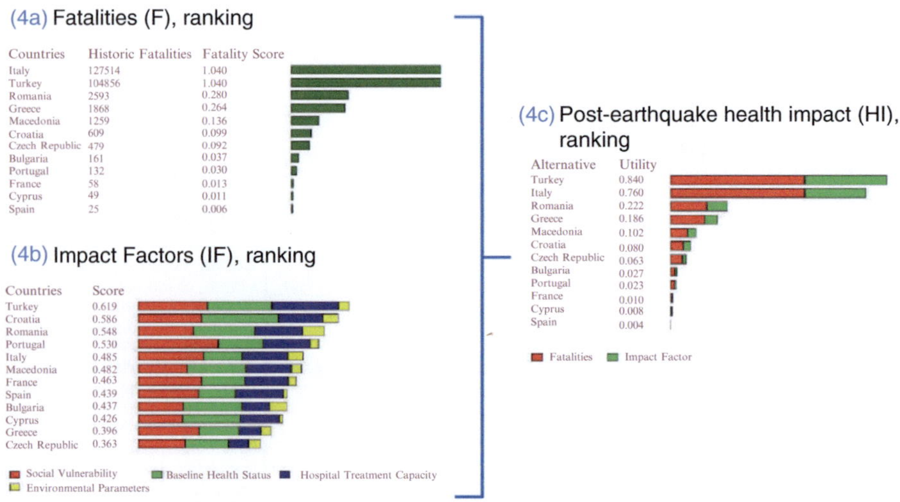

Fig. 4.15 Overall health impact (**c**) obtained as a product of fatalities (**a**) and impact factors (**b**)

estimation model with the hospitality accessibility index and the social vulnerability factors as aggravating factors on a sub-city level. Unfortunately this was not possible due to lack of data in the European Urban Audit on sub-city level for data on baseline health status and health care capacities that could have been combined with the Casualty Estimation for Italy or the transportation accessibility model developed for Thessaloniki, Greece as example. In order to test the approach at least in principle, it was decided to apply it on national level. Based on the literature research presented in Sect. 4.3.2.3, the identified 22 indicators that represent all of the main factors influencing health impact developed in the framework, with the exception of accessibility, were implemented in a multi-criteria hierarchical model (See Fig. 4.14 Multi-Criteria Framework with indicators).

To demonstrate the health impact methodology at the national level of analysis it has been applied to 12 South West European Countries representing the greatest seismic risk in Europe. The mortality ratio was obtained using detailed data on the number of dead persons from 244 events affecting 35 European countries from 1900 till 2012, which were reduced to the top 12 with the highest casualty estimates. Indicators representing the different categories of the impact factor were compiled from various sources including EUROSTAT, UNDP, WHO, World Bank, and CIA. All indicators were derived for the year 2012. It can be observed in Fig. 4.15 that Turkey, for example, has the highest potential for overall health problems following an event due to lack of health care capacities, comparatively lower baseline health status and social vulnerabilities in addition to a very high historic earthquake fatality. Figure 4.15 also shows how impact factor is derived as weighted average of a number of indicators including representing social vulnerability, baseline health status, environmental health and healthcare capacity. The assigning of weights to each of these indicators is a subjective process and their elicitation should be carried out in a participatory approach with a group of experts or other stakeholders using the model. Similarly the weight assigned to each category of the Impact Factor (IF) can be different and should also be determined in a participatory approach. In this analysis equal weights were given to all criteria.

The model is readily extendable to other scales (provincial, municipal, sub-municipal) where individual indicators are selected based on the specific context of the scale and availability of data. The preferred approach for application of models in SYNER-G is the sub-city level of geography. However, based on the principle that only European-wide public available data should be used, application at this scale was not possible as the Urban Audit database does not contain enough relevant factors for criteria identified in this study.

References

Aldrich N, Benson, FW (2008) Disaster preparedness and the chronic disease needs of vulnerable older adults. Prev Chronic Dis 5. http://www.cdc.gov/pcd/issues/2008/jan/07_0135.htm. Accessed 18 June 2013

Alexander D (1982) Disease epidemiology and earthquake disaster: the example of southern Italy after the 23 November 1980 earthquake. Soc Sci Med 16:1959–1969

Alexander D (1996) The health effects of earthquakes in the mid-1990s. Disasters 20:231–247

Alexander D (2011) Mortality and morbidity risk in the L'Aquila, Italy earthquake of 6 April 2009 and lessons to be learned. In: Spence R, So E, Scawthorn C (eds) Human casualties in earthquakes, vol 29, Advances in natural and technological hazards research. Springer, Berlin. ISBN 978-90-481-9455-1

Armenian HK, Melkonian A, Noji EK, Hovanesian AP (1997) Deaths and injuries due to the earthquake in Armenia: a cohort approach. Int J Epidemiol 26:806–813

Bartels SA, VanRooyen MJ (2011) Medical complications associated with earthquakes. Lancet. doi:10.1016/S0140-6736(11)60887-8

Başoğlu M, Şalcioğlu E, Livanou M (2002) Traumatic stress responses in earthquake survivors in Turkey. J Trauma Stress 15:269–276

Berberian M (1978) Tabas-e-Golshan (Iran) catastrophic earthquake of September 16, 1978; a preliminary field report. Disaster 2(4):207–219

Birkmann J (2006) Measuring vulnerability to natural hazards: towards disaster resilient societies. United Nations University Press, New York

Bono F, Eugenio G (2011) A network-based analysis of the impact of structural damage on urban accessibility following a disaster: the case of the seismically damaged Port Au Prince and Carrefour urban road networks. J Transp Geogr 19:1443–1455

Bostrom A, French SP, Gottlieb SJ (2008) Risk assessment, modeling and decision support: strategic directions. Springer, Berlin

Braun J (2011) Post-earthquake vulnerability indicators of shelter and housing in Europe: Masters Thesis. University of Trier

Burton C, Cutter SL (2008) Levee failures and social vulnerability in the Sacramento-San Joaquin Delta area, California. Nat Hazards Rev 9(3):136–149

Cardona OD (2005) Indicators of disaster risk and risk management: program for Latin America and the Caribbean. Summary Report (No. 30718), Inter-American Development Bank

Cavalieri F, Franchin P, Khazai B, Gehl P (2012) Quantitative assessment of buildings habitability based on physical damage and functional interaction with infrastructural systems. Earthq Eng Str Dyn EQE-11-0153

Chadda RK, Malhotra A, Kawa N, Singha J, Sethia H (2007) Mental health problems following the 2005 earthquake in Kashmir: findings of community-run clinics. Prehosp Disaster Med 22:541–545

Chan E (2008) The untold stories of the Sichuan earthquake. Lancet 372:359–362

Chan CL, Wang C-W, Ho AH, Qu Z-Y, Wang X-Y, Ran M-S, Mao W-J, Lu BQ-B, Zhang BQ, Zhang X-L (2012) Symptoms of posttraumatic stress disorder and depression among bereaved and non-bereaved survivors following the 2008 Sichuan earthquake. J Anxiety Disord 26:673–679

Chang H-L, Chang T-C, Lin T-Y, Kuo S-S (2002) Psychiatric morbidity and pregnancy outcome in a disaster area of Taiwan 921 earthquake. Psychiatry Clin Neurosci 56:139–144

Chang SE, Pasion C, Yavari S, Elwood K (2009) Social impacts of lifeline losses: modeling displaced populations and health care functionality. In: Oakland California proceedings of 2009 Technical Council on Lifeline Earthquake Engineering (TCLEE) conference

Chien S, Chen L, Chang S, Cui G, Chu C (2002) Development of an after earthquake disaster shelter evaluation model. J Chin Inst Eng 25(5):2

Chou FHC, Chou P, Su TTP, Ou Yang WC, Chien IC, Lu MK, Huang M-W (2004a) Quality of life and related risk factors in a Taiwanese village population 21 months after an earthquake. Australian and New Zealand. J Psyc 38:358–364

Chou YJ, Huang N, Lee CH, Tsai SL, Chen LS, Chang HJ (2004b) Who is at risk of death in an earthquake. Am J Epidemiol 160:688–695

Christoskov L, Samardjieva E (1984) An approach for estimation of the possible number of casualties during strong earthquakes. Bulg Geophys J X 4:94–106

Coburn AW (1986) Seismic vulnerability and risk reduction strategies for housing in Eastern Turkey. University of Cambridge, PhD Dissertation

Coburn A, Spence R (1992) Earthquake protection. John Wiley & Sons Ltd, Chichester

Coburn A, Spence R (2002) Earthquake protection. Wiley, Chichester

Coburn AW, Ohashi Murakami H, Ohta Y (1987) Factors affecting fatalities and injuries in earthquakes, chair for engineering seismology and earthquake disaster prevention planning. Department of Architectural Engineering, Hokkaido University, Japan

Cutter SL (1996) Societal vulnerability to environmental hazards. Int Soc Sci J 47(4):525–536

Cutter SL, Boruff B, Shirley JWL (2003) Social vulnerability to environmental hazards. Soc Sci Q 84:242–261

Daniell JE, Khazai B, Wenzel F, Vervaeck A (2011) The CATDAT damaging earthquakes database. Nat Hazards Earth Syst Sci 11:2235–2251

Daniell JE, Vervaeck A, Khazai B, Wenzel F (2012) Worldwide CATDAT damaging earth-quakes database in conjunction with Earthquake-report.com – Presenting past and present socio-economic earthquake data. 15th WCEE, Lisbon, Portugal, Paper no. 2025

de Bruycker M, Greco D, Lechat MF (1985) The 1980 earthquake in Southern Italy – morbidity and mortality. Int J Epidemiol 14:113–117

Dobalian A, Stein JA, Heslin KC, Riopelle D, Venkatesh B, Lanto AB, Simon B, Yano EM, Rubenstein LV (2011) Impact of the Northridge earthquake on the mental health of veterans: results from a panel study. Disaster Med Public Health Preparedness 5:S220–S226

Düzgün HSB, Yücemen MS, Kalaycioglu HS, Celik K, Kemec S, Ertugay K, Yilmaz N (2011) An integrated earthquake vulnerability assessment framework for urban areas. Nat Hazards 59:917–947

Elnashai AS, Jefferson T, Friedrich F, Cleveland LJ, Gress T (2009) Impact of new Madrid seismic zone earthquake on the central USA, vol. I. MAE Center report no 09-03

Erdik M, Şeşetyan K, Demircioğlu MB, Hancılar U, Zülfikar C (2011) Rapid earthquake loss assessment after damaging earthquakes. Soil Dynam Earthquake Eng 31(2):247–266

Erek E, Sever M, Serdengeçti K, Vanholder R, Akoğlu E, Yavuz M, Ergin H, Tekçe M, Duman N, Lameire N, Turkish Study Group of Disaster (2002) An over-view of morbidity and mortality in patients with acute renal failure due to crush syndrome: the Marmara earthquake experience. Nephrol Dial Transplant 17(1):33–40

Ertugay K, Duzgun S (2011) GIS-based stochastic modeling of physical accessibility using GPS-based floating Car data and Monte Carlo simulation. Int J Geogr Inform Sci 25(9):1491–1506

FEMA (2003) Chapter 14: Direct social losses: displaced households due to loss of housing habitability and short term shelter needs. HAZUS-MH MR4 Technical Manual, Washington, DC

Federal Emergency Management Agency (FEMA) (2006) HAZUS-MH MR2 technical manual. Federal Emergency Management Agency, Washington, DC

Franchin P, Lupoi A, Pinto PE (2006) On the role of road networks in reducing human losses after earthquakes. J Earthquake Eng 10(2):195–206

Freudenberg M (2003) Composite indicators of country performance: a critical assessment. Organisation for Economic Co-operation and Development (OECD), Paris

Guagliardo MF (2004) Spatial accessibility of primary care: concepts, methods and challenges. Int J Health Geogr 3:1–13

Guha-Sapir D, van Panhuis WG (2009) Health Impact of the 2004 Andaman Nicobar earth-quake and Tsunami in Indonesia. Prehosp Disaster Med 24:493–499

Gunn S (1995) Health effects of earthquakes. Disaster Prev Manag 4:6–10

Harrald JR, Fouladi B, Al-Hajj SF (1992) Estimates of demand for mass care services in future earthquakes affecting the San Francisco bay region. Prepared by George Washington University for the American Red Cross Northern California Earthquake Relief and Preparedness Project (NCERPP)

Hewitt K, Burton I (1971) The hazardousness of a place: a regional ecology of damaging events. Research publication 6. University of Toronto Press, Toronto

Hibino Y, Takaki J, Kambayashi Y, Hitomi Y, Sakai A, Sekizuka N, Ogino K, Nakamura H (2009) Health impact of disaster-related stress on pregnant women living in the affected area of the Noto Peninsula earthquake in Japan. Psychiatry Clin Neurosci 63:107–115. Wiley Online Library

Higgs G (2009) The role of GIS for health utilization studies: literature review. Health Serv Outcom Res Methodol 9:119–139

Horner MW, Widener MJ (2011) The effects of transportation network failure on people's accessibility to hurricane disaster relief goods: a modeling approach and application to a Florida case study. Nat Hazards 59:1619–1634

Howard MJ, Brillmann JC, Burkle FM (1996) Infectious disease emergencies in disasters. Emerg Med Clin North Am 14:413–428

Jaiswal K, Wald D (2010) An empirical model for global earthquake fatality estimation. Earthquake Spectra 26(4):1017–1038

Jibson RW (2002) A public health issue related to collateral seismic hazards: the valley fever outbreak triggered by the 1994 Northridge, California earthquake. Surv Geophys 23:511–523

Karaca E (2005) Regional earthquake loss estimation: role of transportation network, sensitivity and uncertainty, and risk mitigation. PhD dissertation. MIT, Cambridge, MA

Karmakar S, Rathore AS, Kadri SM, Dutt S, Khare S, Lal S (2008) Post-earthquake outbreak of rotavirus gastroenteritis in Kashmir (India): an epidemiological analysis. Public Health 122:981–989

Kawasumi H (1954) Intensity and magnitude of shallow earthquakes. Bureau Central Seism Intern, Scr A Trav Sci 19:99–114

Khazai B, Braun J, Vangelsten BV (2011a) Definition of a group of output indicators representing socio-economic impact from displacement in emergency/temporary shelter. Tech report D4.1 of the SYNER-G project. http://www.syner-g.eu

Khazai B, Daniell JE, Wenzel F (2011b) The March 2011 Japan earthquake – analysis of losses impacts, and implications for the understanding of risks posed by extreme events. Technikfolgenabschätzung – Theorie und Praxis, 20. Jg., Heft, 3 Nov 2011

Khazai B, Vangelsten B, Duzgun S, Braun J, Daniell J (2011c) Social impacts of emergency shelter provision in the aftermath of earthquakes. Geophys Res Abs 13: EGU 2011-7374

Khazai B, Daniell JE, Franchin P, Cavalieri F, Vangelsten B, Iervolino I, Esposito S (2012a) A new approach to modeling post-earthquake shelter demand in the aftermath of earthquakes: integrating social vulnerability in systemic seismic vulnerability analysis. In Proceedings of the 15th WCEE. Paper no. 2105, Lisbon, Portugal, 24–28 Sept 2012

Khazai B, Schaefer D, Mueller T (2012b) MCDA software tool. Tech report D7.1 of the SYNER-G project. http://www.syner-g.eu

Khazai B, Vangelsten B, Duzgun S, Daniell J, Braun J, Plapp-Kunz T, Iervolino I, Esposito S, Elefante L (2012c) Framework for integrating physical and socio-economic models for estimating shelter needs. Tech report D4.7 of the SYNER-G project. http://www.syner-g.eu

Kun P, Wang Z, Chen X, Le H, Gong X, Zahng L, Yao L (2010) Public health status and influence factors after 2008 Wenchuan earthquake among survivors in Sichuan province, China: cross-sectional trial. Public Health 124:573–580

Kuwagata Y, Oda J, Tanaka H, Iwai A, Matsuoka T, Takaoka M, Kishi M, Morimoto F, Ishikawa K, MizushimaY MY, Yamamura H, Hi-raide A, Shimazu T, Yoshioka T (1997) Analysis of 2,702 traumatized patients in the 1995 Hanshin-Awaji earthquake. J Trauma 43:427–432

Liang N-J, Shihs Y-T, Shihs F-Y, Wu H-M, Wang H-J, Shi S-F, Liu M-Y, Wang BB (2001) Disaster epidemiology and medical response in the Chi-Chi earthquake in Taiwan. Ann Emerg Med 38:549–555

Litman T (2011) Evaluating accessibility for transportation planning. Victoria Transport Policy Institute. http://wwwvtpiorg. A shorter version of this paper was presented at the 87th Transportation Research Board annual meeting (www.trb.org) on January 2008

Liu S, Han J, Xiao D, Ma C, Chen B (2010) A report on the reproductive health of women after the massive 2008 Wenchuan earthquake. Int J Gynaecol Obstet 108:161–164

Lupoi G, Franchin P, Lupoi A, Pinto PE, Calvi GM (2008) Probabilistic seismic assessment for hospitals and complex-social systems. IUSS Press, Pavia

Marano KD, Wald DJ, Allen TI (2010) Global earthquake casualties due to secondary effects: a quantitative analysis for improving rapid loss analyses. Nat Hazards 52(2):319–328

Maqsood ST, Schwarz J (2011) Estimation of human casualties from earth-quakes in Pakistan – an engineering approach. Seismol Res Lett 82(1):32–41

Matsuoka T, Yoshioka T, Oda J, Tanaka H, Kuwagata Y, Sugimoto H, Sugimoto T (2000) The impact of a catastrophic earthquake on morbidity rates for various illnesses. Public Health 114:249–253

McCurry J (2011) Japan: the aftermath. Lancet 377:1061–1062

McGrail (2012) Spatial accessibility of primary health care utilising the two step floating catchment area method: an assessment of recent improvements, International J Health Geogr 11:50. doi:10.1186/1476-072X-11-50, http://www.ij-healthgeographics.com/content/11/1/50

Mileti DS (1999) Disasters by design: a reassessment of natural hazards in the United States. National Academies Press, Washington, DC

Mori K, Ugai K, Nonami Y, Kirimura T, Kondo C, Nakamura T, Motoki E, Kaji H (2007) Health needs of patients with chronic diseases who lived through the great Hanshin earthquake. Disaster Manag Response 5:8–13

Muramatsu N, Akiyama H (2011) Japan: Super-Aging Society preparing for the future. Gerontologist 51:425–432

Nakamori Y, Tanaka H, Oda J, Kuwagata Y, Matsuoka T, Yoshioka T (1997) Burn injuries in the 1995 Hanshin-Awaji earthquake. Burns 23:319–322

Nardo N, Saiana M, Saltelli A, Tarantola S, Hoffmann A, Giovannini E (2005) Handbook on constructing composite indicators: methodology and user guide. OECD statistics working paper

Noji EK, Armenian HK, Oganessian A (1993) Issues of rescue and medical care following the 1988 Armenian earthquake international. J Epidemiol 22:1070–1076

Nuti C, Vanzi I (1998) Assessment of post-earthquake availability of hospital system and upgrading strategies. Earthquake Eng Struct Dynam 27(12):1403–1423

Oda J, Tanaka H, Yoshioka T, Iwai A, Yamamura H, Ishikawa K, Sugimoto H (1997) Analysis of 372 patients with crush syndrome caused by the Hanshin-Awaji earthquake. J Trauma-Injury Infect Crit Care 42(3):470–476

Ohashi H, Ohta Y (1983) Distribution characteristics of structural damage and casualty at several recent earthquakes in turkey and its factor analysis. In: A comprehensive study on earthquake disasters in Turkey in view of seismic risk reduction. Hokkaido University, Sappro, pp 131–166

Ohta Y, Ohashi H, Tabban A (1983) Macro statistical analysis on earth-quake damage in Turkey. In: A comprehensive study on earthquake disasters in Turkey in view of seismic risk reduction. Hokkaido University, Sapporo, pp 5–21

Osaki Y, Minowa M (2001) Factors associated with the earthquake deaths in the Great Hanshin-Awaji Earthquake, 1995. Am J Epidemiol 153:153–156

Pedigo AS, Odoi A (2010) Investigation of disparities in geographic accessibility to emergency stroke and myocardial infarction care in East Tennessee using geographic information systems and network analysis. Ann Epidemiol 20(12):924–930

Peek Asa C, Ramirez M, Seligson H, Shoaf K (2003) Seismic, structural, and individual factors associated with earthquake related injury. Inj Prev 9:62–66

Phalkey R, Rheinhardt JD, Marx M (2011) Injury epidemiology after the 2001 Gujarat earthquake in India: a retrospective analysis of injuries treated at a rural hospital in the Kutch district immediately. Glob Health Action 4:7169

Ramirez M, Peek-Asa C (2005) Epidemiology of traumatic injuries from earthquakes. Epidemiol Rev 27:47–55

RGELFE (Research Group for Estimating Losses from Future Earthquakes) (1992) Estimating Losses from Earthquakes in China in the forthcoming 50 years. State Seismological Bureau, Seismological Press, Beijing

Rossetto T, Peiris N, Alarcon J, So E, Sargeant S, Free M, Sword-Daniels V, Re DD, Libberton C, Verrucci E, Sammonds P, Walker JF (2011) Field observations from the Aquila, Italy earthquake of April 6, 2009. Bull Earthq Eng 9:11–37

Roy N, Shah H, Patel V, Coughlin RR (2002) The Gujarat earthquake (2001) experience in a seismically unprepared area: community hospital medical response. Prehosp Disaster Med 17:186–195

Roy N, Shah H, Patel V, Bagalkote H (2005) Surgical and psychosocial outcomes in the rural injured – a follow-up study of the 2001 earthquake victims injury. Int J Care Inj 36:927–934

Schneider E, Haijeh RA, Spiegel RA, Jibson RW, Harp EL, Marshall GA, Gunn RA, McNeil MM, Pinner RW, Baron RC, Burger RC, Hurtwanger LC, Crump C, Kaufman L, Reef SE, Feldman GM, Pappagianis D, Werner SB (1997) A coccidioidomycosis outbreak following the Northridge, Calif, earthquake. J Am Med Assoc 277:904–908

Seplaki CL, Goldman N, Weinstein M, Lin YH (2006) Before and after the 1999 Chi-Chi earthquake: traumatic events and depressive symptoms in an older population. Soc Sci Med 62:3121–3132

Shi YK, Wang LI, Lin YD, Pe FX, Kang Y (2010) Challenges for rear hospital care of Wenchuan earthquake casualties: experience from West China Hospital. Chin J Traumatol 13:131–136

So E (2009) The assessment of casualties for earthquake loss estimation. PhD dissertation, University of Cambridge, Cambridge, UK

So E, Spence R (2013) Estimating shaking-induced casualties and building damage for global earthquake events: a proposed modelling approach. Bull Earthquake Eng 11(1):347–363

Spence R (ed) (2007) Earthquake disaster scenario prediction and loss modelling for urban areas. IUSS Press, University of Pavia

Sullivan KM, Hossain SM (2010) Earthquake mortality in Pakistan. Disasters 36:176–183

Surahman A (2000) Earthquake Vulnerability evaluation of buildings in Bandung Municipality In: Proceedings of the 12th WCEE, Auckland, New Zealand, 30 January-4 February 2000.

Suzuki Y, Kim Y (2012) The Great East Japan earthquake in 2011, toward sustainable mental health care system. Epidemiol Psych Sci 21:7–11

Takahashi T, Goto M, Yoshida H, Sumino H, Matsui H (2011) Infectious diseases after the 2011 Great East Japan earthquake. J Exper Clin Med 4:20–23

Takahashi T, Goto M, Yoshida H, Sumino H, Matsui H (2012) Infectious diseases after the 2011 Great East Japan earthquake. J Exper Clin Med 4:20–23

Tanaka H, Iwai A, Oda J, Kuwagata Y, Matsuoka T, Shimazu T, Yoshioka T (1998) Overview of evacuation and transport of patients following the 1995 Hanshin-Awaji earthquake. Disaster Med 16:439–444

Tanaka H, Oda J, Iwai A, Kuwagata Y, Matsuoka T, Takaoka M, Kishi M, Morimoto F, Ishikawa K, Mizushima Y, Nakata Y, Yamamura H, Hiraide A, Shimazu T, Yoshioka T (1999) Morbidity and mortality of hospitalized patients after the 1995 Hanshin-Awaji-earthquake. Am J Emerg Med 17:186–191

Tierney KJ (2006) Social inequality, hazards, and disasters. In: Daniels RJ (ed) On risk and disaster: lessons from Hurricane Katrina. National symposium on risk and disasters. University of Pennsylvania Press, Philadelphia, pp 110–128

Tural Ü, Coşkun B, Önder E, Çoarpçioğlu A, Yildiz M, Kesepara C, Karakaya I, Aydin M, Erol A, Torun F, Aybar G (2004) Psychological consequences of the 1999 earthquake in Turkey. J Trauma Stress 17:451–459

Vahaboglu H, Gundes S, Karandenizli A, Mutlu B, Cetin S, Kolayli F, Coskunkan F, Dündar V (2000) Transient increase in diarrheal diseases after the devastating earthquake in Kocaeli, turkey: results of an infectious disease surveillance study. Clin Infect Dis 31:1386–1389

Vanholder R, vander Tol A, Smet MD, Hoste E, Koç M, Hussain A, Khan S, Sever M (2007) Earthquake and crush syndrome casualties: lessons learned from the Kashmir disaster. Kidney Int 71:17–23

Watson JT, Gayer M, Conolly MA (2007) Epidemics after natural disasters. Emerg Infect Dis 13:1–5

Wen J, Shi YK, Li YP, Wang L, Cheng L, Gao Z, Li L (2009) Risk factors of earthquake inpatient death: a case control study. Crit Care 13:R24

Wenzel F, Daniell, J, Khazai, B, Kunz-Plapp, T (2012) The CEDIM forensic earthquake analysis and the test case of the 2011 Van earthquake. In: Proceedings of the 15th WCEE, 24–28 Sept 2012, Lisbon, Portugal

Whitman RV, Biggs JM, Brennan JE III, Cornell CA, De Neufville RL, Vanmarcke EH (1975) Seismic design analysis. J Struct Div, ST5, ASCE:1067–1084

Wiggins HH (1977) National losses and mitigation effects for air, earth and waterborne natural hazards. In: Proceedings of the second conference designing to survive severe hazards, IIT Research Institute, Chicago, pp 47–97

Wisner B, Blaikie P, Cannon T, Davies I (2004) At risk – natural hazards, people's vulnerability and disasters, 2nd edn. Routledge, London

Wood NJ, Burton CG, Cutter SL (2010) Community variations in social vulnerability to Cascadia-related tsunamis in the US Pacific Northwest. Nat Hazards 52(2):369–389

Wright KC, Johnston DM (2010) Post-earthquake sheltering needs, how loss of structures and services affects decision making for evacuation. NZSCE conference, Wellington, New Zealand

Yavari S, Chang SE, Elwood K (2010) Modeling post-earthquake functionality of regional health care facilities. Earthquake Spectra 26(3):869–892

Yiannakoulias N, Bland W, Svenson LW (2013) Estimating the effect of turn penalties and traffic congestion on measuring spatial accessibility to primary health care. Appl Geog 39:172–182

Zhang Z, Shi Z, Wang L, Liu M (2011) One year later: mental health problems among survivors in hard-hit areas of the Wenchuan earthquake. Public Health 125:293–300

Zuccaro G, Cacace F (2010) Seismic casualty evaluation: the Italian model, an application to the L'Aquila 2009 event. In: Spence R, So E, Scawthorn C (eds) Human casualties in earthquakes, vol 29, Advances in natural and technological hazards research. Springer, Berlin. ISBN 978-90-481-9455-1

Chapter 5
Specification of the Vulnerability of Physical Systems

Hormoz Modaressi, Nicolas Desramaut, and Pierre Gehl

Abstract The general methodology presented in Chap. 2 of this book, has been conceived in order to be general enough to be adequate for each system. The purpose of this chapter is to decline this methodology to the specificity of each physical system considered: i.e., Buildings, Water Supply System, Waste Water Network, Electrical Power Network, Oil and Gas Network, Transportation Network, Health Care System and Harbours. Each system is described based on its structure and taxonomy, on the dependencies it shares with the other systems, on the available methods to describe its systemic vulnerability and, finally, on the existing indicators to evaluate its performance, but also its functionality according to the societal needs.

5.1 Introduction

The scope of the SYNER-G project includes the definition of some of the systems that comprise the *Infrastructure*, namely the inhabited areas (common buildings), the utility networks (water, waste-water, power, and gas supply networks), the transportation networks (roadways) and some critical facilities/systems (harbours and health-care system). While Chap. 2 of this book has exposed the general methodological framework that allows the representation of all these systems in

H. Modaressi (✉) • P. Gehl
Formerly at Bureau de Recherches Géologiques et Minières (BRGM), 3 avenue Claude-Guillemin, BP 36009, 45060 Orléans Cedex 2, France
e-mail: hormoz.modaressi@gmail.com; p.gehl@ucl.ac.uk

N. Desramaut
Bureau de Recherches Géologiques et Minières (BRGM), 3 avenue Claude-Guillemin, BP 36009, 45060 Orléans Cedex 2, France
e-mail: n.desramaut@brgm.fr

K. Pitilakis et al. (eds.), *SYNER-G: Systemic Seismic Vulnerability and Risk Assessment of Complex Urban, Utility, Lifeline Systems and Critical Facilities*, Geotechnical, Geological and Earthquake Engineering 31, DOI 10.1007/978-94-017-8835-9_5,

an object-oriented architecture, the objective of the present Chapter is to describe each system through various angles. First of all, the object-oriented architecture of the whole Infrastructure is further developed for each considered system, in order to specify the system down to its lowest-level component. Then, the **specific** *attributes* of each class or sub-class are defined: these properties are complementary to the ones already defined in Chap. 2 for **abstract** classes, and they are required to accurately describe each particular system. Interactions between systems are also investigated for each system, based on their type and the modelling assumptions used. Then, various possible approaches to "solve" each system (based on the complexity level of the analysis) are reviewed: the ones that have been implemented in the SYNER-G project are further detailed in terms of *methods*, in the sense of the UML (Unified Modeling Language) terminology (i.e. basic functions or routines used in the solving algorithm). Finally, relevant performance indicators at component- and system-level, whose applicability depends on the level of analysis chosen, are defined.

This specification work is necessary in order to characterize most of the particular features and behaviours of each system, while still staying within the methodological framework defined in Chap. 2. The computation of carefully selected performance indicators serves the purpose of assessing indirect losses (i.e. uninhabitable areas, utility or accessibility losses) that, if combined with direct losses from physical damages, can yield a first partial estimate of the overall socio-economic impact of an earthquake. This key point is enhanced here by the need to feed the socio-economic models defined in Chap. 4 (e.g. emergency shelter needs, health-care capacity), thanks to the various outputs from the specified systems.

5.2 General Specifications

Each system is described according to three main characteristics: (i) the lists of its elements, which is given through the taxonomy of the systems, (ii) the support it provides for the society, which is provided through the system evaluation and the selection of appropriate performance indicators and (iii) the treatment of interactions with the other interconnected systems.

5.2.1 Taxonomy of Systems and Their Components

Following the framework of the general SYNER-G methodology, each class of systems is composed of sub-classes that are used to describe the various types of components, based on the geographical extent and their function within the system:

- **Cell classes** are used to define inhabited areas (i.e., Buildings System) and contain information on buildings' typologies, population or soil land use policy.

- All **network-like systems** (i.e., Water Supply, Electric Power, Gas Network and Road Network) contain two types of sub-classes (Edges and Nodes), which are further sub-divided in specific classes, according to the role played by the component within the system: network nodes can be stations, pumps, reservoirs, sources, distribution nodes, etc.
- **Critical facilities** such as components of the Health-Care System are modelled as point-like objects.

Each of the sub-classes is specified with its characteristic attributes and UML-methods, depending on the type of system considered. For instance, initial properties of the objects may include geographic location, area, length, soil type, typology, associated fragility, capacity, connectivity with other components (for networks), etc. Once the simulation is running, the specific UML-methods update the object properties, such as damage states, losses within each cell or remaining connectivity.

5.2.2 System Evaluation and Performance Indicators

Four main types of system evaluations are considered in the SYNER-G approach:

- **Vulnerability analysis:** This level of analysis (also called Level 0) considers only the potential physical damages of the components of the systems, with no consideration of functionality of either the elements or the whole system.
- **Connectivity analysis:** This level of analysis (Level I) analyzes the probability of the demand nodes to be connected to functioning supply nodes through undamaged paths. In this approach the damaged components are removed from the network and the adjacency matrix is updated accordingly, thus pointing out the nodes or areas that are disconnected from the rest of the system. This qualitative approach is used for all utility networks (water, electricity, gas) and the road transportation system.
- **Capacity analysis**: This level of analysis (Level II) considers the ability of the system to provide to the users the required functionality. This type of approaches is quantitative. For utility networks, graph algorithms and flow equations can be used to estimate capacitive flows from sources (e.g. generators, reservoirs) to sinks (i.e., distribution nodes), based on the damages sustained by the network components (from total destruction to slight damages reducing the capacity).
- **Fault-tree analysis**: This level of analysis concerns critical infrastructures, where multiple conditions are necessary for the systems to ensure its task. This type of approach aims to evaluate the remaining operating capacity of objects such as health-care facilities. The system is broken down into structural, non-structural or human components, each one of them being connected with logic operators.

Performance indicators, at the component or the system level, depend on the type of analysis that is performed. Connectivity analysis gives access to indices such as the connectivity loss (measure of the reduction of the number of possible paths from sources to sinks). Capacitive modelling yields more elaborate performance indicators at the distribution nodes (e.g. head ratio for water system, voltage ratio for electric buses) or for the whole system (e.g. system serviceability index comparing the customer demand satisfaction before and after the seismic event). The fault tree analysis method is generally used for the derivation of fragility curves for specific components that comprise a set of sub-components (e.g. health care facilities, water treatment plants).

5.2.3 Treatment of Interdependencies

The systems are impacted by the other systems by a set of different dependencies. These dependencies can work in one way, i.e. the state of one system conditions the functioning of another system, or two ways, where two systems are mutually dependent, or interdependent. Rinaldi et al. (2001) defined four types of interdependencies: "cyber, logical, geographic and physical".

Out of these four interdependencies, three types of interactions between systems are considered for the present SYNER-G specification:

- **"Demand" interactions**: they correspond to a supply demand from a given component to another system. For instance, the presence of densely populated cells in the vicinity of a given distribution node (e.g. from a water supply or electric power system) will generate a substantial demand on the supply system. Another example could be the number of casualties that will put a strain on the treatment capacity of health-care facilities.
- **Physical interactions**: they are associated with exchanges of services or supplies between systems, like the supply of potable water to inhabited cells, the supply of transportation capacities by roads or the supply of power to various network facilities (e.g. water pumps) by electric generators.
- **Geographical interactions**: they are involved when two components are located in the same area and when the damage of one of them is directly influencing the physical integrity of the second one. For instance, the collapse of buildings in city centres can induce the blockage of adjacent roads due the debris accumulation.

The interactions between systems that are treated in the frame of SYNER-G are listed in Table 5.1: ***D*** stands for Demand, ***P*** for Physical and ***G*** for Geographical interactions.

It should be noted that the "demand" interactions are considered as static, since they are estimated only once, in order to avoid the presence of any feedback loops that would introduce dynamic systems, which are left out of the SYNER-G scope. As a result, this table of interdependencies governs the order in which each system has to be computed during the simulation runs, in order to maintain a straightforward analysis scheme.

Table 5.1 Main interdependencies between systems that have been implemented in SYNER-G

		BDG	**EPN**	**WSS**	**GAS**	**OIL**	**RDN**	**HBR**	**HCS**
Buildings	**BDG**	/	**D**	**D**	**D**		**D/G**		**D**
Power	**EPN**	**P**	/	**P**	**P**	**P**		**P**	**P**
Water	**WSS**	**P**		/					
Gas	**GAS**	**P**			/				
Oil	**OIL**					/			
Roads	**RDN**	**P**					/	**P**	**P**
Harbour	**HBR**						**D**	/	
Hospitals	**HCS**						**D**		/

5.3 Specification of Each System

This section is devoted to the description of each considered system, in terms of internal structure, interactions with other systems, solving algorithms and performance indicators.

5.3.1 Inhabited Systems

Inhabited buildings are somehow different when compared with the other networks and critical infrastructures considered in SYNER-G. Indeed, they do not experience direct intra-dependencies (i.e. each part of the system can function independently of the other states of the systems) and are mainly dependent of the other systems. However, they are still considered as a system, because each component contributes to the general capacity of the system: to shelter people.

5.3.1.1 Structure of the System and Input Attributes

Buildings are the basic point-like component of building aggregates/agglomerates/blocks (where buildings may or may not be in contact, with the ensuing interactions), which are delimited by roads. The description of the vulnerability of an urbanised area at a large scale (e.g. a census tract, where several such building agglomerates are present) implies fragility analyses for each building typology and the use of statistical data on the incidence of each typology in the building population.

There are different levels to consider the inhabited zones. The region encompasses the whole study area, which is then divided in cells, and in sub-cells, according to the localisation of buildings, and their characteristics (see Fig. 5.1 and Table 5.2). This sub-division allows the refinement of the computation of impacts of the earthquake in the area where the density of assets and population is higher.

The various attributes from the Cell are then aggregated at the Region level, in order to generate indicators at the urban or regional level.

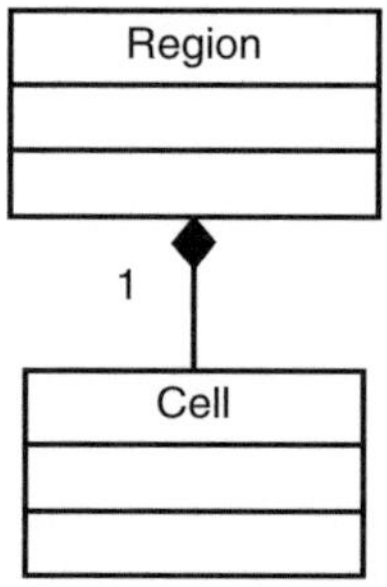

Fig. 5.1 UML class-diagram of the buildings system (BDG)

Table 5.2 Attributes/properties of the cell class

Group	Attribute	Description
Geometry	vertices	Points on a diagonal of the cell
	centroid	Average of vertices, where seismic intensity is predicted
	adjacentCells	Pointers to grid cells sharing a border with the current cell
Physical Damageability	buildingTypologies	$n_T \times 1$ vector with percentages of buildings in each of the n_T typologies (BC)
	fragilitySets	$n_T \times n_{DS} \times 3$ matrix with n_T fragility curves for each of the n_{DS} damage states, specified in terms of IM, median and logarithmic standard deviation
Socio-economic	population	(EAU)
	households	(EAU)
	income	(EAU)
	unemployment	(EAU)
	bldgUsage	4×1 vector of percentages of usage of cell area in use types Green, Residential, Commercial, Industrial (LUP)
Interdependence modeling	refNodes	Pointers to reference nodes in each of the other systems (WSS, EPN, RDN, GAS, OIL, etc.)
State variables recording cell state	states	$n_E \times 1$ collection of properties that describe the current state for each of the n_E events (fields: buildingDamage, utilityLoss, with subfields for each utility, buildingUsability, displacedPopulation, casualties, supplyRequirements, etc.)

BC, EAU, LUP: projected from Building Census, European Urban Audit, Land Use Plan, or other relevant data bases

5.3.1.2 Interdependencies

Interactions of the built areas with other systems are manifold; the main ones are defined in the following list:

- **EPN → BDG [Physical]:**
  Damage to the EPN can reduce the service level in the struck area, possibly below tolerance thresholds, thus leading to population displacement and demand on the Shelter model described in chapter 4.

- **WSS → BDG [Physical]:**
  Damage to the WSS can lower the service level in the struck area, possibly below tolerance thresholds, thus leading to population displacement and demand on the Shelter model.
- **GAS → BDG [Physical]:**
  Damage to the GAS system can reduce the service level in the struck area, possibly below tolerance thresholds, especially in adverse weather conditions, thus leading to population displacement and demand on the Shelter model. Also, leakage of the gas system can induce fires, which could damage buildings.
- **RDN → BDG [Physical]:**
  Damage to the transportation network can block access to damaged buildings hindering emergency response.
- **BDG → WSS [Demand]:**
  Fires in buildings can be triggered by earthquake induced damage thus raising the water-supply demand on the WSS (when this is not independent of the FFS).
- **BDG → RDN [Geographical]:**
  In an urban setting, structural damage to buildings produces debris that can cause road blockages.
- **BDG → HCS [Demand]:**
  Structural and non-structural damage to buildings may result in casualties that need to be treated in a health-care facility and hence determine the demand on this system.

5.3.1.3 Methods for Systemic Analysis

In practice, while the vulnerability assessment of a single building of special interest is based on a detailed and specific structural analysis, the global evaluation of vulnerability (i.e., for several hundreds or thousands of buildings at urban or regional scale) relies mostly on the use of statistical or probabilistic vulnerability functions. These functions represent the "typical" behaviour of a group of buildings characterized by a limited number of similar physical parameters.

Whatever the procedure used, a vulnerability assessment study of common buildings at urban or regional scale is based on the following elements:

- A building typology and its census within the studied area: while the seismic behaviour of buildings cannot be specified one by one, it is required to define a building typology based on structural criteria (i.e. material used, height, bracing system), that can be more or less accurate.
- A damage probability matrix or fragility curves that correspond to the chosen typology: for a given building typology, they represent the percentage of buildings that exceed a given damage state, for a given level of seismic intensity. Extensive details on fragility functions for different building typologies are provided in the Chaps. 3 and 4 of Pitilakis et al. (2014).

The very large number of buildings at regional or even at an urban scale can then be treated according to two possible approaches:

- The first one considers each building individually. Goda and Hong (2008) developed a methodological framework to estimate the influence of the degree of spatial correlations and simultaneous occurrence of the seismic solicitations on the assessment of seismic risk for buildings. Four classes of correlation are considered, namely, no correlation, full correlation, and partial correlation with/without intra-event components. Estimations of damage are performed on sets of buildings subject to a series of earthquakes for a given period of time, according to the four classes of correlation. The authors conclude that a bad estimation (either over-estimation or under-estimation) of the real correlation has big influence on the distributions of damages to buildings but do not influence the average value of this damage.
- The second approach is carried out when a detailed individual analysis of all buildings is not feasible. In this case, buildings are modelled in 'statistical terms' as populations for which information is given at the level of the buildings group (group size depending on the refinement of the analysis and varying from a single block to a larger extent of the urban territory). Information includes percentage of each building typology within the group, with associated fragility models, population, income, education, and other urban and social features.

Due to the diversity of scales and existing inventories, both approaches have been used in the SYNER-G project. The study of Vienna has been conducted with both individual and statistical approaches (Chap. 8), while the buildings of Thessaloniki have been analysed with the second, i.e. statistical approach (Chap. 7).

But due to the fact that most of the areas of concern will be more similar to Thessaloniki than Vienna, regarding the size and the existence of detailed inventories, the second approach has been adopted in the SYNER-G methodology (Cavalieri et al. 2012), where improvement of spatial resolution is adopted for areas with high population density, in order to reduce the variability of buildings' typology existing in each cell. The influence of spatial refinement (i.e. higher resolution grid) on the variability of the mean damage ratio for large groups of buildings has been shown by Bal et al. (2010).

Then, the analysis of the built-up areas within the SYNER-G framework is achieved through a set of UML-methods (Table 5.3).

5.3.1.4 Performance Indicators

Performance indicators of the built area system are expressed at the region- and cell-level, depending on the requirements of the systemic analysis. In particular in SYNER-G the performance indicators have been selected to require the need of the socio-economic analyses (Chap. 4). The corresponding performance indicators are the following

Table 5.3 Most relevant UML-methods of the Region class

Method	Description
evaluateBuildingDamage	Evaluates the damage state for each typology of buildings in the cell employing the corresponding set of fragility curves and the current intensity at the centroid
retrieveUtilityLoss	Reads from the reference node of each utility system the corresponding service level and computes a total (weighted sum) utility loss
evaluateBuildingUsability	Determines usability based on physical damage and empirically derived *usability ratios*
evaluateBuildingOccupancy	Determines building occupancy based on the number of households per building and the land use plan
evaluateBuildingHabitability	Determines building habitability based on usability and residual utility service level
evaluateCasualties	Determines number of dead and injured people based on physical damage and empirically derived *lethality ratios*
evaluateDisplacedPopulation	Determines number of people displaced from home based on building habitability
evaluateSupplyRequirements	Determines required amount of good/service based on cell population and demand model. Called by reference node in each system to aggregate demands from tributary cells

- **Building damage/collapse**
  It estimates the physical damage of the buildings after an earthquake. It is strongly dependent on the type of structure analysed and it describes the probability of a structure to exceed different limit states (such as different level of damage) given a level of ground shaking;
- **Building usability**
  It identifies the extent to which a building can be used by the inhabitants, and depends mainly on the physical damage to the structure (building damage/collapse).
- **Building habitability**
  It identifies whether the occupants can inhabit the building, and depends on the building usability and the utility loss (to the building).
- **Casualty model.**
  This model leads to indicators that estimate the number of deaths and injuries after an earthquake. It depends both on the type of building and on the number of people that live or reside temporarily in the damaged structure.
- **Debris model**
  This model leads to indicators that estimate the amount of debris following an earthquake. It depends on the building type and on the structural and non-structural damage.

5.3.2 Network Systems

This section includes the systems that can be defined as an organized set of edges and nodes: as defined in Chap. 2, networks can be subdivided in two abstract classes, i.e. *Undirected* graphs for utility networks (i.e. water, electric power and gas supply) and *Directed* graphs for transportation networks (i.e. roadways and railways).

5.3.2.1 Water Supply and Waste-Water Systems

The following paragraphs are devoted to the specification of the Water Supply System (WSS), through the description of its class structure, interactions with other systems and possible approaches for performance assessment.

Structure of the System and Input Attributes

The water-supply system as a whole is composed of a number of point-like critical facilities (Water sources, Treatment plants, Pumping stations, Storage tanks) and of the Water distribution network itself. The internal logic of the critical facilities and their function in the management of the whole system should be modelled explicitly. The network portion of the system is made of pipelines, tunnels and canals and the supervisory control and data acquisition (SCADA) sub-system.

The identified system components are:

WSS01:	Source (Springs, shallow or deep wells, rivers, natural lakes, and impounding reservoirs)	[Points]
WSS02:	Treatment Plant	[Points, critical facility]
WSS03:	Pumping station	[Points, critical facility]
WSS04:	Storage Tank	[Points]
WSS05:	Pipe	[Edges]
WSS06:	Tunnel	[Edges]
WSS07:	Canal	[Edges]
WSS08:	SCADA system	[System]

The structure of the Water supply system and the main input attributes are detailed in Fig. 5.2 and Table 5.4.

The waste-water system as a whole is composed of a number of point-like critical facilities (Treatment Plants, Pumping Stations) and of the distribution network itself (Pipelines, Tunnels). The internal logic of the critical facilities and their function in the management of the whole system should be modelled explicitly.

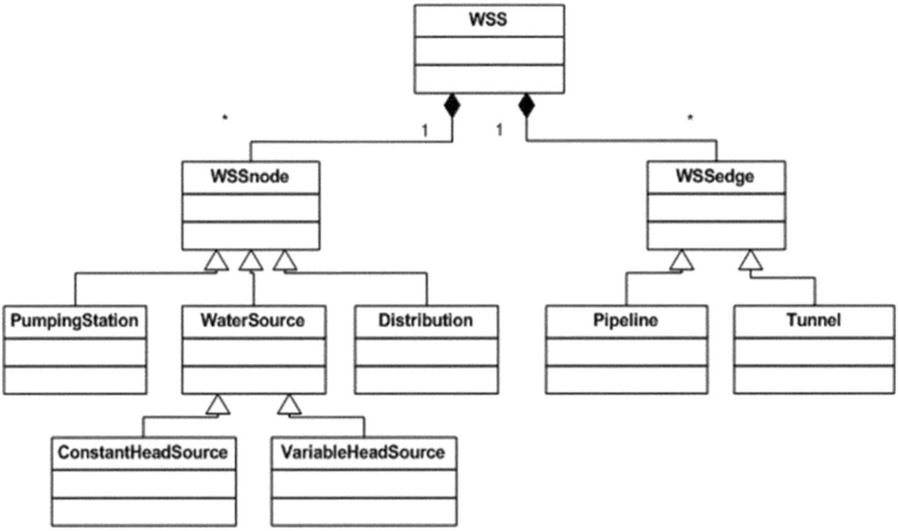

Fig. 5.2 UML class-diagram of the Water Supply System

The identified system components are:

WWN01:	Waste-water treatment plant	[Points, critical facility]
WWN02:	Pumping (lift) station	[Points, critical facility]
WWN03:	Pipelines	[Edges]
WWN04:	Tunnels	[Edges]
WWN05:	SCADA system	[System]

5.3.2.2 Interdependencies

The water supply system is strongly interconnected with the other systems. Most components of the WSS are dependent on the power supply and they are in turn used to feed inhabited areas and health-care centers.

- **EPN ➔ WSS [Physical]:**
  Damage to the EPN can reduce the functionality of pumping stations in the WSS.
- **WSS ➔ BDG [Physical]:**
  Damage to the WSS can lower the service level in the struck area, possibly below tolerance thresholds thus leading to population displacement and demand on the Shelter model.

Table 5.4 Main attributes/properties of the Water Supply System class

Group	Attribute(s)	Description
Global properties	sourceHead	Water head at source nodes
	endUserDemand, hydricEquipment	Required water flow at demand nodes, either assigned or evaluated by aggregating over tributary cells, employing population and hydricEquipment for the region (expressed in [l/inhab./day])
	refEPNnode	Pointers to EPN node(s) feeding power to pumping stations (for inter-dependence modelling)
Pointers	pipe	Pointers to all the pipes in the system, objects from the pipe class
	demand	Pointers to all end-user nodes, objects from the DistributionNode class
	source	Pointers to all sources in the system, in general objects from the ConstantHeadSource and VariableHeadSource (for finite reservoirs) classes
	pump	Pointers to all pumping station nodes, objects from the PumpStation class
Edge properties stored at WSS level	edgeMaterial, edgeDiameter, edgeRoughness, edgeDepth	Length, centroid, etc. are attributes inherited from the Network class. Here the network-specific properties are listed (roughness, diameter, laying depth, etc.)
Node properties stored at WSS level	nodeMinimalHead	Minimal head required at nodes for delivery of the assigned demand water flow; this property is a function of the average building elevation in the region of interest
	nodeDepth	–
State variables recording WSS state	states	$n_E \times 1$ collection of properties that describe the current state for each of the n_E events (fields: demandFlow, outFlow, average head ratio, system serviceability index, number of leaks, number of breaks, etc.)

- **WSS → HCS [Physical]:**
  Damage to the WSS can prevent water from being delivered to the health-care facilities, hindering emergency response over time in case backup reservoirs are depleted.

5.3.2.3 Methods for Systemic Analysis

The seismic reliability of water networks can be assessed using different indices of physical nature like vulnerability, connectivity, serviceability, maximum flow, redundancy, or immaterial estimates like economic loss (ATC 1991).

Connectivity analyses measure the post-earthquake integrity of the system, i.e., the extent to which links and nodes are still connected. Serviceability analyses estimate the post-earthquake capacity between selected source-to-sink nodes. Serviceability is a performance measure that considers mostly the hydraulic behaviour of the networks and less the robustness of the network in terms of its layout.

Closely related to reliability is redundancy, a characteristic of the overall system performance that is often neglected. Redundancy in a water supply network determinates the existence of backup capacities and of alternatives for routing, i.e. the existence of additional paths from supply to demand nodes in case of breaks in the main supply links (Awumah et al. 1991). The redundancy of a water supply system can be evaluated along with reliability to assess system performance under earthquake stimulations, in order to design a new network, and for efficient seismic mitigation of the existing network.

Reliability assessment could be performed with the mitigation prioritization procedure. Multi-criteria analysis (MCA) is probably more efficient than traditional cost-benefit analysis, as it copes with the uncertainties inherent to the judgment of experts. Moreover, the model has to consider customers importance, pipeline properties and hazard factors. Hence, to prioritize renewing of the lifeline systems, a fuzzy analytic hierarchy process (FAHP) to support the MCA has advantages. This optimized fuzzy prioritization method can be applied as an evaluation tool (Alexoudi et al. 2009), where uncertain and imprecise judgments of experts are translated into fuzzy numbers.

Seismic risk of water system has been investigated extensively (Ballantyne et al. 1990; Taylor 1991; Shinozuka et al. 1992; Hwang et al. 1998; Shi et al. 2006; Wang 2006). Chang et al. (2002) assessed the seismic risk of the WSS in Memphis, whereas the impacts of the 1906 San Francisco and 1989 Loma Prieta earthquakes on the system reliability of the auxiliary WSS of the city of San Francisco are detailed by Scawthorn et al. (2006). In this paper, the authors also described the induced consequences of the inoperability of some part of the Water system on earthquake-triggered fires.

The methods to address the risk of WSS can be classified as follows:

- **Level 0 (Vulnerability Analysis)**:
  - The scope is to estimate the percentage of the physical damages to the WSS elements in a specific region based on the vulnerability analysis of water network components, which can be estimated through appropriate fragility curves or/and Monte-Carlo technique.
  - The majority of the studies performed for water systems can be categorized as Level 0, implying simple physical vulnerability studies of water system components (ATC 1985; ATC 1991; NIBS 2004). The performance index used in Level 0 studies is the "Damage Ratio" that describes the expected number of failures per unit of length (for pipes), per link or per node of the system. Moreover, the "Damage Ratio" can be considered as a percentage of the damaged nodes/links.
- **Level I (Connectivity Analysis):**
  - In this level of systemic analysis, the concern is the connectivity between functioning supplying nodes (water sources and pumps) to demand nodes (linked to other systems), through undamaged pipes. In order to do that,

the damaged components are removed from the network. Then, some of the remaining nodes which are completely isolated from all supply nodes must be removed from the original network.

- A simple connectivity analysis (Level I) of the network can be accomplished using Graph Theory (clustering coefficient of a graph, Redundancy Ratio, Service Ratio Reachability Ratio) and Statistical Methods (Simple Level I Analysis). Illustration of Advanced Level I studies can be found in Shinozuka et al. (1977) and O'Rourke et al. (1985), which use minimal cut set paths in reliability evaluation of lifeline networks. Moreover, available techniques to identify the minimal paths and minimal cut sets have mainly been presented in literatures as connectivity analyses of the network (Jasmon and Kai 1985; Fotuhi-Firuzabad et al. 2004). Another example of Level I analysis is the study performed by Kawakami (1990), which uses the "Damage Ratio" (Level 0) and "Service Ratio" (Level I) as performance indices. Service Ratio indicates the ratio of normally supplied houses over the total number in the system. Dueñas-Osorio et al. (2007a) propose the concept of "Connectivity Loss" in order to quantify the average decrease of the ability of distribution vertices to receive flow from the generation vertices. Dueñas-Osorio et al. (2007b) introduce "Redundancy Ratio" as the appropriate parameter to measure the performance of water system. Moghtaderi-Zadeh et al. (1982) propose "*Reachability*" of water as performance index, indicating the probability that a certain amount of water flow would reach key locations (nodes). Conclusively, "*Damage Ratio*", "*Service Ratio*", "*Connectivity Loss*", "*Redundancy Ratio*" and "*Reachability*" are the performance indicators used in such level of analysis (Level I).

- **Level II (Flow Analysis/Serviceability Analysis):**
  - At this level, the concern is the quantity of the water provided to the user and the ability of the system to meet the needs. The physical-based indicators such as water head, flow rate and amount of leakage at each demand node are calculated under intact (pre-earthquake) conditions. Equivalent physical-estimates are assessed for pipes, like the quantity of flow and head loss. After the evaluation of the physical damages to the pipes (break, leak), a flow analysis is performed involving the newly formed "damaged" network. It is assumed that, when a pipe is broken, a shutdown device is automatically activated to prevent water leakage in pipe. Another underlying hypothesis is the unchanged capacity of the supplying nodes. Vulnerability and damage estimations of water system components, with the resulting flow analysis can be repeated for different seismic intensities using Monte-Carlo simulations. Average values of the flow rate and water pressure are then calculated at each node, and these values are compared to the measures for normal (pre-earthquake) conditions. The results are generally returned as ratios of post- to pre- earthquakes measures, and given in percentage of reduction of functionality.

– Many researchers have contributed to the improvement of seismic reliability methods for water supply systems from the flow and serviceability analysis viewpoint (Level II) (e.g. Shinozuka et al. 1981; Isoyama and Katayama 1981; O'Rourke et al. 1985; Javanbarg and Takada 2009). One example of Level II study is the one performed by Shinozuka et al. (1981). Their methodology allowed assessing the seismic reliability of water supply system in Los Angeles regarding the serviceability. The condition to consider the system serviceable is the remaining intact capabilities of fire-fighting systems after the earthquake. Monte Carlo simulations were carried out in order to estimate the probability of serviceability levels on the basis of estimated physical damage states of the elements of system at the aftermath of an earthquake. O'Rourke et al. (1985) also simulated the consequences of an earthquake on the serviceability of the water supply system in city of San Francisco through a flow analysis. The considered performance indicator for their analysis was the ratio of available water flow over the required one at a given and requested pressure at proximity to the fire outbreak (Level II). Potential performance indices used in Level II analyses include the probabilistic distribution of the percentage of customers who would lose their service after a specific earthquake.

These approaches require complex hydraulic analyses, which are time consuming and require expertise and availability of extensive data. For this reasons, a number of researchers have developed simplified models to assess the serviceability of pipeline networks under various amounts of pipe damages (Markov et al. 1994; Hwang et al. 1998; Javanbarg et al. 2006 and Shi et al. 2006). HAZUS (NIBS 2004) methodology proposes a diagram correlating the Serviceability Index (SI) (see Sect. 5.3.2.1 for definition) to average break rate (i.e. the number of complete failure of the pipeline section per unit of length).

Besides the models classified in the above three categories (Level 0 to II), other models have been also proposed, such as redundancy approaches (Awumah et al. 1991; Kalungi and Tanyimboh 2003; Hoshiya and Yamamoto 2002; Hoshiya et al. 2004) and studies for the identification of critical links of water supply systems under earthquakes (Wang et al. 2010).

The main UML-methods corresponding to these 3 levels of analysis used in the SYNER-G approach are described in Table 5.5 and their applications are detailed in Chaps. 7 and 8.

Over the last twenty years, waste-water systems worldwide have also been heavily damaged by natural disasters such as earthquakes. The societal and economic disruption caused by waste-water network damages is important, as for example, the impact on public health and environment due to the discharge of raw/inadequately treated sewage.

The required effort to assess the performance of waste-water systems varies with the level of analysis and the complexity of the system. Most of the available methodologies used for waste-water systems, estimate the physical damages, the replacement cost and the restoration time of the system's component without

Table 5.5 Most relevant UML-methods of the Water Supply System class

Method	Description
computeDemand	Aggregates demand from tributary cells in demand nodes
isBreakAndLeaksNumber	Evaluates the damage state of each pipeline segment employing the corresponding set of fragility functions and the current intensity at the centroid
computeLeakageArea	Computes the amount of leakage from the numbers of leaks in each pipe segment
updateConnectivity	Updates the adjacency matrix based on the pipe breaks and/or the failure of the nodes (e.g. pumping stations, reservoirs)
computeFlow	Computes the actual flow from the sources to the demand nodes based on an optimization algorithm, using the demand level and the leakage amount
computePerformanceIndicator	Computes the different PIs at component- and system-level

considering the overall performance of the network (ATC 1991; ATC 1985; NIBS 2004). While this system has been fully addressed in the SYNER-G project, its systemic evaluation should follow the same approach as for the water supply network.

5.3.2.4 Performance Indicators

Component-Level Performance Indicators (PIs)

- *Junctions/Nodes:* **Head Ratio**, or HR. ***[Level II]***

For each node, this index is defined as the ratio of the water head in seismically damaged network (H_{si}) over the reference value for the non-seismic, normal operations conditions (H_{0i}):

$$HR_i = H_{si} / H_{0i} \tag{5.1}$$

The determination of the water head requires a flow analysis of the network. Hence this index expresses the functional consequence in the i-th component of the physical damage to all system components (within the WSS). When interactions with other systems are modeled, HR_i expresses the functional consequence in the i-th component of the physical damage to components of the other systems (WSS, EPN, etc.), i.e., it is the value of the index that changes due to the inter- and intra-dependencies, not its definition.

- *Pipes:* the **Damage Consequence Index**, or DCI. ***[Level II]*** (Wang et al. 2010)

This index measure the impact of each pipe on the overall system serviceability and identify critical links that significantly affect the system's seismic performance. The index is defined at the component level in terms of a system-level PI that measures serviceability; the System Serviceability Index (SSI) is defined afterwards. Thus, as for the HR index, this is a PI that reflects at component-level the functional consequence of damage to all systems' components and incorporates the effect of the inter- and intra-dependencies, when modelled. The *DCI* for the *i*-th pipe is defined to reflect the consequence from damaging the pipe, including pipe breaks and leaks. It is expressed as:

$$DCI_i = \frac{E\left[SSI\right] - E\left[SSI \,|\, L_i\right]}{1 - E\left[SSI\right]} \tag{5.2}$$

in which *E[SSI]* is the (unconditional) expected value of *SSI* from a set of simulations in which the *i*-th pipe might or might not be damaged; and *E[SSI|Li]* is the conditional expectation of *SSI* from another set of simulations under the same seismic hazard, but given that the *i*-th pipe is damaged.

- *Pipes:* **Upgrade Benefit Index**, or UBI. ***[Level II]*** (Wang et al. 2010)

Similarly to the *DCI*, this index measures the impact of an upgrade of an individual pipe on the overall system serviceability, and reflects at the component level the systemic functional consequence of damage to the whole system(s). It is defined as:

$$UBI_i = \frac{E_{upgrade}\left[SSI\right] - E\left[SSI\right]}{1 - E\left[SSI\right]} \tag{5.3}$$

in which $E_{upgrade}[SSI]$ is the expected value of *SSI* given that the *i*-th pipe is "upgraded." By "upgrade", it is meant that the probability of pipe damage given an earthquake is significantly smaller than its value before upgrade. UBI_i is the percent increase of *SSI* given that the *i*-th pipe is upgraded, and its relative value is a measure of the pipe impact on the overall system serviceability.

System-Level Performance Indicators (PIs)

- **Average Head Ratio**, or AHR. ***[Level II]***

This index is defined as the average over the network nodes of the *HR* index:

$$AHR = \overline{HR} = \frac{1}{n_N}\sum\nolimits_{i=1}^{n_N} HR_i \tag{5.4}$$

where n_N is the number of nodes in the WSS.

- **System Serviceability Index**, or SSI. ***[Level II]*** (Wang et al. 2010).

The System Serviceability Index is defined as the ratio of the sum of the satisfied customer demands after an earthquake over the ones before the earthquake:

$$SSI = \frac{\sum\nolimits_{i=1}^{n} Q_i}{\sum\nolimits_{i=1}^{n_0} Q_i} \tag{5.5}$$

where n and n_0 are the number of satisfied demand nodes after and before the earthquake, and Q_i is the demand at the i-th node. The *SSI* varies between 0 and 1.

A single value can be determined for a given condition of the network. Its probabilistic characterization, in terms of either its full distribution or its expected value *E[SSI]* that enters in the definitions of *DCI* and *UBI*, requires running multiple simulations for different earthquake realizations. The above definition from Wang et al. (2010) assumes that the demand remains fixed before and after the earthquake, since it looks only at a single system, without considering the interactions of the WSS with the other systems.

Finally, regarding waste-water systems, ALA (2004) proposes different performance indicators like (i) capacity measures (e.g. flow of waste-water at selected points); (ii) measures of reliability (such as frequency and magnitude of sanitary sewer overflows (SSOs) or combined sewer overflows (CSOs), and the frequency and magnitude of discharge of inadequately treated sewage, percentage treated, etc.); (ii) measures of safety and health (backup of any raw sewage into buildings-not acceptable, overflow of raw sewage into streets-acceptable in localized areas for less than 24 h); or (iv) financial measures. The Environmental Protection Agency National Pollution Discharge Elimination System (EPA NPDES) permit requirements incorporate relevant performance measures such as discharge volume and water quality.

5.3.2.5 Gas and Oil Networks

This section focuses on the systems that are in charge of the delivery of natural gas and oil from production/gathering facilities to inhabited areas, especially the transmission/distribution network and the related support stations (i.e. for compression/reduction of the hydrocarbon flows).

Structure of the System and Input Attributes

The natural gas or oil system as a whole is composed of a number of point-like critical facilities (Production and gathering facilities, Treatment plants, Storage facilities, Intermediate stations where gas is pressurized/depressurized or simply metered) and of the transmission/distribution network itself. The internal logic of the critical facilities and their function in the management of the whole system should be modelled explicitly. The network portion of the system is made of edges (i.e. pipelines) and of the supervisory control and data acquisition (SCADA) sub-system.

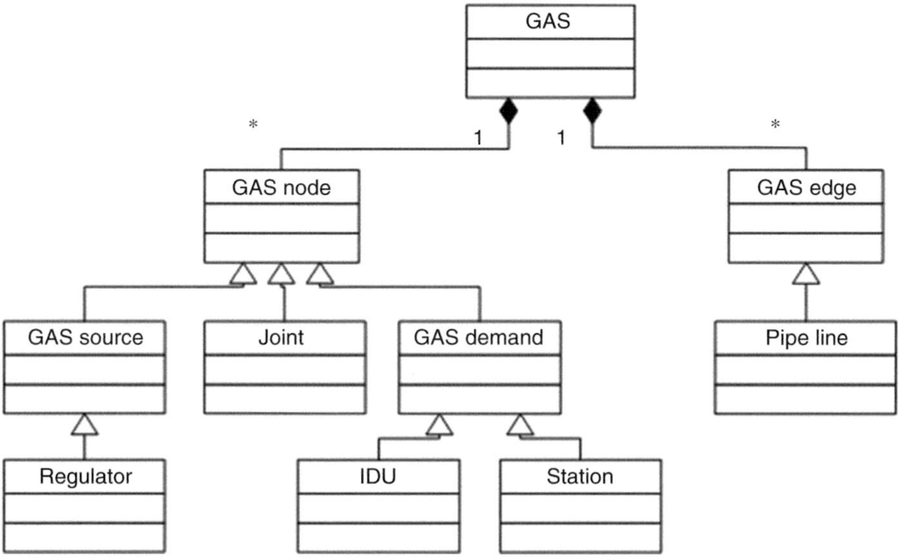

Fig. 5.3 UML class-diagram of the Gas Network System (Esposito and Iervolino 2012). IDU class represents the end-user nodes (i.e. final node for low-pressure networks)

For the GAS system the identified components are:

GAS01:	Production and gathering facility (Onshore, Offshore)	[Points, critical facility]
GAS02:	Treatment plant	[Points, critical facility]
GAS03:	Storage tanks	[Points]
GAS04:	Station (Compression, Metering Compression/ metering, Regulator/metering)	[Points, critical facility]
GAS05:	Pipelines	[Edge]
GAS06:	SCADA	[Systems]

For the OIL system the identified components are:

OIL01:	Production and gathering facility (Onshore, Offshore)	[Points, critical facility]
OIL02:	Refinery	[Points, critical facility]
OIL03:	Storage tank farm	[Points]
OIL04:	Pumping Station	[Points, critical facility]
OIL05:	Pipelines	[Edges]
OIL06:	SCADA	[Systems]

The object-oriented structure and the main class attributes of the Gas network system are presented in Fig. 5.3 and Table 5.6.

Table 5.6 Main attributes/properties of the Gas Network System class

Group	Attribute(s)	Description
Global properties	sourceHead	Gas pressure at source nodes
	endUserDemand	Required gas pressure at demand nodes, either assigned or evaluated by aggregating over tributary cells, employing population for the region
	refEPNnode	Pointers to EPN node(s) feeding power to compression/reduction stations (for inter-dependence modelling)
Pointers	pipe	Pointers to all the pipes in the system, objects from the Pipe class
	demand	Pointers to all end-user nodes, in general objects from the IDU (low-pressure network) or the Station (medium-pressure network) classes
	source	Pointers to all sources in the system, objects from the GASsource class
	station	Pointers to all compression/reduction stations in the system, objects from the Station class
	joint	Pointers to all joint nodes in the system, to reproduce the geometry of the network system, objects from the Joint class
Edge properties stored at GAS level	edgeMaterial, edgeDiameter, edgeRoughness, edgePressure	Length, centroid, etc. are attributes inherited from the Network class. Here the network-specific properties are listed (roughness, diameter, operating pressure, etc.)
Node properties stored at GAS level	fragility	Fragility type for Station objects
State variables recording GAS state	states	$n_E \times 1$ collection of properties that describe the current state for each of the n_E events (fields: demandFlow, outFlow, average pressure ratio, system serviceability index, number of leaks, number of breaks, etc.)

5.3.2.6 Interdependencies

Interactions of gas and oil networks with other systems are mainly of physical nature, since it is based on the supply of hydrocarbons to the customers.

+ Natural Gas System

- **EPN → GAS [Physical]:**
  Damage to the EPN can hinder the proper operation of re-gasification and regulation/metering stations in the GAS system.

- **GAS → BDG [Physical]:**
  Damage to the GAS system can lower the service level in the struck area, possibly below tolerance thresholds, especially in adverse weather conditions, thus potentially leading to population displacement.

- **GAS → HCS [Physical]:**
  Damage to the GAS system can prevent natural gas to be fed to the health-care facilities hindering emergency response in case backup power sources depend on gas fuel.

+ Oil System

- **EPN → OIL [Physical]:**
  Damage to the EPN can prevent functioning of stations in the OIL system.
- **OIL → EPN [Physical]:**
  Damage to the OIL system can stop production in generators within the EPN inducing power shortages.

5.3.2.7 Methods for Systemic Analysis

The selected works referenced in this section can be classified based on the different goals that the network is expected to meet and the approach used for the network analysis.

+ Natural Gas System

Depending on the purpose of the study, three levels of analysis can be used for the evaluation of seismic performance of gas networks.

- **Level 0 (Vulnerability analysis):**
  - Level 0 analysis is a basic vulnerability analysis and it is related to the physical performance of a single component of the network (e.g. for a gas system it could involve the number of breaks per kilometre for the pipeline system).
- **Level I (Connectivity analysis):**
  - Level I analysis is related to the existence of a path connecting sources and the demand nodes, when the links and the nodes may fail (Ching and Hsu 2007), allowing the assessment of *serviceability*, for example in terms of the number of distribution nodes which remain accessible from at least one supply node after the earthquake (Adachi and Ellingwood 2008; Poljanšek et al. 2012).
- Connectivity analysis requires a simple description of the network in terms of a graph, defined as a collection of nodes (i.e., stations) and links (i.e., pipes) connecting nodes. Moreover, in order to perform the connectivity analysis, nodes should be distinguished considering their functionality. Connectivity analysis tools are limited to those of graph theory (e.g. Ching and Hsu 2007). These algorithms are applied on the network after removing the parts of the system that are failed after the seismic event.
- **Level II (Flow analysis):**
  - Flow analysis includes consideration of the network's capacity, for example maintaining minimum head pressure related to leakages from two particular

Table 5.7 Most relevant methods of the Gas Network System class

Method	Description
computeDemand	Aggregates demand from tributary cells in demand nodes
isBreakAndLeaksNumber	Evaluates the damage state of each pipeline segment employing the corresponding set of fragility functions and the current intensity at the centroid
computeLeakageArea	Computes the amount of leakage from the numbers of leaks in each pipe segment
updateConnectivity	Updates the adjacency matrix based on the pipe breaks and/or the failure of the nodes (e.g. pumping stations, sources)
computePressure	Computes the actual pressure from the sources to the demand nodes based on an optimization algorithm, using the demand level and the leakage amount
computePerformanceIndicator	Computes the different PIs at component- and system-level

points of the network or related to a demand node (Li et al. 2006; Helseth and Holen 2006). In flow analysis, the network's performance is measured evaluating the satisfied end user demand, in terms of flow, after the earthquake event with respect to that before the earthquake. For the purpose of calculating pipe flow and nodal pressure before and after the seismic event, it is necessary to consider flow equations and a method to solve the network analysis problem (Osiadacz 1987).

In the SYNER-G framework, the main UML[1]-methods used to solve the gas network system are described in Table 5.7. An application of these functions is described in Chap. 9.

+ Oil system

The UML-methods used in the vulnerability assessment of natural gas systems can be also applied to oil systems. The classification mentioned above can therefore be used here too.

5.3.2.8 Performance Indicators

Performance indicators are defined to estimate the performance of the system at component or system-level, for the different level of analysis.

Component-Level Performance Indicators (PIs)

- *Pipelines:* **Damage Consequence Index** or DCI. ***[Level II]*** (Wang et al. 2010)

This index is defined at the component level in terms of a system-level PI that measures serviceability; the System Serviceability Index (*SSI*) is defined afterwards.

[1]UML: Unified Modeling Language.

Thus, as for the HR index, this is a PI that reflects at component-level the functional consequence of damage to all systems' components and incorporates the effect of the inter- and intra-dependencies, when modelled. The *DCI* for the *i*-th pipe is defined to reflect the consequence from damaging the pipe, including pipe breaks and leaks. It is expressed as:

$$DCI_i = \frac{E\,[SSI] - E\,[SSI\,|\,L_i\,]}{1 - E\,[SSI]} \tag{5.2}$$

in which *E[SSI]* is the (unconditional) expected value of *SSI* from a set of simulations in which the *i*-th pipe might or might not be damaged; and *E[SSI|Li]* is the conditional expectation of *SSI* from another set of simulations under the same seismic hazard, but given that the *i*-th pipe is damaged.

- Nodes: Pressure Ratio or PR. *[Level II]*

The Pressure Ratio is defined, for each node, as the ratio between the gas pressure at the *i*-th node, in the seismically damaged network, P_{si}, and the reference value of the pressure P_{oi}, for normal operating conditions. The determination of the gas pressure in the seismically damaged network, P_{si} requires a flow analysis of the network.

$$PR_i = \frac{P_{si}}{P_{0i}} \tag{5.6}$$

- *Demand Nodes*: **Customer Connectivity** or CC. ***[Level I]***

The Customer Connectivity evaluates the capacity of demand nodes (or stations) in the gas distribution network to satisfy customers receiving flow from supply nodes (stations or plants). *CC* counts the number of customers satisfied by the *i*-th demand node $N^i_{customer,s}$ if the *i*-th demand node is accessible from at least one supply node, with respect to the number of customers in the undamaged network $N^i_{customer,0}$ Moreover this index can be evaluated for each type of customer, i.e. residential, industrial or strategic:

$$CC_i = \frac{\mathrm{N}^{\mathrm{i}}_{\mathrm{customer,s}}}{\mathrm{N}^{\mathrm{i}}_{\mathrm{customer,0}}} \tag{5.7}$$

System-Level Performance Indicators (PIs)

- **System Serviceability Index** or SSI. ***[Level II]*** (Wang et al. 2010)

Originally defined by Wang et al. (2010) for a Water Supply System, the System Serviceability Index is proposed as a system performance indicator for the Gas and Oil networks. The *SSI* is a relative index that compares the serviceability of the utility network, in terms of customer demand satisfaction, before and after the earthquake. The description of this indicator can be found in Sect. 5.3.2.1.

- **Connectivity Loss** or CL. ***[Level I]*** (Poljanšek et al. 2012)

Connectivity loss (CL) measures the average reduction in the ability of sinks (e.g. gas-fired power plants or distribution nodes for inhabited areas) to receive flow from sources (gas fields and LNG terminals) by counting the number of the sources connected to the *i*-th sink in the original (undamaged) network $N^i_{source,0}$ and then in the damaged network $N^i_{source,s}$.

$$CL = 1 - \left\langle \frac{N^i_{source,0}}{N^i_{source,s}} \right\rangle_i \tag{5.8}$$

- **Serviceability Ratio** or SR. ***[Level II]*** (Adachi and Ellingwood 2008)

The Serviceability ratio, originally defined by Adachi and Ellingwood (2008) for Water Supply Systems, is also proposed for Gas and Oil systems. This index is directly related to the number of distribution nodes in the utility network, which remain accessible from at least one supply facility following the earthquake. It is computed as:

$$P\left[SR \leq s\right] = P\left[SR \leq \frac{\sum_{i=1}^{N} w_i X_i}{\sum_{i=1}^{N} w_i}\right] \tag{5.9}$$

Where SR is the serviceability ratio of the system defined on the domain [0;1], w_i is a weighting factor assigned to the distribution node *i* and X_i represents the functionality of the node *i*, which is modeled as the outcome of a Bernoulli trial ($X_i = 1$ if the node is accessible from at least one supply facility) and N is the number of distribution nodes.

- **Average Pressure Ratio** or APR. ***[Level II]***

The Average Pressure Ratio is defined as the average ratio of the gas pressure in the seismically damaged network over the reference value for non-seismic, normal operations conditions considering n_N nodes.

$$APR = \frac{1}{n_N} \sum_{i=1}^{n_N} PR_i \tag{5.10}$$

Where PR_i is the pressure ratio defined above.

- **Utility Customer density.** ***[Level I]***

The Utility Customer Density measures the average number of customers connected to the utility services per square kilometre at the certain time. It can be evaluated considering the type of customer, i.e. residential, industrial or strategic.

5.3.2.9 Electric Power Network

The focus is put here on the electric power transmission network, from the power generators to the distribution substations.

Structure of the System and Input Attributes

The Electric-power system as a whole is composed of a number of point-like critical facilities (Power generation facilities, Transformation substations, Maintenance facilities) and of the Electric power transmission network itself. The internal logic of the critical facilities and their function in the management of the whole system should be modelled explicitly. The network portion of the system is made of lines and of the supervisory control and data acquisition (SCADA) subsystem.

The identified system components are:

EPN01:	Electric power grid	[System]
EPN02:	Power plant	[Points, critical facilities]
EPN03:	Sub-station (distribution, transformation-distribution)	[Points, critical facilities]
EPN04:	Distribution circuits	[Points, critical facilities]
EPN05-23:	Substation Components	[Points, critical facilities]
EPN24:	Transmission or distribution line	[Edges]

The Electric Power Network is described in terms of objected-oriented structures (Fig. 5.4) and the corresponding attributes (Table 5.8).

5.3.2.10 Interdependencies

The Electric Power Network is a key component of critical infrastructures and it is at the basis of the operation conditions of almost all other systems.

- **EPN → BDG [Physical]:**
  Damage to the EPN can lower the service level in the struck area, possibly below tolerance thresholds thus potentially leading to population displacement.
- **EPN → WSS [Physical]:**
  Damage to the EPN can prevent functioning of pumping stations in the WSS.
- **EPN → GAS [Physical]:**
  Damage to the EPN can prevent functioning of re-gasification and regulation/metering stations in the GAS system.
- **EPN → OIL [Physical]:**
  Damage to the EPN can prevent functioning of stations in the OIL system.

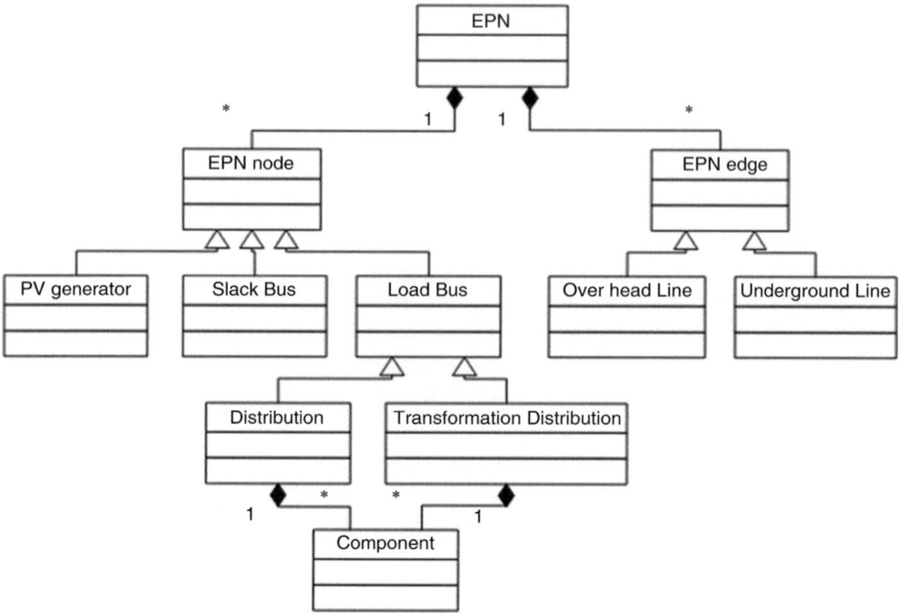

Fig. 5.4 UML class-diagram of the Electrical Power Network (EPN)

- **EPN ➔ HBR [Physical]:**
  Damage to the EPN can prevent functioning of critical components in the HBR system.
- **EPN ➔ HCS [Physical]:**
  Damage to the EPN can prevent power to be fed to the health-care facilities hindering emergency response in case a joint failure of backup power sources occur.

5.3.2.11 Methods for Systemic Analysis

Damages following recent earthquakes revealed that electric power supply, one of the most important services that need to be guaranteed after an earthquake, is maybe the least reliable function. Examples include earthquakes in many countries worldwide: after the earthquake of Kocaeli, Turkey, in 1999, the half of the region's hospitals were not supplied with electricity; about the same happened in Kobe, Japan, 1995, when the whole area was isolated for a period from three to five days; in Northridge, U.S.A., 1994, the electric isolation lasted a day; other earthquakes, even of moderate intensity, caused severe damage either to the entire network, preventing power flow, or to single stations, isolating single nodes.

There are many reasons for carrying out a seismic vulnerability analysis of an EPN. First, the construction of electric networks in industrialized countries, dates

Table 5.8 Main attributes/properties of the Electric Power Network class

Group	Attribute(s)	Description
Global properties	endUserDemand	Required power level at demand nodes, either assigned or evaluated by aggregating over tributary cells, employing population for the region
	admittanceMatrix	Admittance matrix of the EPN, containing the self and mutual bus admittances
Pointers	line	Pointers to all the transmission lines in the system, objects from the EPNLink class
	slack	Pointers to the slack bus, one object from the SlackBus class
	generator	Pointers to all power generators (excluded the slack bus) in the system, objects from the PVGenerator class
	transdistr	Pointers to all transformation/distribution substations in the system, objects from the TransformationDistribution class
	distribution	Pointers to all distribution substations in the system, objects from the Distribution class
Edge properties stored at EPN level	voltage, resistance, voltageRatio	Length, centroid, etc. are attributes inherited from the Network class. Here the network- specific properties are listed (voltage range, etc.)
Node properties stored at EPN level	busType	Typology of the bus that is used to assign a given fragility function, as well as the role in the network
State variables recording EPN state	states	$n_E \times 1$ collection of properties that describe the current state for each of the n_E events (fields: busDown, isolatedBus, shortCircuitIn, shortCircuitOut, VoltageRatio, etc.)

back to a period when earthquake engineering was not at an advanced stage: priority was naturally given to electrical issues when designing components, and thus the equipment currently in place within the stations is not designed for seismic forces. Further, for many types of equipment, the most effective electrical configuration (a slender vertical beam, with steel below, ceramic above and heavy equipment on top) happened to be the least effective structural configuration. Moreover, short-circuits may spread from one station to another, thus isolating large parts of the network.

It should be noted, however, that for a widely distributed and redundant network, damage to a few of the network components will not necessarily lead to a widespread power black-out as a result of alternative paths within the system. Also, as a result of its redundancy, the seismic performance and reliability of an electric power transmission system may be enhanced by upgrading just a few of the network components (Shumuta 2007). Quantitative (probabilistic) information on the likelihood of different levels of damage and extent of affected areas under different earthquake intensities would, therefore, be worthwhile for determining the necessary upgrading of an existing system and for emergency planning and disaster reduction preparedness, including restoration of power.

Economic and social consequences arising from direct and indirect losses due to seismic failures in the EPN are huge, since post-emergency civil protection operations, hospitals, telecommunications, industries and other functions are all affected.

Seismic behaviour of electric power network thus appears a rewarding field of research; however, the efforts on this subject have been limited as compared to other topics. This is probably due to the fact that electric networks more naturally fall within the expertise of electrical engineers, and also to intrinsic complexity in modelling, requiring advanced mathematical tools and interdisciplinary knowledge.

The analysis of an EPN in a seismically active environment can be carried out, as for other lifeline systems, at three different levels.

- **Level 0 (Vulnerability analysis):**
  - Level 0 analysis is a basic vulnerability analysis and it is related to the physical performance of a single component of the network (e.g. power plants, substations, lines . . .)
- **Level I: Connectivity analysis.**
  - Connectivity-based methods focus on finding connected components within the network so that supply and demand can be connected. In their basic form, the methods only lead to a binary statement on whether any given node is connected with another node, specifically a source node, through the network.
  - Li and He (2002) and Kim and Kang (2013) used a non-simulation-based network reliability method, the Recursive Decomposition Algorithm, for risk assessment of generic networks whose operation is defined by the connections of multiple initial and terminal node pairs. Kang et al. (2008) proposed another non-simulation-based method, the Matrix-based System Reliability method, which is able to compute the probability of general system events (at the connectivity level) with correlated system components based on efficient matrix manipulations and minimal set identification.

    Dueñas-Osorio and Rojo (2011) introduced a closed form technique to obtain the entire probability distribution of a reliability metric of customer service availability (CSA) for generic radial lifeline systems. Further works falling within the framework of complex system theory are those by Arianos et al. (2009) and Bompard et al. (2011 The work by Buritica et al. (2012) also relies on a hierarchical representation of networks, the Markov Clustering Algorithm (Gomez et al. 2011), which uses the affinity matrix and random walks to simulate flow through the network and identify communities.
- **Level II (Capacity Analysis)**
  - Capacity analysis is based on the power flow analysis and the point that the actual electrical quantities (voltages, currents, powers) in the network nodes and lines must be determined to make any meaningful statement on the

satisfaction of the power demand at the node, not just its state of continued connectivity. The latter is an intrinsically systemic problem since it depends on the determination of the flows on the entire (damaged) network. Further, before being able to evaluate flows it is necessary to determine which EPN portion remains functional after an event.

– Dueñas-Osorio and Vemuru (2009) included in their reliability assessment study the analysis of flow dynamics, thus allowing to capture the possibility that the system undergoes large-scale cascading failures, the latter being caused by flow redistribution after the occurrence of disruptive events. Pires et al. (1996) presented a simulation-based model to evaluate the seismic reliability of electric power transmission systems, allowing estimating the probability of disconnection of substations from supply nodes, as well as the probability of abnormal power flow in substations. These latter facilities are considered as series systems of a number of electrical components, characterized each by a fragility function.

Some authors (Vanzi 1995, 1996, 2000; Giannini and Vanzi 2000; Nuti et al. 2007) did not simply consider the network nodes (buses) as points characterized by a unique fragility function; rather, they modelled the substations' internal logic. In this model, seismically-induced damage to the components of a substation can have non-local consequences, leading to a short-circuit that may or may not propagate within the substation and eventually further away from that substation to adjacent others, generating in extreme cases very large black-outs. In the analysis of short-circuit propagation, circuit breakers are the only active components playing a key role in arresting the short-circuit spreading. This model allows for intermediate non-binary states to be captured.

Among the "probability-based" vulnerability assessment methods, Ma et al. (2010) proposed a method to evaluate the power system vulnerability in terms of voltage magnitudes and transmission lines passing their limits; a probabilistic technique is applied to obtain the PDF and CDF of the voltage magnitude and transmission line power flows. Xingbin and Singh (2004) employed the power flow computation within an integrated scheme to study the power system vulnerability considering protection system failures.

The three types of level analysis are implemented in SYNER-G, and used according to the levels of the available data and requested details. The corresponding UML-methods implemented are detailed in Table 5.9 and an application of an analysis of the EPN is provided in Chap. 10.

5.3.2.12 Performance Indicators

Performance indicators are defined to estimate the performance of the Electrical Power Network at component or system-level, for the aforementioned different level of analysis.

Table 5.9 Most relevant UML-methods of the Electric Power Network class

Method	Description
computeDemand	Aggregates demand from tributary cells in demand nodes
computeDamage	Evaluates the damage state of each station/bus employing the corresponding set of fragility functions and power loss
spreadShortCircuitsInStation	Computes the short-circuit propagation
checkStationDamage	Deletes the transmission lines affected by short circuits
computePerformanceIndicator	Computes the different PIs at component- and system-level

Component-Level Performance Indicators (PIs)

- **Damage Consequence Index**, or DCI [Level II] (Wang et al. 2010)

  Same as for WSS (Sect. 5.3.2.4)

  This index is defined at the component level in terms of a system-level PI that measures serviceability, the System Serviceability Index (SSI), defined afterwards. Thus, as for the HR index, this is a PI that reflects at component-level the functional consequence of damage to all systems' components (and incorporates the effect of the inter- and intra-dependencies, when modelled). The DCI for the *i*-th element is defined to reflect the consequence from damaging the element *i*. It is expressed as:

$$DCI_i = \frac{E\,[SSI] - E\,[SSI\,|L_i\,]}{1 - E\,[SSI]} \tag{5.11}$$

in which *E[SSI]* is the (unconditional) expected value of *SSI* from a set of simulations in which the *i*-th pipe might or might not be damaged; and *E[SSI|Li]* is the conditional expectation of *SSI* from another set of simulations under the same seismic hazard, but given that the *i*-th pipe is damaged.

- **Upgrade Benefit Index**, or UBI [Level II] (Wang et al. 2010)

  Same as for WSS (Sect. 5.3.2.4)

  Similarly to the *DCI*, this index measures the impact of an upgrade of an individual pipe on the overall system serviceability, and reflects at the component level the systemic functional consequence of damage to the whole system(s). It is defined as:

$$UBI_i = \frac{E_{upgrade}\,[SSI] - E\,[SSI]}{1 - E\,[SSI]} \tag{5.12}$$

in which $E_{upgrade}[SSI]$ is the expected value of *SSI* given that the *i*-th pipe is "upgraded." By "upgrade", it is meant that the probability of pipe damage given an earthquake is significantly smaller than its value before upgrade. UBI_i is the percent increase of SSI given that the *i*-th pipe is upgraded, and its relative value is a measure of the pipe impact on the overall system serviceability.

- **Voltage Ratio**, or VR [Level II]

For each bus inside the substations, this index is defined as the ratio of the voltage magnitude in the seismically damaged network ($V_{i,s}$) to the reference value for non-seismic, normal conditions $V_{i,0}$:

$$VR_i = V_{i,s}/V_{i,0} \tag{5.13}$$

The voltage computation requires a power-flow analysis on the network. Hence this index expresses a functional consequence in the i-th component of the physical damage to all system components. When interactions with other systems are modelled, VR_i expresses the functional consequence in the i-th component of the physical damage to components of all the interacting systems, i.e. it is the value of the index that changes due to the inter- and intra-dependencies, not its definition.

System-Level Performance Indicators (PIs)

- **Simple Connectivity Loss** or SCL [Level I] (Poljanšek et al. 2012)

Same as for GAS (Sect. 5.3.2.8),

Connectivity loss (CL) measures the average reduction in the ability of sinks (e.g. load buses) to receive flow from sources (power plants) by counting the number of the sources connected to the i-th sink in the original (undamaged) network $N^i_{source,orig}$ and then in the damaged network $N^i_{source,dam}$.

$$CL = 1 - \left\langle \frac{N^i_{source,dam}}{N^i_{source,orig}} \right\rangle_i \tag{5.14}$$

- **System Serviceability Index**, or SSI [Level II]

The System Serviceability Index can be defined for EPN as in Vanzi (1995), by the ratio of the sum of the real power delivered from load buses after an earthquake, to that before the earthquake:

$$SSI = \frac{\sum_{i=1}^{N_D} P_{i,0}\,(1 - R_i)\,w_i}{\sum_{i=1}^{N_D} P_{i,0}} \tag{5.15}$$

Where $P_{i,0}$ is the real power delivered from the i-th load bus in non-seismic condition. In order to compute the eventually reduced power delivered in seismic conditions, two factors are considered. The first one, $R_i = \frac{|V_{i,s} - V_{i,0}|}{V_{i,0}}$, with $V_{i,s}$ and $V_{i,0}$ the voltage magnitudes in seismic and non-seismic conditions, is the percent reduction of voltage in the i-th load bus and if $Vi,s < Vi$, one has $1 - R_i = VR_i$ The second factor, w_i, is a weight function accounting for the small tolerance on voltage reduction: in particular, its value is 1 for $R_i < 10\ \%$ and 0 otherwise. The *SSI* index varies between 0 and 1, assuming the value 0 when there is no solution for the power-flow analysis and 1 when the EPN remains undamaged after the earthquake.

The above definition assumes that the demand remains fixed before and after the earthquake, since the index looks only at a single system, without considering the interactions of the EPN with the other infrastructure systems. It can be improved upon and redefined as the *ESSI* that follows.

- **Enhanced System Serviceability Index**, or ESSI [Level II]

The Enhanced System Serviceability Index is an enhancement of the *SSI*, defined to capture the interaction of the EPN with the built area of the study region. In order to model this interaction, the power demand is eventually reduced of the fraction corresponding to collapsed buildings. The *ESSI* is defined as:

$$ESSI = \frac{\sum_{i=1}^{N_D} P_{i,0}\,(1 - R_i)\, w_i \dfrac{\sum_{j \in I_i} N_{j,\overline{CO}}}{\sum_{j \in I_i} N_j}}{\sum_{i=1}^{N_D} P_{i,0} \dfrac{\sum_{j \in I_i} N_{j,\overline{CO}}}{\sum_{j \in I_i} N_j}} \tag{5.16}$$

where I_i is the set of tributary cells for the i-th load bus, N_j is the total number of buildings inside the *j*-th tributary cell and $N_{j,\overline{CO}}$ is the number of not collapsed buildings inside the *j*-th tributary cell. As the *SSI*, the *ESSI* index also varies between 0 and 1, assuming the value 0 when there is no solution for the power-flow analysis or all buildings in the study region are collapsed and 1 when the EPN remains undamaged after the earthquake.

5.3.2.13 Transportation Networks

This chapter concerns mostly the road system. The railway system will be introduced, but no details will be provided. The damage to the network causes traffic congestion, resulting in increased travel time, which is in turn translated into monetary terms.

Structure of the System and Input Attributes

The Road Network is a directed graph composed of a number of nodes and edges.. In a directed graph, one-way edges are usually referred to as arcs and a two-way edge can then be virtually decomposed in two arcs (one for each opposite direction).

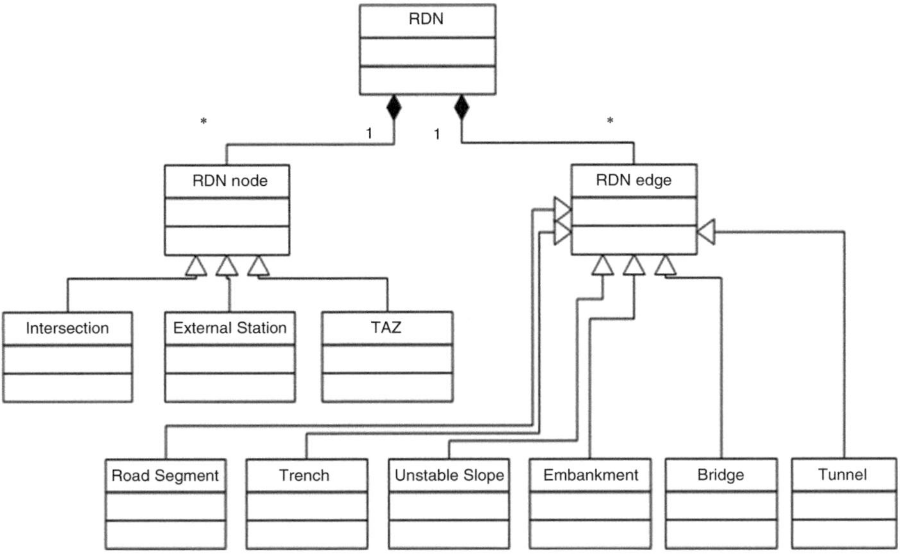

Fig. 5.5 UML class-diagram of the Road Transportation Network (RDN)

All edges are in general vulnerable to seismic shaking or geotechnical hazards (i.e., ground failure due to liquefaction, landslides and fault rupture). Some types of edges or road segments, like those identified below, have specific types of response to seismic action and associated vulnerability (Fig. 5.5 and Table 5.10).

The identified system components are:

RDN01:	Bridge	[Points or edges]
RDN02:	Tunnel	[Edges]
RDN03:	Embankment (road on)	[Edges]
RDN04:	Trench (road in a)	[Edges]
RDN05:	Unstable slope (road on, or running along)	[Edges]
RDN06:	Road pavements	[Edges]
RDN07:	Bridges abutments	[Points or edges]

The Railway system as a whole is composed of a number of point-like critical facilities (Stations) and of the Railway Network itself. The internal logic of the stations and their function in the traffic management of the whole system should be modelled explicitly. The network portion of the system has the same components as a Road network, plus a supervisory control and data acquisition – SCADA – subsystem. The difference is in the fragility models as the railway tracks present lower tolerance to damage compared to roadways.

Table 5.10 Main attributes/properties of the Road Transportation Network class

Group	Attribute(s)	Description
Global properties	tripDemand	Origin-destination matrix built from the TAZ nodes
	roadBlockageModel	Road blockage model to be used
	roadBlockageCoefficients	Coefficients of road blockage model to be used
Pointers	road, trench, embank, unstSlope, tunnel, bridge	Pointers to all road pavements, trenches, embankments, unstable slopes, tunnels, bridges, which are assigned specific fragility functions
	intersection	Pointers to all intersections, objects from the Intersection class
	external	Pointers to all external stations, objects from the ExternalStation class
	taz	Pointers to all Traffic Analysis Zones, objects from the TAZ class
Edge properties stored at RDN level	speed, lanes, dependency, hierarchy	Length, centroid, etc. are attributes inherited from the Directed Network class. Here the network-specific properties are listed (free-flow speed, number of lanes, classification, etc.)
Node properties stored at RDN level	tazType	Type of Traffic Analysis Zones (type of trip demand)
State variables recording RDN state	states	$n_E \times 1$ collection of properties that describe the current state for each of the n_E events (fields: damage state, isBroken, isBlocked, SCL, WCL, isolatedTAZ, etc.)

The identified system components are:

RWN01:	Bridge, same as per RDN	[Points or edges]
RWN02:	Tunnel, same as per RDN	[Edges]
RWN03:	Embankment (road on), same as per RDN	[Edges]
RWN04:	Trench (road in a), same as per RDN	[Edges]
RWN05:	Unstable slope, same as per RDN	[Edges]
RWN06:	Tracks	[Edges]
RWN07:	Bridges abutments	[Points or edges]
RWN08:	Station	[Points]

5.3.2.14 Interdependencies

Road transportation networks play a central part in the analysis of the system of systems, as they must connect all the strategic facilities and the inhabited areas.

- **BDG → RDN [Geographical]:**
  In a urban setting, structural damage to buildings produces debris that can cause road blockages.
- **HBR → RDN [Demand]:**
  Demand for transportation (which concur to the determination of the origin-destination matrix that drives traffic flows) of goods is generated in HBR (HBR is an origin).
- **HCS → RDN [Demand]:**
  Demand for transportation is generated in HCS (as origins for ambulances searching for victims and destination for returning ambulances).
- **RDN → BDG [Physical]:**
  Damage to the transportation network can block access to damaged buildings hindering emergency response.
- **RDN → HBR [Physical]:**
  Damage to the transportation network can block access to the HBR preventing goods to be dispatched and causing large economic loss.
- **RDN → HCS [Physical]:**
  Damage to the transportation network can block access to health-care facilities hindering emergency response.
- **WSS → RDN [Geographical]:**
  Damage and leakage of water pipes underneath the roadways can cause disruption of traffic.

5.3.2.15 Methods for Systemic Analysis

The selected works referenced in this section can be classified according to the level of analysis of the functionality of the transportation network. In a way of classification, available studies can be assigned to the following three levels:

- **Level 0 (Vulnerability analysis):**
  - Level 0 analysis is a basic vulnerability analysis and it is related to the physical performance of a single component of the network (e.g. damages to roads, tunnel or bridges
- **Level I (Connectivity analysis):**
  - Level I analyses are studying the integrity of the network in terms of pure connectivity focussing on the services provided by the network, most typically the rescue function immediately after the earthquake. They may be of interest in identifying portions of the network that are critical to keep the connectivity between most of the points of the networks.
  - Two similar examples of Level I studies can be found in Franchin et al. (2006) and in Nuti and Vanzi (1998). In the latter study the road network serves the purpose of connecting the hospitals to a regional health-care system. A further

example of Level I study is given by Kang et al. (2008). The authors apply a matrix-based system reliability (MSR) method to a transportation network where bridge structures are considered as vulnerable, in order to evaluate the probability of disconnection between each city/county and a critical facility.

- **Level IIa (Capacity analysis):**
  - The scope of the Capacity analysis is widened to include consideration of the network capacity to accommodate traffic flows.
  - Examples of Level II studies are those in Shinozuka et al. (2003) and Chang et al. (2011). The approach in Shinozuka et al. (2003) aims at determining the direct and indirect economic loss due to damage to a transportation network. Direct loss is related to physical damage to vulnerable components, while indirect loss is related to functionality of the transportation system, whose degradation is measured in terms of a system-level performance index called Driver's delay (DD), i.e., the increase in total daily travel time for all travellers. This study is extended in Zhou et al. (2004), to consider the effect of retrofit strategies in improving the performance in future events. The work by Chang et al. (2011) advances a proposal for going beyond the use of the pre-earthquake (static) origin-destination matrix as an input for traffic flow analysis. The post-quake travel demand is complicated and the change of traffic pattern after the event is coupled with the damage of transportation infrastructures.

- **Level IIb (Serviceability analysis):**
  - This more general approach aims at obtaining a realistic estimate of total loss, inclusive of direct physical damage to the built environment (residential and industrial buildings as well as network components), loss due to reduced activity in the economic sectors (industry, services), and losses due to (increased travel time). Economic interdependencies are accounted for, such as the reduction in demand and supply of commodities (due to damaged factories, etc.), hence in the demand for travel, and due to the increased travel costs. At this level the relevance and the complexity of the economic models become dominant over that of the transportation network. This is a full systemic study requiring important inputs from the economic disciplines.
  - Among the few available Level IIb studies, an example is the work by Karaca (2005). The work reports a regional earthquake loss methodology that emphasizes economic interdependencies at regional and national scales and the mediating role of the transportation network. The effectiveness of alternative mitigation strategies is also considered. The loss assessment methodology includes spatial interactions (through the transportation network) and business interaction (through an input-output model). The losses reflect damage to buildings and transportation components, reduced functionality, changes in the level of economic activity in different economic sectors and geographical regions, and the speed of the reconstruction/recovery process.

Table 5.11 Most relevant UML-methods of the Road Transportation Network class

Method	Description
evaluateRDNdamage	Evaluates the damage state of each component
addSecondEdge	Adds a second edge in the model from end to start node, if a two-way travel is requested
discretizeEdges	Subdivides all the links with a length greater than a threshold into smaller segments
updateConnectivity	Sets to 0 the elements in the adjacency matrix corresponding to broken edges and checks if TAZ's are isolated from each other
setRoadBlockageModel	Computes and Samples the road blockage probability based on the collapsed buildings in each cell
computePerformanceIndicator	Computes the different PIs at component- and system-level

The connectivity approach used in the SYNER-G framework uses the set of UML-methods presented in Table 5.11 and is applied to a road network in Italy (Chap. 10).

5.3.2.16 Performance Indicators

Performance indicators are defined to estimate the performance of the Road Network at component or system-level, for the aforementioned different level of analysis.

Component-Level Performance Indicators (PIs)

- *Nodes:* **Connectivity reliability** ***[Level I]***

Connectivity reliability estimates the probability that the network nodes remain connected. A special case of connectivity reliability is the terminal reliability (Iida and Wakabayashi 1989), which concerns the existence of a path between a specific origin-destination (OD) pair. For each node, the network is considered successful if at least one path is operational. A path consists of a set of components (roadways, also called arcs), which are characterized by a binary variable denoting their state (operating or failed). Capacity constraints on the arcs are not accounted for.

- *Nodes:* **Travel time reliability** ***[Level IIa]***

This indicator is defined as the probability that a trip between a given OD pair can be made successful within a specified interval of time (Asakura and Kashiwadani 1991). This measure is useful to evaluate network performances under both normal daily flow variations and seismic conditions. Let C and C_0 be the vectors of damaged and undamaged states of the arcs along the paths and the corresponding travel times between the OD pair w in these two states be denoted as $t_w(C)$ and $t_w(C_0)$.

The travel time reliability is the defined as the probability $\tau_w(\theta)$ of the ratio of $t_w(C)$ to $t_w(C_0)$ being lower than an acceptable level θ.

$$\tau_w(\theta) = P\left(\frac{t_w(C)}{t_w(C_0)} \leq \theta\right) \tag{5.17}$$

The value θ can be interpreted as the level of service that should be maintained despite the capacity reduction that has occurred on some arcs in the network. This index expresses a functional consequence for OD pair of the physical damage to $\tau_w(\theta)$ expresses the functional consequence for OD pair of the physical damage to components of all the interacting systems, i.e. it is the value of the index that changes due to the inter-and intra-dependencies, not its definition.

- Nodes: **Minimum travel time** ***[Level II or III]***

It is the time needed to reach a critical facility, for example a hospital, computed for each TAZ centroid.

System-Level Performance Indicators (PIs)

- **Simple Connectivity Loss**, or SCL. ***[Level I]***

This definition of this index is based on the concept of connectivity (Poljanšek et al. 2012); for a generic system it measures the average reduction in the ability of sink nodes (i.e. destination points in this case) to receive flow from source nodes (i.e. origin nodes in this case):

$$SCL = 1 - \left\langle \frac{N_s^i}{N_0^i} \right\rangle_i \tag{5.18}$$

Where < > denotes averaging over all sink vertices, while N_s^i and N_0^i are the number of sources connected to the i-th sink in the seismically damaged network and in non-seismic conditions, respectively. With reference to a RDN, all the single TAZ's, taken one at a time, are considered sinks, whereas all the remaining TAZ's are sources.

- **Weighted Connectivity Loss**, or WCL. ***[Level I]***

This index upgrades the simple connectivity loss by weighting the number of sources (i.e. origin point) connected to the i-th sink (i.e. destination point), in the seismically damaged network and in non-seismic conditions, respectively:

$$WCL = 1 - \left\langle \frac{N_s^i W_s^i}{N_0^i W_0^i} \right\rangle_i \tag{5.19}$$

Where the weights W_s^i and W_0^i can be defined in different ways. The authors here defined them as the sum of the inverse of the number of edges composing the single paths between the i-th sink and the sources, in the seismically damaged network and in non-seismic conditions, respectively:

$$W^i = \sum_{j,j \neq i} I_{ij} \frac{1}{TT_{ij}} \tag{5.20}$$

where I_{ij} is the indicator function (indicating the existence of a path between the i-th sink and the j-th source, TT_{ij} is the travel time of the path between the i-th sink and the j-th source and j spans all the source nodes, i.e. all TAZ's excluded the i-th one.

- **Driver's delay**, or DD. ***[Level IIa]***

This system-level performance index one of the most classical (Shinozuka et al. 2003); it is defined as the increase in total daily travel time (hours/day) for all travellers, not distinguishing between commuters and commercial vehicles:

$$DD = \sum_a x'_a t'_a (x'_a) - \sum_a x_a t_a (x_a) \tag{5.21}$$

Where x_a and x'_a denote the traffic flows (in PCU[2]/day) on the a-th link in the prevent undamaged and the damaged conditions, respectively, while $t_a(x_a)$ and $t'_a(x'_a)$ denote the corresponding travel times (hours/PCU), which depend on the congestion level through the model:

$$t_a = t_a^0 \left[1 + \alpha \left(\frac{x_a}{c_a} \right)^\beta \right] \tag{5.22}$$

Where c_a is the practical capacity of the link (in PCU/day), t_a^0 the travel time at "zero" flow in the link, α and β are model parameters (frequently assigned values for α and β are 0.15 and 4.0, respectively).

- **Capacity reliability** ***[Level IIb]***

This quantity is defined as the probability that the network can accommodate a certain traffic demand at a required service level, while accounting for drivers' route choice behaviour (Chen et al. 1999). Travel time reliability can also be obtained as a side product. This measure provides important information for efficient flow control, capacity expansion and other relevant works to enhance the reliability of a road network. The maximum capacity of the network, μ, can be computed from the capacities of all the arcs:

$$\mu = g(c_1, c_2, ..., c_a) \tag{5.23}$$

[2]Passenger Car Unit.

Let μ_r denote a required demand level the capacity reliability is given as the probability of μ exceeding μ_r:

$$R(\mu_r) = P(\mu \geq \mu_r) \tag{5.24}$$

This probability predicts how reliably the existing network with damaged arcs can accommodate a given level of required demand. It is easy to see that the boundary conditions must satisfy the following cases:

$$-R(\mu_r = 0) = 1 \tag{5.25}$$

$$-R(\mu_r = \infty) = 0 \tag{5.26}$$

It should be noted that connectivity reliability (level I) is actually a specific case of capacity reliability (level III), where only binary damage states are used and the arcs are either functional or not (i.e. no capacity constraint).

- **Overall travel time reliability *[Level IIb]***

It is sometimes more convenient to use a single index to describe the overall performance of the system and this OD travel time reliability satisfies this need for a reliability measure of the whole road network (Chen et al 2002). However, it is difficult to define such an index because of the interdependence of the individual OD travel times. In the literature, three possible indices representing the overall travel time reliability of the system are provided:

$$\tau_{\min}(\theta) = \min_w \{\tau_w(\theta)\} \tag{5.27}$$

$$\tau_{avg}(\theta) = \frac{1}{W}\sum_{w=1}^{W} \tau_w(\theta) \tag{5.28}$$

$$\tau_{wgt}(\theta) = \frac{\sum_{w=1}^{W} \tau_w(\theta)\, q_w}{\sum_{w=1}^{W} q_w} \tag{5.29}$$

$\tau_{min}(\theta)$ takes the minimum of all OD travel time reliabilities as the overall travel time reliability for a given level of service θ . It is a conservative measure and may not truly reflect the performance of the system. $\tau_{avg}(\theta)$ is a simple arithmetic average of all OD travel time reliabilities and $\tau_{wgt}(\theta)$ is a weighted average of all OD travel time reliabilities by weighing the contribution of each OD pair by its travel demand q_w.

5.3.3 *Critical Facilities*

In SYNER-G, only two critical facilities are considered: harbours and health-care facilities. They are two examples on how critical facilities can be integrated in the global assessment of a system of systems, and play important roles in society in normal times, and an even exacerbated role during crisis.

5.3.3.1 Harbour System

Port transportation systems are critical facilities whose function is to transport cargos and people. They contain a wide variety of facilities for passenger operations and transport, cargo handling and storage, rail and road transport of facility users and cargoes, communication, guidance, maintenance, administration, utilities, and various supporting operations. Ports offer wide-open areas that can be used for emergency or refuge activities after a damaging earthquake. Moreover, ports can play an important role during the recovery period, as they contribute to the reconstruction assistance and the transportation of goods for homeless citizens.

Harbours are part of the general transportation system, often either as an entrance or an exit to close continental, terrestrial systems.

Structure of the System

Harbours are complex systems comprising all the activities related to the transfer of goods/passengers between the maritime transportation and the earth-bound transportation systems. Often they have important storage facilities as well (oil reservoirs, tanks, silos, etc.). They are serviced by a number of other systems including: EPN, WSS, WWN, FFS, GAS, RDN, RWN. The identified system components are:

- *HBR01:* Waterfront components (wharves, breakwaters, etc.)
- *HBR02:* Earthen embankments (backfills, some time hydraulic fills, and native soil material)
- *HBR03:* Cargo handling and storage components (cranes, tanks, etc.)
- *HBR04:* Buildings (sheds, warehouse, offices, control towers etc.)
- *HBR05:* Liquid fuel system (components as per the OIL system)

Also almost all other utility and transportation systems are present within port facilities, like water and waste-water systems, electric power networks, gas supplying systems, road and railway networks. The ports' functionality is dependent on the functioning of each system/component, taking also into consideration the interactions between them.

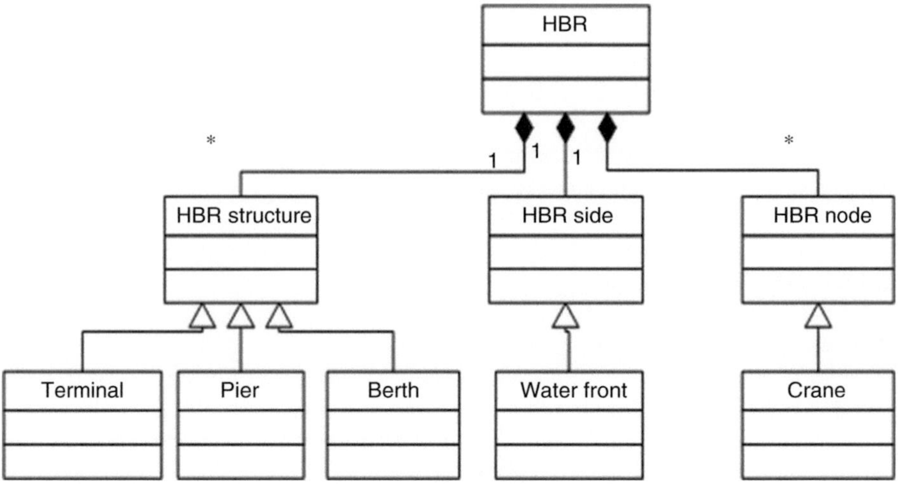

Fig. 5.6 UML class-diagram of the Harbour System (HBR)

Table 5.12 Main attributes/properties of the Harbour System class according to SYNER-G

Group	Attribute	Description
Geometry	nodePosition	Coordinates of the component vertices
Physical damageability	vulnSites	List of vulnerable sites of the HBR, containing their location and IM types
	isVulnerable	Boolean variable determining whether or not the physical damage of the component has to be computed
	typeFragility	Fragility model to be assigned to each of the vulnerablecomponents
System functionality	typeFunctionality	Functionality model, relating physical damages to functional damages, for each vulnerable component
	crane, waterfront, berth, pier, terminal	Pointers to all the cranes, waterfronts, berths, piers and terminals in the system
Interdependence modeling	EPNlinks, RDNlinks	Pointers to the EPN and RDN links connecting the different HBR components
State variables recording system/component state	states	$n_E \times 1$ collection of properties that describe the current state for each of the n_E events (fields: damage state, isolated EPN and RDN nodes, TCoH, TCoM, etc.)

The objet-oriented structure of the Harbour System as defined in SYNER-G as well as the main attributes of the class are presented in Fig. 5.6 and Table 5.12. An illustration of potential applications is described for the Thessaloniki harbor in Chap. 12 of this book.

5.3.3.2 Interdependencies

As Harbours are critical facilities composed of the different systems considered in SYNER-G, interdependencies exist with all the systems. However the main ones are the following ones:

- **EPN ➔HBR [Physical]:**
  Damage to the EPN can prevent functioning of critical components in the HBR system (e.g., cargo handling equipment).
- **RDN ➔ HBR [Physical]:**
  Damage to the transportation network can block access to the HBR preventing goods to be dispatched and causing large economic loss.
- **HBR ➔ RDN [Demand]:**
  Demand for transportation (which concur to the determination of the origin-destination matrix that drives traffic flows) of goods is generated in HBR (HBR is an origin).

5.3.3.3 Methods for Systemic Analysis

Current engineering practice for seismic risk reduction of port facilities is typically based on design or retrofit criteria for individual physical components (e.g. wharf structures) expressed as prescribed levels of displacement, strain, etc. However, the resilience and continuity of shipping operations at a port after an earthquake depend not only on the performance of these individual components, but on their locations, redundancy, and physical and operational connectivity to utility networks as well; that is, on the port system as a whole.

- **Level 0 (Vulnerability analysis)**
  - In most of the post-seismic studies, the performances of the harbours are analysed regarding the physical integrity of the different elements constituting the port systems. The loss of functionality and the recovery are only seldom considered.
  - Hence, almost all the available literature on seismic risk evaluation for port systems focus on the direct physical damages, sometimes with the estimation of the associated cost (NIBS 2004). Shinozuka (2009) developed a model to estimate the physical vulnerability of harbour systems to earthquakes, and the corresponding uncertainties using fragility curves.
  - The economic consequences, caused by losses of incomes, interruption of business or other induced effects for other economic sectors, are estimated only in few studies (Pachakis and Kiremidjian 2003, 2004; Na et al. 2007, 2008).
- **Level II (Capacity analysis)**
  - The integration of indirect costs and functionality losses in seismic risk evaluation is more recent.

Table 5.13 Most relevant UML-methods of the Harbour System class

Method	Description
evaluateHBRdamage	Evaluates the damage state of each Harbour component
evaluateHBRfunctionality	Evaluates the functionality of each Harbour component
retrieveEPNandRDNstates	Checks the state of the EPN and RDN components to ensure the functionality of the Harbour components
computePerformanceIndicator	Computes the different PIs at component- and system-level

– Combining a model estimating the physical damages with a model to estimate losses of revenues caused by the induced closure of the ports, Pachakis and Kiremidjian (2003, 2004) developed a methodology to simulate the response of a harbour system to earthquakes. Two classes of losses are defined: direct losses due to physical damages and indirect losses caused by the modification of conditions of operability of the port systems. Hence, the methodology simulates physical damages, but also takes into account planning and management of the risk. More recent studies also estimate the seismic risk of harbours through the evaluations of both physical damages and revenue losses (Na et al. 2007, 2008; Na and Shinozuka 2009).

The main UML-methods used in the SYNER-G approach to analyse the Harbour System are described in Table 5.13.

5.3.3.4 Performance Indicators

Performance indicators of harbours can be estimated in terms of either quantities of inputs handled or number of boats taken care of.

Container terminals:

- *Terminal:* **Total number of containers handled or** TCoH
  TCoH = total number of containers handled (loaded and unloaded) per day, in Twentyfoot Equivalent Units (TEU)[3]
- *Gate:* **Total number of containers' movements or** TCoM
  TCoM = total number of containers' movements per day, in Twenty-foot Equivalent Units (TEU) (in the whole harbor facility)

Bulk cargo terminals:

- *Terminal:* **Total cargo handled** or TCaH
  TCaH = total cargo handled (loaded and unloaded) per day, in tones
- *Gate:* **Total cargo movements** or TCaM
  TCoM = total cargo movements per DAY, in tones (in the whole harbor facility)

[3]TEU is not a standardized unit. It corresponds to the volume of a 20-foot-long (6.1m) intermodal container, and is often used to estimate the capacity of transportation systems (e.g. boats or terminals).

5.3.3.5 Health-Care System

Hospital facilities are critical infrastructures of the health-care systems From an engineering point of view these systems are made of many components of different types that jointly contribute to provide an output, which are the medical services in the case of hospital. From a social point of view, hospitals provide a fundamental assistance to citizens in every-day life and their function becomes of paramount importance in the case of a disaster. Therefore they are classified as critical facilities.

Structure of the System

The health-care system is made up of health-care facilities (HCF): hospitals, clinics, and all buildings providing medical cares. Hospitals are systems whose function is delivering medical services, which consist of standardized procedures to guarantee an adequate treatment of patients. These procedures are delivered to patients by a joint contribution of the three "active" components of the system:

- The *operators* (human component) namely medical personnel, doctors, nurses and in general whoever plays an active role in providing medical care;
- The *facility* (physical component) i.e. buildings and other sub-components and facilities where medical services are delivered;
- The *organization* (organizational component), which consists of the hospital management, responsible of setting up adequate conditions (standardized procedures for ordinary and emergency conditions) so that the medical services can be delivered.

The identified system components are:

- *HCS01:* Organizational component
- *HCS02:* Human component
- *HCS03:* Physical Component
  - *HCS03-1:* Structural elements (of the buildings within the complex/facility)
  - *HCS03-2:* Non-structural elements/Architectural
  - *HCS03-3:* Non-structural elements/Basic installations/Medical gases
  - *HCS03-4:* Non-structural elements/Basic installations/Power system
  - *HCS03-5:* Non-structural elements/Basic installations/Water system
  - *HCS03-6:* Non-structural elements/Basic installations/Conveying system
  - *HCS03-7:* Non-structural elements/Content-Equipment

The structure of the Health-Care System and the corresponding attributes are detailed in Fig. 5.7 and Table 5.14.

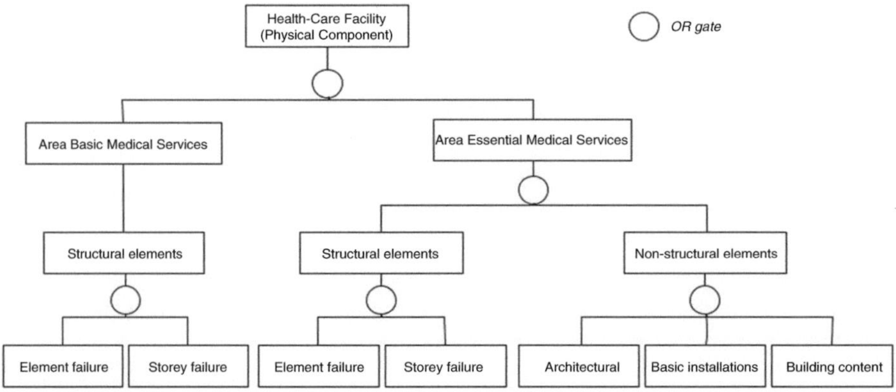

Fig. 5.7 Fault-tree structure of the physical component of the Health-Care System with a description of the main sub-components

Table 5.14 Main attributes/properties of the Health-Care System class

Group	Attribute	Description
Geometry	nodePosition	Coordinates of the network vertices (i.e. health-care centers)
	connectivity	Connectivity matrix listing the start and end nodes of each RDN link
System Functionality	HTC	Health-care facility's treatment capacity
Interdependence modeling	accessibility	Accessibility to the health-care facilities through the rod network
	utilityLoss	Level of service in basic utilities for the health-care facilities
State variables recording system/component state	states	$n_E \times 1$ collection of properties that describe the current state for each of the n_E events (fields: damage state, isolated health-care facilities, HTC, Nb of available operating theatres, Nb of beds, etc.)

5.3.3.6 Interdependencies

Health-care facilities form "high-end" systems, in the sense that they are located at downstream of the global system, needing inputs from almost all other systems to operate.

- **BDG ➔ HCS [Demand]:**
  Structural and non-structural damage to buildings may result in casualties that need to be treated in a health-care facility and hence determine the demand on this system.
- **EPN➔ HCS [Physical]:**
  Damage to the EPN can prevent power to be fed to the health-care facilities hindering emergency response in case a joint failure of backup power sources occur.

- **WSS→ HCS [Physical]:**
  Damage to the WSS can prevent water to be delivered to the health-care facilities hindering emergency response over time in case backup reservoirs are depleted.
- **GAS→ HCS [Physical]:**
  Damage to the GAS system can prevent natural gas to be fed to the health-care facilities hindering emergency response in case backup power sources depend on gas fuel.
- **RDN→ HCS [Physical]:**
  Damage to the transportation network can block access to damaged buildings hindering emergency response.
- **HCS → RDN [Demand]:**
  Demand for transportation is generated in HCS (as a destination).

5.3.3.7 Methods for Systemic Analysis

The health-care system is made up of health-care facilities, collectively serving a region, city or part of a city and coping with the earthquake induced surge in treatment demand in the aftermath of an event. Notwithstanding the criticality of the function of the HCS, the technical literature on the matter is all but abundant. Few studies can be found, some with a focus on the assessment of the capacity of a single facility to remain operational, even if partially, under emergency conditions with possible damage to the facility structural and non-structural components. The remaining few studies deal with the entire system at the regional level and try to evaluate so-called community impact.

For instance, in Monti and Nuti (1996) a reliability-based (FORM, SORM and bounds) procedure to evaluate the functional vulnerability of the surgical function of a hospital system is presented. In Nuti and Vanzi (1998) the regional system of hospitals is studied with the aim of setting up a model for their availability. Such a model is proposed to assess the best retrofit strategies from a systemic point of view, as well as emergency measures such as the use of camp hospitals. Another study which deals with the system as a collection of facilities is Menoni et al. (2002), where the capacity of public facilities can continue providing their service under stressful conditions, even when a certain degree of physical damage has been suffered by structures or by medical equipment, is investigated.

Recent studies try to look at the resilience of the hospital system, as in Cimellaro et al. (2010, 2011). The latter introduces an organizational model, a metamodel, describing the response of the Hospital Emergency Department (ED), which is able to estimate the hospital capacity and the dynamic response in real time and to incorporate the influence of the damage of structural and non-structural components on the organizational ones. The performance indicator chosen to assess the structure is the waiting time. The metamodel covers a large range of hospital configurations and takes into account hospital resources, in terms of staff and infrastructures, operational efficiency and existence of an emergency plan, maximum capacity and behaviour both in saturated and over-capacitated conditions.

Table 5.15 Most relevant UML-methods of the Health-Care System class

UML-Method	Description
evaluateHCSdamage	Evaluates the physical damage state of each health-care center
evaluateHTC	Evaluate the Hospital Treatment Capacity of each health-care center
evaluateHCSaccessibility	Evaluates the accessibility of each health-care center based on the functional state of the road network
performCasualtiesTransportation	Assign each casualty to a health-care center based on the HTC and the accessibility, through an iterative algorithm
computePerformanceIndicator	Computes the different PIs at component- and system-level

Similarly, in Lupoi et al. (2008), a methodology is given to compare treatment demand and capacity for a facility under emergency conditions. Performance is measured in terms of the mean annual rate of demand exceeding a random treatment capacity:

The capacity is measured in terms of number of surgical operations that can be carried out per hour. The demand is evaluated starting from the total number of casualties and using severity classes to find the subset of those requiring surgical treatment.

The capacity term is the result of three contributions, coming from the three macro-components (m/c) making up the hospital system: the physical m/c (structural and non-structural element of the facility), the organizational m/c (the procedure in the emergency plans) and the human m/c (skill and training of the operators using the facilities and equipment according to the procedures).

Some of the UML-methods used is the SYNER-G approach are presented in Table 5.15.

5.3.3.8 Performance Indicators

System-Level Performance Indicators (PIs)

- **Hospital Treatment Capacity**, or HTC

This system-level index expresses the number of patients that can be given surgical treatment per hour (Lupoi et al. 2008). It is defined as:

$$HTC = \frac{\alpha \cdot \beta \cdot \gamma_1 \gamma_2}{t_m} \tag{5.30}$$

where α and β are factors accounting for organizational and human macro-components of the hospital system, γ_1 is the number of undamaged operating theatres, γ_2 a Boolean variable that takes upon the value of one when essential utilities needed for the functioning of the operating theatres are properly working, zero otherwise, and t_m is the average duration of surgical treatment. The performance of the system relative to its pre-earthquake state can be measured through HTC either by taking its ratio to the pre-earthquake value HTCR = HTC/HTC$_0$, or by taking its ratio to the corresponding demand HTC/HTD.[4]

5.4 Synthesis

The several systems that have been described in this chapter are summarized in Table 5.16, where the analysis levels and corresponding performance indicators are outlined.

Table 5.16 Summary of the possible analysis levels and performance indicators for each of the systems studied within the SYNER-G project

System	Analysis levels	Main performance indicators
Buildings	+ Level 0 (vulnerability analysis): + Level II (serviceability)	**Component level** *Building:* **Building damage *[Level 0]*** *Building:* **Building usability *[Level 0]*** *Building:* **Casualties *[Level 0]*** *Building:* **Building habitability *[Level II]*** **System level** **Repartition of Building damages *[Level 0]*** **Repartition of Building usability *[Level 0]*** **Repartition of Casualties *[Level 0]*** **Repartition of Building habitability *[Level II]***
Electric power	+ Level 0 (vulnerability analysis): + Level I (connectivity analysis) + Level II (capacity analysis)	**Component level** *Lines/Nodes:* **Damage Consequence Index**, or DCI. ***[Level II]*** (Wang et al. 2010) *Lines/nodes:* **Upgrade Benefit Index**, or UBI. ***[Level II]*** (Wang et al 2010) *Substations:* **Voltage Ratio**, or VR ***[Level II]*** **System level** **Average Head Ratio**, or AHR. ***[Level II]*** **Simple Connectivity Loss** or SCL ***[Level I]*** (Poljanšek et al. 2012) **System Serviceability Index**, or SSI ***[Level II]*** **Enhanced System Serviceability Index**, or ESSI ***[Level II]***

(continued)

[4]HTD = Hospital Treatment Demand.

Table 5.16 (continued)

System	Analysis levels	Main performance indicators
Water supply and waste-water	+ Level 0 (vulnerability analysis): + Level I (connectivity analysis) + Level II (flow analysis/ serviceability analysis)	**Component level** Junctions/Nodes: **Head Ratio**, or HR. ***[Level II]*** Pipes: the **Damage Consequence Index**, or DCI. ***[Level II]*** (Wang et al 2010) Pipes: **Upgrade Benefit Index**, or UBI. ***[Level II]*** (Wang et al. 2010) **System level** **Average Head Ratio**, or AHR. ***[Level II]*** **System Serviceability Index**, or SSI. ***[Level II]*** (Wang et al. 2010).
Gas and oil	+ Level 0 (vulnerability analysis): + Level I (connectivity analysis) + Level II (flow-performance reliability analysis)	**Component level** Demand Nodes: **Customer Connectivity** or CC. ***[Level I]*** Pipelines: **Damage Consequence Index** or DCI. ***[Level II]*** (Wang et al. 2010) Nodes: **Pressure Ratio** or PR. ***[Level II]*** **System level** **Utility customer density.** ***[Level I]*** **System Serviceability Index** or SSI. ***[Level II]*** (Wang et al. 2010) **Connectivity Loss** or CL. ***[Level I]*** (Poljanšek et al. 2012) **Serviceability ratio** or S. ***[Level II]*** (Adachi and Ellingwood 2008) **Average Pressure Ratio** or APR. ***[Level II]***
Road transportation	+ Level 0 (vulnerability analysis): + Level I (connectivity analysis) + Level IIa (capacity analysis) + Level IIb (serviceability analysis)	**Component level** Nodes: **Connectivity reliability** ***[Level I]*** Nodes: **Travel time reliability** ***[Level II]*** (Asakura and Kashiwadani 1991) Nodes: **Minimum travel time** [***Level IIa or IIb]*** **System level** **Simple Connectivity Loss**, or SCL. ***[Level I]*** (Poljanšek et al. 2012) **Weighted Connectivity Loss**, or WCL. ***[Level I]*** **Driver's delay**, or DD. ***[Level IIa]*** (Shinozuka et al. 2003) **Capacity reliability** ***[Level IIb]*** (Chen et al. 1999) **Overall travel time reliability** ***[Level IIb]*** (Chen et al. 2002) **Capacity reliability** ***[Level IIb]*** (Chen et al. 1999) **Overall travel time reliability** ***[Level IIb]*** (Chen et al. 2002)

(continued)

Table 5.16 (continued)

System	Analysis levels	Main performance indicators
Harbour		***Container terminals:***
		Container Terminal: **Total number of containers handled or** TCoH
		Gate: **Total number of containers' movements or** TCoM
		Bulk cargo terminals:
		Bulk Terminal: **Total cargo handled** or TCaH
		Gate: **Total cargo movements** or TCaM
Health-care		**System level**
		Hospital Treatment Capacity, or HTC (Lupoi et al. 2008)

References

Adachi T, Ellingwood BR (2008) Serviceability of earthquake-damaged water systems: effects of electrical power availability and power backup systems on system vulnerability. Reliab Eng Sys Saf 93:78–88

ALA (2004) Wastewater system performance assessment guideline. American Lifelines Association/FEMA and NIBS

Alexoudi M, Kakderi K, Pitilakis K (2009) Seismic risk and hierarchy importance of the interdependent lifeline systems using fuzzy reasoning. In: 10th international conference on structural safety and reliability, Osaka, Japan

Arianos S, Bompard E, Carbone A, Xue F (2009) Power grids vulnerability: a complex network approach. Chaos 19(1):013–119

Asakura Y, Kashiwadani M (1991) Road network reliability caused by daily fluctuation of traffic flow. In: Proceedings of the 19th PTRC summer annual meeting, Sussex, 9–13 Sept 1991

ATC (1985) ATC-13, earthquake damage evaluation data for California. Applied Technology Council, Redwood City

ATC (1991) ATC-25, seismic vulnerability and impact of disruption on conterminous United States. Applied Technology Council, Redwood City

Awumah K, Goulter I, Bhatt S (1991) Entropy-based redundancy measures in water distribution network design. J Hydrol Eng 117(3):595–614

Bal IE, Bommer JJ, Stafford PJ, Crowley H, Pinho R (2010) The influence of geographical resolution of urban exposure data in an earthquake loss model for Istanbul. Struct Saf 26(3):619–634

Ballantyne DB, Berg E, Kennedy J, Reneau R, Wu D (1990) Earthquake loss estimation modeling of Seattle water system. Technical report: US Geological Survey Grant Award No 14-08-0001-G1526. Kennedy/Jenks/Chilton, Federal Way WA

Bompard E, Wu D, Xue F (2011) Structural vulnerability of power systems: a topological approach. Elect Power Sys Res 81(7):1334–1340

Buritica J, Tesfamariam S, Sanchez-Silva M (2012) Seismic vulnerability assessment of power transmission networks using complex-systems based methodologies. In: Proceedings of the 15th world conference on earthquake engineering, Lisbon, 24–28 Sept 2012

Cavalieri F, Franchin P, Gehl P, Khazai B (2012) Quantitative assessment of social losses based on physical damage and interaction with infrastructural systems. Earthq Eng Struct Dyn 41(11):1569–1589

Chang SE, Svekla WD, Shinozuka M (2002) Linking infrastructure and urban economy: simulation of water disruption impacts in earthquakes. Environ Plan B 29(2):281–301

Chang L, Elnashai AS, Spencer BF Jr, Song J, Ouyang Y (2011) Transportation system modeling and applications in earthquake engineering. Report 10-03, Mid-America Earthquake (MAE) Center

Chen A, Yang H, Lo HK, Tang WH (1999) A capacity related reliability for transportation networks. J Adv Transport 33(2):183–200

Chen A, Yang H, Lo HK, Tang WH (2002) Capacity reliability of a road network: an assessment methodology and numerical results. Trans Res Part B-Methodol 36(3):225–252

Ching J, Hsu W-C (2007) An efficient method for evaluating origin-destination connectivity reliability of real-world lifeline networks. Comput-Aid Civil Infra Eng 22:584–496

Cimellaro GP, Reinhorn AM, Bruneau M (2010) Seismic resilience of a hospital system. Struct Infrastr Eng 6(1):127–144

Cimellaro GP, Reinhorn AM, Bruneau M (2011) Performance-based metamodel for healthcare facilities. Earthq Eng Struct Dyn 40(11):1197–1217

Dueñas-Osorio L, Rojo J (2011) Reliability assessment of radial lifeline systems. Comput-Aid Civil Infra Eng 26(2):111–128

Dueñas-Osorio L, Vemuru MV (2009) Cascading failures in complex infrastructure systems. Struct Saf 313(2):157–167

Dueñas-Osorio L, Craig JI, Goodno BJ (2007a) Seismic response of critical interdependent networks. Earthq Eng Struct Dyn 36:285–306

Dueñas-Osorio L, Craig JI, Goodno BJ (2007b) Interdependent response of networked systems. J Infrastr Sys 13:185–194

Esposito S, Iervolino I (2012) Application and validation study to a gas pipeline network. SYNER-G Deliverable D65

Fotuhi-Firuzabad M, Billinton R, Munian TS, Vinayagam B (2004) A novel approach to determine minimal tie-sets of complex network. IEEE Trans Reliab 53(1):61–70

Franchin P, Lupoi A, Pinto PE (2006) On the role of road networks in reducing human losses after earthquakes. J Earthq Eng 10(2):195–206

Giannini R, Vanzi A (2000) Seismic reliability of electric networks and interaction with other damage indicators. In: Proceedings of 12th world conference on earthquake engineering, New Zealand Society for Earthquake Engineering, Upper Hutt, 30 Jan–4 Feb 2000

Goda K, Hong HP (2008) Estimation of seismic loss for spatially distributed buildings. Earthq Spectra 24(4):889–910

Gomez C, Buritica J, Sanchez-Silva M, Duenas-Osorio L (2011) Optimisation-based decision-making for complex networks in disastrous events. Int J Risk Assess and Manage 15(5–6):417–436

Helseth A, Holen AT (2006) Reliability modelling of gas and electric power distribution systems; similarities and differences. In: Proceedings of 9th international conference on probabilistic methods applied to power systems, New Zealand Society for Earthquake Engineering, Stockholm, 11–15 June 2006

Hoshiya M, Yamamoto K (2002) Redundancy index of lifeline systems. J Eng Mech – ASCE 128(9):961–968

Hoshiya M, Yamamoto K, Ohno H (2004) Redundancy index of lifeline for mitigation measures against seismic risk. Prob Eng Mech 19:205–210

Hwang H, Lin H, Shinozuka M (1998) Seismic performance assessment of water distribution systems. J Infrastr Sys 4(3):118–125

Iida Y, Wakabayashi H (1989) An approximation method of terminal reliability of a road network using partial minimal path and cut set. In: Proceedings of the fifth world conference on transport research, Yokohama, 10–14 July 1989

Isoyama R, Katayama T (1981) Reliability evaluation of water supply systems during earth-quakes. Report of Institute of Industrial Science, University of Tokyo, 30(1)

Jasmon GB, Kai OS (1985) A new technique in minimal path and cutset evaluation. IEEE Trans Reliab 34(2):136–143

Javanbarg MB, Takada S (2009) Seismic reliability assessment of water supply systems. In: Proceedings 10th international conference on structural safety and reliability, Osaka, 13–17 Sept 2009
Javanbarg MB, Takada S, Kuwata Y (2006) Seismic vulnerability evaluation of water delivery system. In: Proceedings of the 12th Japan earthquake engineering symposium, Tokyo
Kalungi P, Tanyimboh TT (2003) Redundancy model for water distribution system. Reliab Eng Sys Saf 82:275–286
Kang WH, Song J, Gardoni P (2008) Matrix-based system reliability method and applications to bridge networks. Reliab Eng Sys Saf 93:1584–1593
Karaca E (2005) Regional earthquake loss estimation: role of transportation network, sensitivity and uncertainty, and risk mitigation., PhD dissertation, MIT, Cambridge
Kawakami H (1990) Earthquake physical damage and functional serviceability of lifeline network models. Earthq Eng 19:1153–1165
Kim Y, Kang W-H (2013) Network reliability analysis of complex systems using a non-simulation-based method. Reliab Eng Sys Saf 110:80–88
Li J, He J (2002) A recursive decomposition algorithm for network seismic reliability evaluation. Earthq Eng Struct Dyn 31(8):1525–1539
Li J, Shulin W, Wei L (2006) Seismic reliability analysis of urban water distribution network. Earthq Eng Eng Vib 5(1):71–77
Lupoi G, Franchin P, Lupoi A, Pinto PE, Calvi GM (2008) Probabilistic seismic assessment for hospitals and complex-social systems. Rose School Tech Rep 2008/02. IUSS Press, Pavia
Ma J, Huang Z, Wong PC, Ferryman T (2010) Probabilistic vulnerability assessment based on power flow and voltage distribution. In: Proceedings of transmission and distribution conference and exposition, 2010 IEEE PES, New Orleans, 19–22 Apr 2010
Markov IJ, Grigoriu M, O'Rourke TD (1994) An evaluation of seismic serviceability of water supply networks with application to San Francisco Auxiliary Water Supply System. Technical report MCEER-94-0001. Multidisciplinary Center for earthquake Engineering Research, Buffalo
Menoni S, Pergalani BMP, Petrini V (2002) Lifelines earthquake vulnerability assessment: a systemic approach. Soil Dy Earthq Eng 22:1199–1208
Moghtaderi-Zadeh M, Wood K, Der Kiureghian A, Barlow RE (1982) Seismic reliability of lifeline networks. J Tech Council – ASCE 108:60–78
Monti G, Nuti C (1996) A procedure for assessing the functional reliability of hospital systems. Struct Saf 18(4):277–292
Na UJ, Shinozuka M (2009) Simulation-based seismic loss estimation of seaport transportation system. Reliab Eng Sys Saf 94(3):722–731
Na UJ, Chaudhuri SR, Shinozuka M (2007) Scenario-based simulation of revenue loss at seismically damaged seaports. In: Proceedings of the 2nd international conference on urban disaster reduction, Taipei, 27–29 Nov 2007
Na UJ, Chaudhuri SR, Shinozuka M (2008) Probabilistic assessment for seismic performance of port structures. Soil Dyn Earthq Eng 28(2):147–158
NIBS (2004) HAZUS-MH: user's manual and technical manuals. Report prepared for the federal emergency management agency. National Institute of Building Sciences, Washington, DC
Nuti C, Vanzi I (1998) Assessment of post-earthquake availability of hospital system and upgrading strategies. Earthq Eng Struct Dyn 27(12):1403–1423
Nuti C, Rasulo A, Vanzi I (2007) Seismic safety evaluation of electric power supply at urban level. Earthq Eng Struct Dyn 36(2):245–263
O'Rourke TD, Grigoriu MD, Khater MM (1985) Seismic response of buried pipes. In: Sundrarajan C (ed) Pressure vessel and piping technology – a decade of progress. ASME, New York, pp 281–323
Osiadacz AJ (1987) Simulation and analysis of gas network, 1st edn. D and F.N. Spon Ltd., London
Pachakis D, Kiremidjian A (2003) The use of simulation in disaster response planning and risk management of ports and harbors. In Proceedings of the 6th U.S. conference on lifeline earthquake engineering. ASCE's Technical Council on lifeline earthquake engineering, Long Beach, 10–13 Aug 2003

Pachakis D, Kiremidjian A (2004) Estimation of downtime-related revenue losses in seaports following scenario earthquakes. Earthq Spectra 20(2):427–449

Pires JA, Ang AH-S, Villaverde R (1996) Seismic reliability of electrical power transmission systems. Nucl Eng Desig 160(3):427–439

Pitilakis K, Crowley H, Kaynia H (2014) Fragility functions for physical elements at seismic risk: buildings, lifelines, transportation networks and critical facilities: the SYNER-G approach. Springer

Poljanšek K, Bono F, Gutiérrez E (2012) Seismic risk assessment of interdependent critical infrastructure systems: the case of european gas and electricity networks. Earthq Eng Struct Dyn 41(1):61–79

Rinaldi RM, Peerenboom JP, Kelly TK (2001) Identifying, understanding and analyzing critical infrastructures interdependencies. IEEE Cont Sys Mag 21(6):11–25

Scawthorn C, O'Rourke TD, Balckburn FT (2006) The San Francisco earthquake and fire of 1906 – enduring lesson for fire protection and water supply. Earthq Spectra 22(S2):135–158

Shi P, O'Rourke TD, Wang Y (2006) Simulation of earthquake water supply performance. In: Proceedings of 8th U.S. national conference on earthquake engineering, San Francisco, 18–22 Apr 2006

Shinozuka M (2009) Performance reliability of port facilities. In: Proceedings of 10th international conference on structural safety and reliability, ICOSSAR, Osaka, 13–17 Sept 2009

Shinozuka M, Takada S, Kawakami, H (1977) Risk analysis of underground network systems. In: Proceedings of the U.S. Southeast Asia symposium on engineering for natural hazards protection, Manila, 26–30 Sept 1977

Shinozuka M, Tan RY, Koike T (1981) Serviceability of water transmission systems under seismic risk. In: Smith DJ Jr (ed) Proceedings of the ASCE specialty conference on lifeline earthquake engineering: the current state of knowledge, Oakland, 20–21 Aug 1981

Shinozuka M, Hwang H, Murata M (1992) Impact on water supply of a seismically damaged water delivery system. Lifeline Earthquake Engineering in the Central and Eastern US. Tech Council Lifel Earthq Eng Monogr 5:43–57

Shinozuka M, Murachi Y, Dong X, Zhou Y, Orlikowski MJ (2003). Seismic performance of highway transportation networks. In: Proceedings of China-US workshop on protection of urban infrastructure and public buildings against earthquakes and man-made disasters, 21–22 Feb 2003

Shumuta Y (2007) Practical seismic upgrade strategy for substation equipment based on performance indices. Earthq Eng Struct Dyn 36(2):209–226

Taylor CE (1991) Seismic loss estimation for a hypothetical water system. Technical council on lifeline earthquake engineering monograph no. 2. ASCE, Reston

Vanzi I (1995) Seismic reliability of electric power networks. PhD dissertation, Sapienza University of Rome

Vanzi I (1996) Seismic reliability of electric power networks: methodology and application. Struct Saf 18(4):311–327

Vanzi I (2000) Structural upgrading strategy for electric power networks under seismic action. Earthq Eng Struct Dyn 29(7):1053–1073

Wang Y (2006). Seismic performance evaluation of water supply systems. PhD dissertation. Cornell University, Ithaca

Wang Y, Siu-Kui A, Qiang F (2010) Seismic risk assessment and mitigation of water supply systems. Earthq Spectra 26(1):257–274

Xingbin Y, Singh C (2004) Probabilistic power system security analysis considering protection failures. Int J Comput Math Elect Electron Eng 23(1):35–47

Zhou Y, Murachi Y, Kim S-H, Shinozuka M (2004) Seismic risk assessment of retrofitted transportation systems. In: Proceedings of 13th world conference on earthquake engineering, Vancouver, 1–6 Aug 2004

Chapter 6
Introduction to the Applications of the SYNER-G Methodology and Tools

Kyriazis Pitilakis and Sotiris Argyroudis

Abstract The objective of this Chapter is to introduce the application case studies that are presented in the following Chapters. At first, the general framework of the SYNER-G methodology is outlined, including the seismic hazard, the physical vulnerability and the systemic (functional and socio-economic) models, in order to summarize the basic concepts and methods that are applied and facilitate the understanding of the following applications. The SYNER-G approach is presented in detail in Part I of this book (Chaps. 2, 3, 4, and 5). The case studies aim to demonstrate and test the applicability of the developed methodology and tools in different systems and levels of analysis, accounting for inter and intra-dependencies among infrastructural components and systems: at urban level, in the city of Thessaloniki in Greece and a district in the city of Vienna in Austria; at system level, in the gas system of L'Aquila in Italy, the road network of Calabria region in Southern Italy and the electric power network of Sicily again in Italy; in complex infrastructures: a network of hospitals at regional scale and the harbor of Thessaloniki in Greece. A brief description of the case studies is given and their key elements are outlined, including the seismic and geotechnical hazard assessment, the selection of fragility curves and performance indicators, as well as the systemic analysis, the sources of uncertainties, the estimation of socio-economic impacts and finally the interpretation of results.

K. Pitilakis (✉) • S. Argyroudis
Department of Civil Engineering, Aristotle University, 54124 Thessaloniki, Greece
e-mail: kpitilak@civil.auth.gr; sarg@civil.auth.gr

K. Pitilakis et al. (eds.), *SYNER-G: Systemic Seismic Vulnerability and Risk Assessment of Complex Urban, Utility, Lifeline Systems and Critical Facilities*, Geotechnical, Geological and Earthquake Engineering 31, DOI 10.1007/978-94-017-8835-9_6,

6.1 Introduction

The goal of the SYNER-G methodology is to assess the seismic vulnerability and losses of the *Infrastructure* (i.e., set of interconnected civil infrastructural systems) of urban or regional extension, accounting for inter and intra-dependencies among infrastructural components and systems, as well as for the uncertainties characterizing the problem. The goal has been achieved setting up a model of the *Infrastructure* in urban or regional scale and of the hazard acting upon them, and then enhancing it with the introduction of the uncertainty and of the analysis methods that can evaluate the individual system's or system of systems' performance accounting for such uncertainty. In its general form the SYNER-G methodology is based on a sequence of three models: (a) seismic hazard model, (b) components' physical vulnerability model, and (c) system (functional and socio-economic) model.

The *hazard model* provides a tool for sampling events in terms of location (epicentre) and possibly extension, magnitude and faulting style according to the seismicity of the study region and predicting maps of seismic intensities considering spatial variability of ground motion and associated geotechnical hazards (see Chap. 3). These maps, conditional on earthquake magnitude, epicentre, etc. describe the variability and spatial correlation of intensities at different sites where the different infrastructures are located. Further, when different vulnerable components exist at the same location, which are sensitive to different seismic intensity measures (e.g. acceleration, velocity or permanent displacement), the hazard model should predict intensity measures that are consistent at the same site.

Development of the physical model involves first the definition of an extensive *taxonomy of systems* and *components* and requires: (a) for each system, a description of the functioning of the system under both *undisturbed* and *disturbed* conditions (i.e. in the damaged state following an earthquake); (b) a model for the physical and functional (seismic) damageability of each component within each system; (c) identification of all dependencies between the systems; (d) definition of adequate performance indicators for components and systems, and the *Infrastructure* as a whole (see Chap. 5). The SYNER-G typology of components and fragility models for each element at risk are presented in detail in Pitilakis et al. (2014).

Development of the *socio-economic model* starts with an interface to outputs from the physical model in each of the four domains of SYNER-G (i.e., buildings, transportation systems, utility systems and critical facilities). Thus, in the present applications, four main performance indicators are used to determine both direct and indirect impacts on society: *Building Usability, Transportation Accessibility, Utility Functionality and Healthcare Treatment Capacity*. Direct social losses are computed in terms of casualties and displaced populations. Indirect social losses are considered in two models, *Shelter Needs and Health Impact*, which employ the multi-criteria decision analysis (MCDA) theory for combining performance indicators from the physical and social vulnerability models (see Chap. 4).

In order to tackle the complexity of the described problem the object-oriented paradigm (OOP) has been adopted, which encompasses abstraction, encapsulation, modularity, hierarchy, typing, concurrency, and persistence brought together in a synergistic way (see Chap. 2). In abstract terms, within such a paradigm, the problem is described as a set of *objects*, characterized in terms of *attributes* and *methods*, interacting with each other. *Objects* are *instances* (concrete realizations) of *classes* (abstract models, or *templates* for all objects with the same set of properties and methods).

The object model developed within SYNER-G is described by means of class-diagrams, used to show the existence of classes and their relationships. The object of the analysis is the *Environment*, which is composed of three classes, the *Infrastructure*, the *Hazard* and the *Weather*. The *Infrastructure* class is the composition of the Critical facility class, the Network class and the BDG class. The *Network and Critical facility* classes are abstract ones, and are the generalizations of all types of networks and of critical facilities. The evaluation of interactions between them is performed through the establishment of an order in the evaluation of states of the objects, described with a state diagram.

The SYNER-G prototype framework has been implemented in MATHWORKS-MATLAB® for each network and infrastructure. In addition, EQvis, an advanced seismic loss assessment and risk management software based on the Mid-America Earthquake Center software tool MAEviz (MAEviz 2013), was further developed and adapted in SYNER-G, as a platform for performing deterministic earthquake simulations as well as various other tools for pre and post processing of the input and output data of both the deterministic and probabilistic (prototype software) analyses.

6.2 Applications

The SYNER-G methodology and software tools have been applied and tested in the following case studies at urban and regional level as well as at complex infrastructure level.

6.2.1 City of Thessaloniki

The city of Thessaloniki in Northern Greece is located in a high seismicity area. The study area covers the municipality of Thessaloniki, which is divided in 20 Sub City Districts as defined by EUROSTAT and Urban Audit approach (Fig. 6.1). The case study presented in Chap. 7 includes the following elements: building stock (BDG), road network (RDN), water supply system (WSS) and electric power network (EPN). The networks are comprised by the main lines and components and cover the wider Metropolitan area. The internal functioning of each network is simulated and a connectivity analysis is performed. Moreover, specific interdependencies between systems are considered: EPN with WSS (electric power supply to pumping stations), RDN with BDG (road blockage due to building collapses), BDG with EPN and WSS

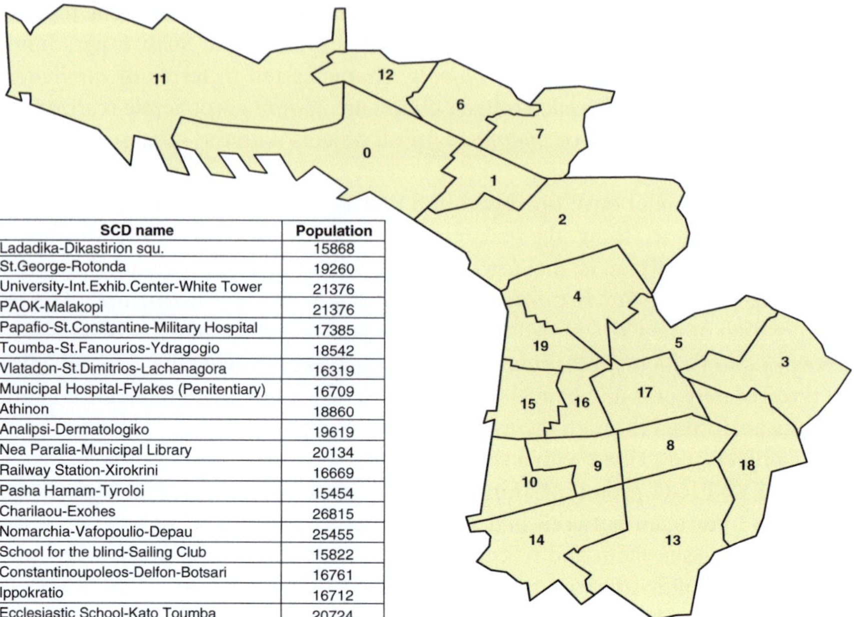

ID	SCD name	Population
0	Ladadika-Dikastirion squ.	15868
1	St.George-Rotonda	19260
2	University-Int.Exhib.Center-White Tower	21376
3	PAOK-Malakopi	21376
4	Papafio-St.Constantine-Military Hospital	17385
5	Toumba-St.Fanourios-Ydragogio	18542
6	Vlatadon-St.Dimitrios-Lachanagora	16319
7	Municipal Hospital-Fylakes (Penitentiary)	16709
8	Athinon	18860
9	Analipsi-Dermatologiko	19619
10	Nea Paralia-Municipal Library	20134
11	Railway Station-Xirokrini	16669
12	Pasha Hamam-Tyroloi	15454
13	Charilaou-Exohes	26815
14	Nomarchia-Vafopoulio-Depau	25455
15	School for the blind-Sailing Club	15822
16	Constantinoupoleos-Delfon-Botsari	16761
17	Ippokratio	16712
18	Ecclesiastic School-Kato Toumba	20724
19	Makedonia Palace-Paedagogiki Academia	16729

Fig. 6.1 Sub-city districts (SCD) of Thessaloniki study area as defined by Urban Audit

(displaced people due to utility loss). An accessibility analysis to hospital facilities considering the damages in RDN is also performed (see Chap. 4) and a shelter demand analysis based on a multi-criteria approach and using the loss assessments for buildings and utilities is applied (Chap. 7).

6.2.2 City of Vienna

The city of Vienna in Austria is located in a low seismicity area. The majority of seismic risk in Austria is associated with the Vienna transform fault zone, which runs through the eastern part of Austria beneath the city of Vienna and surrounding areas. The region of interest selected for the case study in Chap. 8 is the Brigittenau district, which is the 20th district of Vienna. It is located north of the central district, north of Leopoldstadt on the same island area between the Danube and the Danube Canal. Brigittenau is a heavily populated urban area with many residential buildings and several networks and infrastructures (Fig. 6.2). The present test case is mainly an attempt to look at SYNER-G methods at the building level, using high-resolution data. This data is not commonly available, thus a specific building identification procedure has been formulated to identify and inventory buildings that were considered in the case study. A deterministic and a probabilistic analysis have

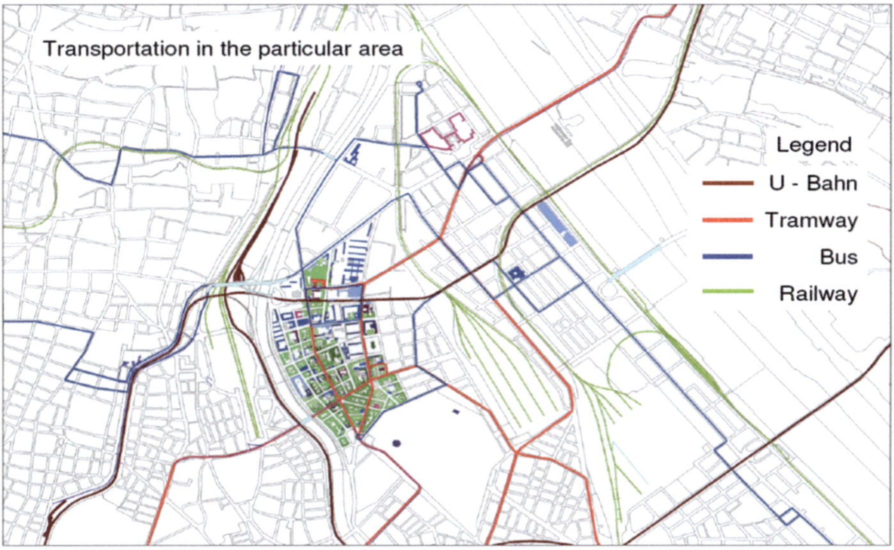

Fig. 6.2 Brigittenau district in the city of Vienna and overview of transportation networks in the considered area

been performed in the area of interest in order to assess the expected losses for buildings and lifelines and to highlight the benefits of both approaches. The EQvis software which is based on MAEviz tool (MAEviz 2013) has been used for the deterministic analysis, while the SYNER-G prototype software was used for the probabilistic analysis.

6.2.3 L'Aquila Gas System

The medium-pressure part of the L'Aquila (central Italy) gas distribution system is considered (Fig. 6.3) in Chap. 9. The selected network is characterized by three reduction stations connecting the network to the high-pressure nationwide network, more than 200 km of pipelines, either made of steel or high density polyethylene (HDPE) pipes, and about 200 Reduction Groups (M/R stations). The network is composed of 602 nodes (3 sources, 209 RGs and 390 joints) and 608 links. The study employs probabilistic seismic and geotechnical hazard analysis, empirical relations to estimate pipeline response, fragility curves for the evaluation of M/R stations' vulnerability, and connectivity performance indicators (serviceability ratio and connectivity loss) to characterize the functionality of the network. Earthquakes are generated using as source the Paganica fault and considering characteristic earthquakes of moment magnitude *Mw* 6.3 while the landslide potential of L'Aquila region is also considered.

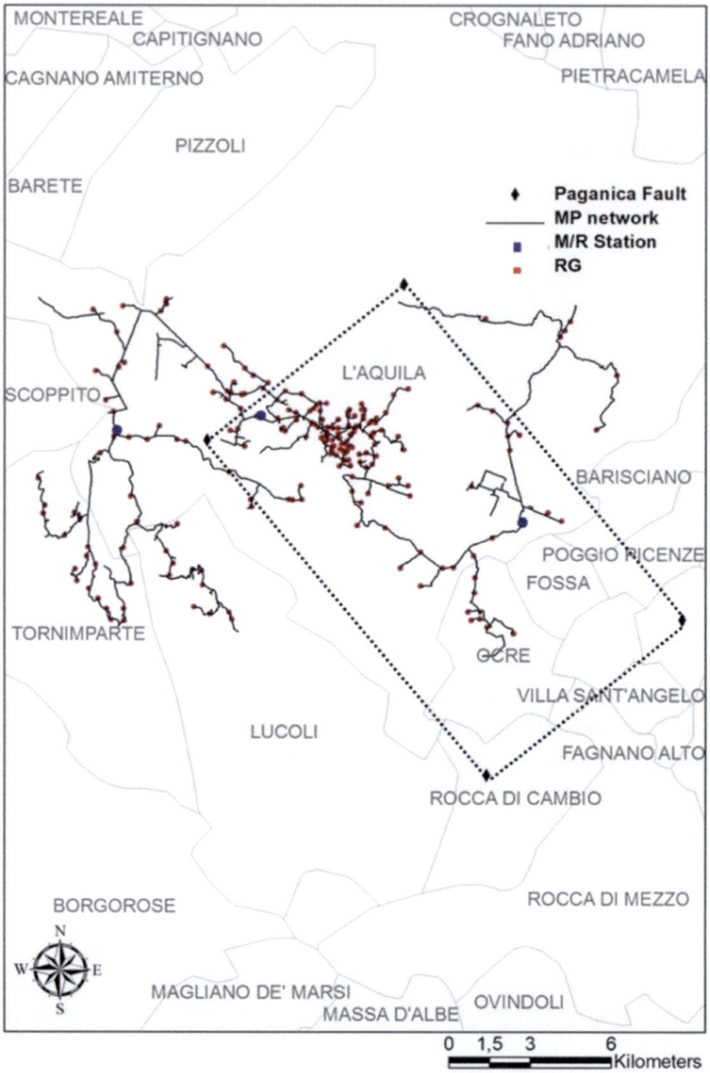

Fig. 6.3 Application gas system of L'Aquila (central Italy)

6.2.4 Selected Transportation and Electric Networks in Italy

The road network in Calabria region, Southern Italy has been chosen as the case study in Chap. 10. A data reduction process was performed in order to remove the irrelevant components at the regional scale. A pure connectivity analysis is performed considering 2,861 nodes and 5,970 edges of the network (Fig. 6.4).

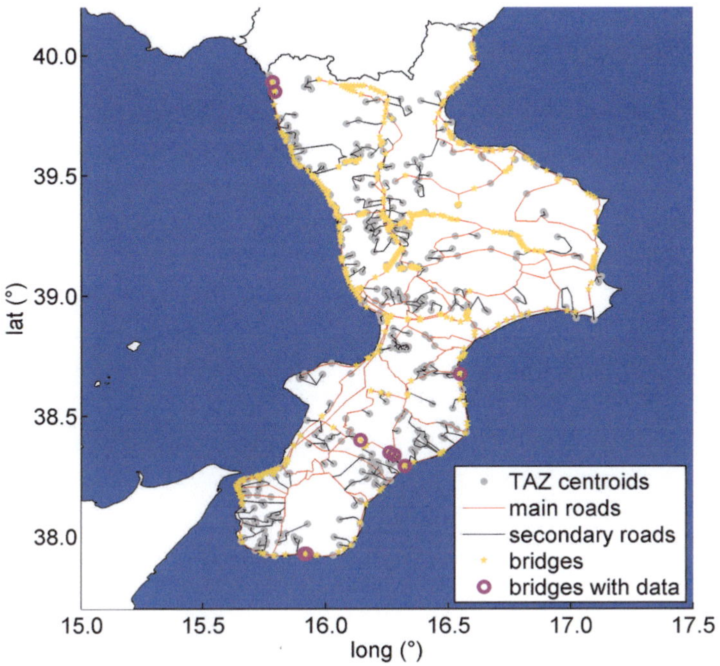

Fig. 6.4 Calabria region road network topology

The seismic hazard is modeled through 20 faults taken from the Italian DISS (Database of Individual Seismogenic Sources) database.

The electric power network of Sicily, Southern Italy has been chosen as a second case study in Chap. 10. A capacitive study is performed, with power flow analysis that follows the analysis of short-circuit propagation, in which circuit breakers are active components playing a key role in arresting the short-circuit spreading. The main vulnerable elements are the micro-components within substations, whereas the vulnerability of lines is neglected being relatively minor. The total number of municipalities served by the network is 390 (Fig. 6.5). The network is composed of 181 nodes and 220 transmission lines. The seismic hazard is modeled through 18 faults taken from the Italian DISS database.

6.2.5 Network of Hospitals at Regional Scale

The seismic performance of a generic regional Health-Care System (HCS) is investigated in Chap. 11. The earthquake effects both on hospitals and on the Road Network (RDN), connecting towns to hospitals, are evaluated and the interaction among them is accounted (Fig. 6.6). Victims move to hospitals until their request for a bed or for a surgical treatment is satisfied, if possible. A novel "dynamic"

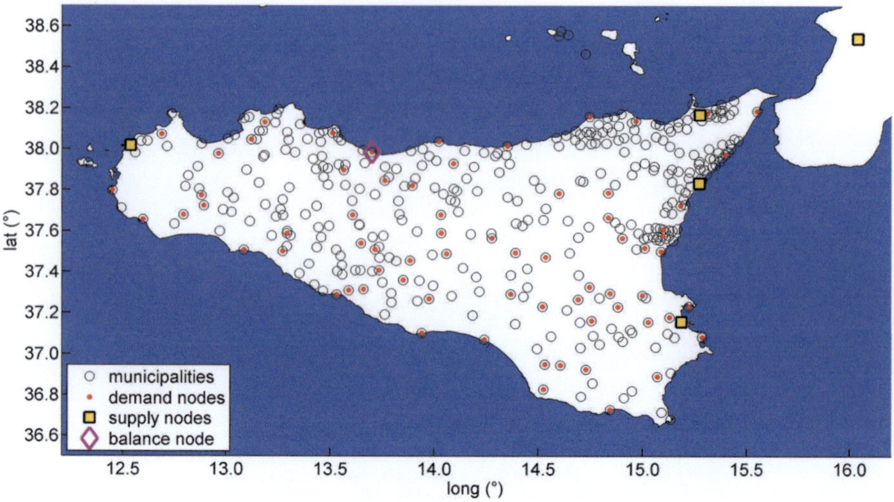

Fig. 6.5 Position of municipalities and EPN nodes in Sicily case study

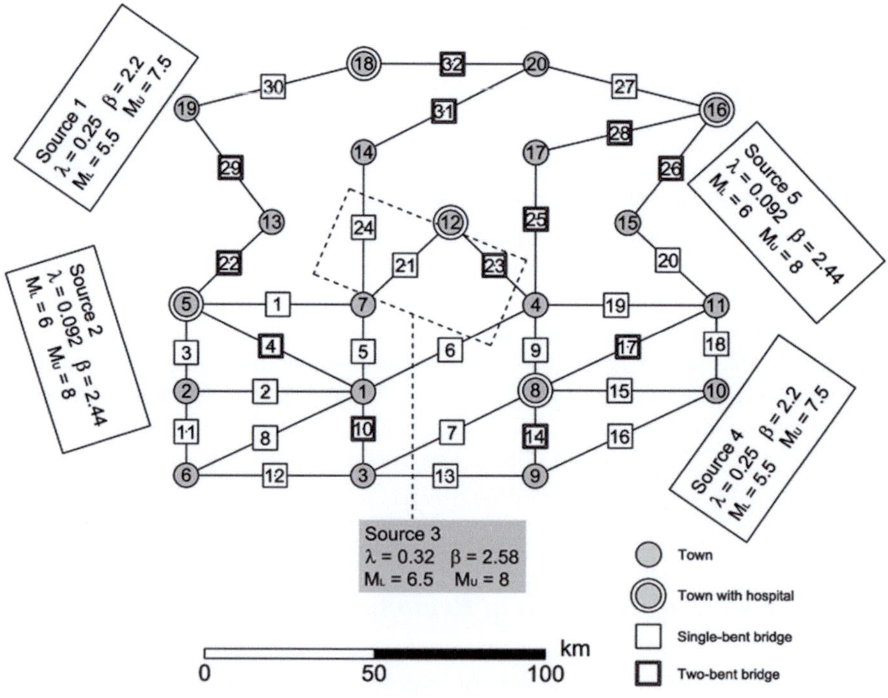

Fig. 6.6 The study area of regional Health-Care System

Fig. 6.7 Aerial view of Thessaloniki's harbor

model for hospitalization is developed and implemented where the path of the victims from the town of origin to the hospital of destination through the road network is considered. Hospitalization, for each homogenous type of victims, is accorded on a "first-come, first-served" criterion. The road network is modelled in connectivity terms. The vulnerability of hospitals and bridges is expressed by pre-evaluated fragility curves relative to existing structures located in Italy. Uncertainties associated to seismic hazard, ground motion intensities and estimation of victims are accounted. In particular, the reliability problem is solved by Monte Carlo simulation. The un-hospitalized victims, the risk that hospitals are unable to provide medical care, the demand of medical care on hospitals and the hospitalization travel time, are among the useful results of the analysis. The methodology is exemplified with reference to a hypothetical region, with population of 877,000, 20 towns, 5 hospitals and 32 bridges.

6.2.6 Harbor of Thessaloniki

The assessment of the systemic vulnerability of Thessaloniki's port is performed in Chap. 12. The port covers an area of 1,550,000 m^2 and trades approximately 16,000,000 t of cargo annually, having a capacity of 370,000 containers and 6 piers with 6,500 m length (Fig. 6.7). In the case study, waterfront structures, cargo handling equipment, power supply system, roadway system and buildings are examined. In particular, for the systemic analysis, waterfront structures of a total 6.5 km length, 48 crane-nodes and 2 terminals (1 container and 1 bulk cargo) are considered. The interactions accounted for in the analysis are the supply of EPN to cranes and the road closures due to building collapses. For seismic hazard, five seismic zones have been selected, based on the recent results of the SHARE European research project (Giardini et al. 2013), as for the Thessaloniki case study. A plain Monte Carlo simulation (MCS) has been carried out sampling earthquake events for these zones and computing selected performance indicators (PIs) based on the estimated damages and functionality loss of the different components.

6.3 Main Features of the Applications

6.3.1 Seismic and Geotechnical Hazard

A probabilistic approach is followed which samples earthquake events based on the hazard characterization of each area following the specifications provided in Chap. 3. Each sampled event represents a single earthquake (Shakefields method) and all systems are analyzed for each event. The results are then aggregated over all the sampled events.

The source models provided in SHARE project (Giardini et al. 2013) are applied either as seismic zones (e.g. Thessaloniki case study, harbor of Thessaloniki) or as faults (e.g. transportation network in South Italy and electric power network in Sicily). In other cases, specific faults or/and earthquake magnitudes are used as sources (e.g., Paganica fault, with $M_w = 6.3$ in L'Aquila case study) or a deterministic analysis for specific earthquake events is considered (Vienna case study).

The ground motion prediction equation introduced by Akkar and Bommer (2010) is applied for the estimation of the ground motion parameters on rock, while the spatial variability is modelled using appropriate correlation models. For each site of the grid the averages of primary IM from the specified GMPE are calculated, and the residual is sampled from a random field of spatially correlated Gaussian variables according to the spatial correlation model. The primary IM is then retrieved at the sites of vulnerable components by distance-based interpolation and finally the local IM is sampled conditional on primary IM. In case of L'Aquila gas system the influence of modeling of spatial correlation of intra-event residuals on risk assessment was also investigated showing a relatively small impact on risk evaluation of the selected system and with reference to the considered performance indicators.

To scale the hazard to the site condition different amplification methods are available in the SYNER-G prototype software: Present Eurocode 8 provisions, Eurocode 8 improved amplification factors as modified by Pitilakis et al. (2012), NEHRP, Choi and Stewart (2005), context-specific. Depending on the available information for site characterization an amplification method is selected. In case that no soil classification data are available IM values are computed for stiff soil conditions (e.g. transportation network in South Italy and electric power network in Sicily).

For the geotechnical hazards (liquefaction and landslide) the modelling approach of HAZUS (NIBS 2004) is adopted for the estimation of the permanent ground deformations, PGD, at the vulnerable sites. A detailed description of the entire hazard model adopted in the methodology and hence implemented in the SYNER-G prototype software can be found in Chaps. 2 and 3.

6.3.2 *Fragility Curves*

Appropriate fragility curves are selected for the components of each case study considering the typology/taxonomy of the network or infrastructure and the proposals of SYNER-G, which are presented in another volume of the present book series (Pitilakis et al. 2014). The fragility curves for most elements exposed in ground shaking (i.e., buildings, bridges, electric power substations, gas M/R reduction stations, cranes and waterfront structures) are defined as a function of peak ground acceleration (PGA). Damages for elements that are exposed in ground failure due to liquefaction or landsliding (i.e. cranes or roads) are estimated based on permanent ground displacements (PGD). The pipelines' damages are estimated based on fragility functions that correlate the repair rate (repairs per km) with peak ground velocity (PGV) for ground shaking and PGD for ground failure.

In case of Thessaloniki application, new analytical fragility curves have been developed for bridges and buildings respecting the specific typologies and features of the Thessaloniki structures. In case of Thessaloniki harbour the epistemic uncertainty related to different fragility functions and functionality definitions was investigated performing sensitivity analysis with the use of alternative fragility curves and functionality thresholds for the waterfront structures. In the case of hospitals and medical care infrastructure the fragility is estimated through a fault tree approach considering the various physical components.

6.3.3 *Performance Indicators and Systemic Analysis*

The risk assessment is performed in terms of appropriate performance indicators (PIs) for each system (see Chap. 5). In this way, indicators that are able to quantify the degree to which the system is able to meet established specifications and/or customer requirements following an earthquake event give the quantitative measure of the functionality of each network.

The performance of the networks is generally measured according to two categories of indices: (a) those based on connectivity analysis that allows assessment of serviceability in terms of the aggregate functionality of facilities composing the system (i.e., the number of distribution nodes which remain accessible from at least one supply node after the earthquake), and (b) those based on capacitive analysis that include consideration of graph algorithms and flow equations used to compute flows from sources to sinks (i.e., distribution nodes), based on the damages sustained by the network components.

A connectivity analysis is performed in most networks of the present case studies in order to evaluate the performance of the system under earthquake conditions (e.g. networks in Thessaloniki and Vienna, road network in South Italy, gas system in L'Aquila, regional hospitals system in Italy). This is simply due to lack of all required information or due to large computational demand for a complete flow analysis. However, in case of road network at least, this type of analysis is coherent with the time frame of the study that is limited to rescue operations in the aftermath of the seismic event. The interest is the identification of the portions of the network, which are critical with respect to the continued connectivity of the network. In this type of analysis, the performance indicators are most commonly described through the Serviceability Index or Connectivity Loss, which measure the average reduction in the ability of demand nodes to receive flow from sources of the network. Further indicators applied in case of road network in South Italy, include the Minimum Travel Time to reach hospitals and the Terminal Reliability (i.e., the probability that a path exists between specific nodes).

A flow analysis is performed in case of electric power network in Sicily. The performance here is evaluated both with connectivity-related measures (i.e., Connectivity Loss and Power Loss) and flows-based indicators (i.e., Voltage Ratio and System Serviceability Index).

For complex systems (i.e., Thessaloniki harbor, health-care facilities) their internal logic and functions are simulated. In particular, the performance of the harbor is basically measured with the total cargo/containers handled and/or delivered (to the port's gate) in a pre-defined time frame per terminal and for the whole port system, considering the seismic damages as well as specific interdependencies. In case of hospitals the overall performance of the system is quantified through the un-hospitalized victims at regional and town level, the ability of each hospital to provide surgical treatment (comparison between the hospital capacity and demand) or the maximum travel time for hospitalization.

6.3.4 Results

The overall performance of each network and system is expressed through the moving average μ and moving standard deviation σ (averaged over simulations), as well as the Mean Annual Frequency (MAF) of exceedance of the PIs. The average loss is defined based on the moving average graph. Through the MAF graphs the annual probability of exceeding specific levels of loss can be defined and the loss for specific mean return period of the particular PI can be estimated. The earthquake event(s) that correspond to a particular return period (i.e., 500 years) can be identified in the performance curve and maps with the distribution of damages can be produced for the selected event(s). It is noted that this event is a single sample of a stochastic set, representing a randomly selected epicentre, magnitude as well as a realization of spatially correlated ground motions. Therefore, the spatial distribution of damages and losses for this specific simulation are strongly related not only to the location and the size of the earthquake, but also to the (randomly sampled) spatial distribution of intensities. Thus, they are not meant to be those that are expected in average for the specific seismic source position and magnitude, as they are strongly dependent also on where the positive peaks of intensity cluster with respect to the elements at risk in that specific realization of spatial correlations.

In order to evaluate the contribution of certain components on the overall performance of the network, the correlation between damaged components and system's functionality is defined. In the case of Thessaloniki the correlation of each component (EPN, WSS, RDN) to the system PIs is estimated. This type of analysis is based on the results of each single event, and thus it preserves the information about systems' topology and its behavior in case of spatial correlated damages (related to single earthquakes). Thus, it allows identifying the most critical elements for the functionality of each system (i.e., the damaged components that more closely control the performance of the network).

6.3.5 Uncertainties

Several sources of uncertainties are inherent in the analysis, which are related among others to the seismic hazard and spatial correlation models, the fragility and loss assessment or the functionality thresholds of each component. The SYNER-G methodology incorporates a rather comprehensive representation of uncertainty in the problem, with a refined and effective seismic hazard model (Chap. 3) and vulnerability model (Chap. 5), including epistemic modelling of the uncertainty in a hierarchical fashion.

In case of Thessaloniki harbor, the epistemic uncertainty related to different fragility functions and functionality definitions is investigated through a sensitivity analysis with the use of alternative fragility curves and functionality thresholds for the waterfront structures. Similar results are obtained when different fragility curves are applied. This can be attributed to the small frequency of damage occurrence to the waterfront structures and the fact that the total port performance is mostly prescribed by the cranes functionality.

6.3.6 Socioeconomic Analysis

A socio-economic analysis has been performed in the case of Thessaloniki based on the framework that is described in Chap. 4 and the systemic risk assessments for buildings and networks (Chap. 7). In particular, a GIS-based accessibility modelling has been implemented for shelter and healthcare services. It is a representative example, without considering the whole metropolitan area and related networks. It is shown that the SYNER-G methodology and analysis is an important tool for seismic risk management purposes before, during and after disaster. GIS based accessibility modelling can directly provide a vital support to disaster managers in terms of accessibility, location/allocation of available resources and service/catchment related issues. A shelter needs analysis has been also applied. The shelter model considers households' decision-making features along with physical, socio-economic, climatic, spatial and temporal factors in addition to modelled building damage states and utility loss. From the analysis, different SCDs are identified, as "Hot Spots" for shelter needs. These results can help for the planning of shelter allocation.

References

Akkar S, Bommer JJ (2010) Empirical equations for the prediction of PGA, PGV and spectral accelerations in Europe, the Mediterranean region and the Middle East. Seismol Res Lett 81(2):195–206

Choi Y, Stewart JP (2005) Nonlinear site amplification function of 30 m shear wave velocity. Earthq Spectra 21(1):1–30

Giardini D, Woessner J, Danciu L, Crowley H, Cotton F, Grünthal G, Pinho R, Valensise G, Akkar S, Arvidsson R, Basili R, Cameelbeeck T, Campos-Costa A, Douglas J, Demircioglu MB, Erdik M, Fonseca J, Glavatovic B, Lindholm C, Makropoulos K, Meletti C, Musson R, Pitilakis K, Sesetyan K, Stromeyer D, Stucchi M, Rovida A (2013) Seismic Hazard Harmonization in Europe (SHARE). Online data resource, http://portal.share-eu.org:8080/jetspeed/portal/, doi:10.12686/SED-00000001-SHARE

MAE (2013) MAEviz, developed by the Mid-America Earthquake Center and the National Centre for Supercomputing Applications. http://mharp.ncsa.illinois.edu/

National Institute of Building Sciences (NIBS) (2004) HAZUS-MH: user's manual and technical manuals. Report prepared for the Federal Emergency Management Agency, Washington, DC

Pitilakis K, Riga E, Anastasiadis A (2012) Design spectra and amplification factors for Eurocode 8. Bull Earthq Eng 10(5):1377–1400

Pitilakis K, Crowley H, Kaynia A (eds) (2014) SYNER-G: typology definition and fragility functions for physical elements at seismic risk, vol 27, Geotechnical, geological and earthquake engineering. Springer, Heidelberg. ISBN 978-94-007-7871-9

Chapter 7
Application to the City of Thessaloniki

Sotiris Argyroudis, Jacopo Selva, Kalliopi Kakderi, and Kyriazis Pitilakis

Abstract This chapter presents the application of the SYNER-G general methodology and tools to the case study of Thessaloniki. The application includes the building stock (BDG), the electric power network (EPN), the water supply system (WSS) and the road network (RDN) with specific interdependencies between systems. The seismic hazard model is based on the seismic zones proposed in the SHARE project (Giardini et al., Seismic hazard harmonization in Europe (SHARE). Online data resource, http://portal.share-eu.org:8080/jetspeed/portal/. doi:10.12686/SED-00000001-SHARE, 2013). For each system, the main features for the systemic analysis and the system topology and characteristics are described. Then the analysis results are presented. They include damages, casualties (deaths, injuries) and displaced people for BDG and connectivity-based Performance Indicators (PIs) for EPN, WSS and RDN systems. Apart from the average performance and the Mean Annual Frequency (MAF) of exceedance of the PIs, the distribution of estimated damages and losses for specific events is also given through thematic maps. The significant elements for the functionality of each system are defined through correlation factors to the system PIs. Finally, representative results of the shelter demand analysis are given. They are based on a multi-criteria approach that takes into account the outcomes of the systemic risk analysis for buildings and utility systems as well as other indicators.

S. Argyroudis (✉) • K. Kakderi • K. Pitilakis
Department of Civil Engineering, Aristotle University, 54124 Thessaloniki, Greece
e-mail: sarg@civil.auth.gr; kkakderi@gmail.com; kpitilak@civil.auth.gr

J. Selva
Istituto Nazionale di Geofisica e Vulcanologia, via D. Creti 12, 40128 Bologna, Italy
e-mail: jacopo.selva@bo.ingv.it

K. Pitilakis et al. (eds.), *SYNER-G: Systemic Seismic Vulnerability and Risk Assessment of Complex Urban, Utility, Lifeline Systems and Critical Facilities*, Geotechnical, Geological and Earthquake Engineering 31, DOI 10.1007/978-94-017-8835-9_7,

7.1 Introduction

The objective of the study is to demonstrate the applicability of the methods and tools developed in SYNER-G using the city of Thessaloniki as a case study. The main study area is the municipality of Thessaloniki, which is divided in 20 Sub City Districts as defined by Eurostat and the Urban Audit approach. The case study includes the following elements: building stock (BDG), road network (RDN), water supply system (WSS) and electric power network (EPN). The networks comprise the main lines and components and cover the wider Metropolitan area. The internal functioning of each network is simulated and specific interdependencies between systems are considered according to the SYNER-G methodology (Chap. 2): EPN with WSS (electric power supply to pumping stations), RDN with BDG (road blockage due to building collapses), BDG with EPN and WSS (displaced people due to building damages and utility loss).

For the seismic hazard, five seismic zones with magnitude between $M_{min} = 5.5$ and $M_{max} = 7.5$ are selected based on the results of the SHARE European research project (Giardini et al. 2013; www.share-eu.org). Appropriate fragility curves are applied for the vulnerability assessment of each element at risk. For bridges and buildings (RC and masonry) new analytical fragility curves are applied, which have been developed in the framework of SYNER-G specifically for the Thessaloniki area. A Monte Carlo simulation (MCS) has been carried out with 10,000 runs, which samples earthquake events, damage and functional consequences based on the methods and tools developed in SYNER-G.

For each system analyzed (EPN, WSS, BDG, RDN) we provide the description of the systemic vulnerability methodology and software implementation, the description of the system topology and main characteristics, and finally the results of the application. Note that, in order to analyse all the inter- and intra- dependencies among these systems, each simulation considered simultaneously all the systems in the whole area of the applications (see Chap. 2). This poses severe computational problems that lead us to controlled simplifications of single systems analysis. The selected Performance Indicators (PIs) are calculated based on the estimated damages and functionality losses of the different components, using specific simulation methods (Chaps. 2 and 5). The overall performance of each network is expressed through the moving average μ and moving standard deviation σ (averaged over all simulations), as well as the Mean Annual Frequency (MAF) of exceedance of the PIs values (performance curve). The average loss is defined based on the moving average graph (as assessed for the final run of the MCS). Earthquake events that correspond to a specific return period of the considered PI (e.g. $T_R = 500$ years) can be identified for each system, and maps with the distribution of damages/non-serviceabilities can be produced for this events. It is noted that this event is a single sample of a stochastic set, representing a randomly selected epicentre, magnitude as well as a realization of spatially correlated ground motions (see Chap. 3). Therefore, the spatial distribution of damages and losses for this specific simulation are strongly related not only to the location and the size of the earthquake, but also to the (randomly sampled) spatial distribution of intensities. Thus, they are not meant to be those that are expected in average for the specific seismic source position and magnitude, as they are strongly dependent also on where the positive peaks of intensity cluster with respect to the elements at risk in that specific realization of spatial correlations.

The correlation of each component PI to the system PIs is also estimated. This type of analysis is based on the results of each single event, and thus it preserves the information about systems' topology and its behaviour in case of spatial correlated damages (related to single earthquakes), and it may be applied simultaneously to all the analysed systems. Thus, it allows identifying the most critical elements (in whatever system) for the functionality of each specific system (i.e., the damaged components that more closely control the performance of the network).

To demonstrate the application of the socio-economic methodology in SYNER-G, a shelter needs analysis has been applied. The shelter model simulates households' decision-making process and considers physical, socio-economic, climatic, spatial and temporal factors in addition to modelled building damage states and utility loss.

Finally, a GIS-based transportation accessibility model has been implemented for healthcare services of Thessaloniki considering the results of the road network analysis under earthquake conditions. Example healthcare accessibility results are presented in Chap. 4.

7.2 Seismic Hazard

7.2.1 Seismic Source Model

The study area is characterized by intense seismic activity with strong historical earthquakes of magnitudes larger than 6.0. The most recent destructive earthquake occurred in the broader area of Thessaloniki on the Gerakarou-Stivos fault, along the Mygdonian graben (20 June 1978, $M = 6.5$). The mainshock caused extensive damage and loss of life in the metropolitan area of Thessaloniki and the surrounding villages (Papazachos and Papazachou 1997).

For the seismic hazard input of the present application, five seismic zones with $M_{min} = 5.5$ and $M_{max} = 7.5$ are selected based on the results of the SHARE European research project (Giardini et al. 2013) (Fig. 7.1). A Monte Carlo simulation (MCS) is carried out (10,000 runs) by sampling seismic events in these zones.

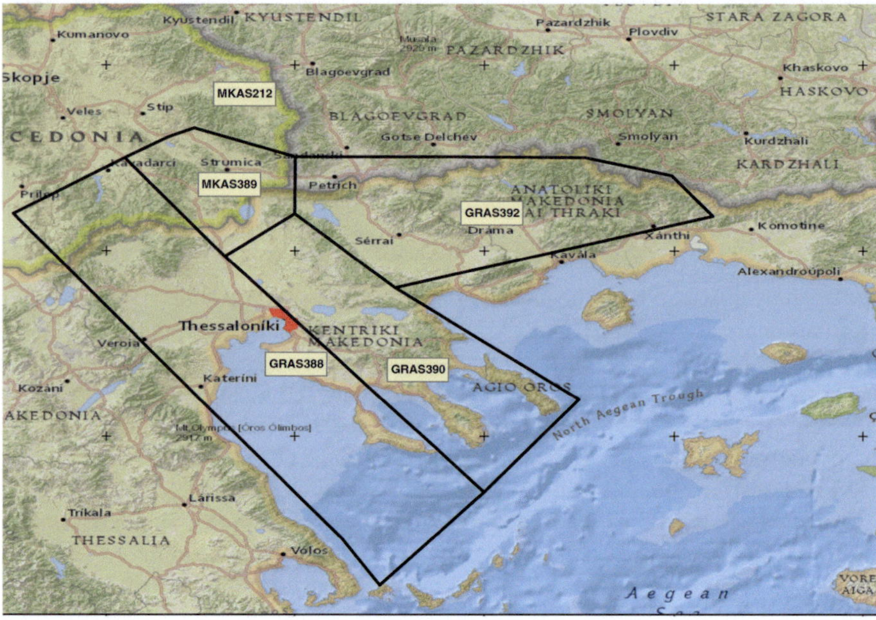

Fig. 7.1 Seismic zones considered in the case study, obtained by SHARE project

7.2.2 Geotechnical Maps

A detailed microzonation study has been conducted for Thessaloniki during the last years. A detailed model of the surface geology and geotechnical characteristics for site effect studies was generated. The resulting geotechnical map (Anastasiadis et al. 2001) was based on numerous data provided by geotechnical investigations, geophysical surveys, microtremor measurements, classical geotechnical and special soil dynamic tests (Pitilakis et al. 1992; Pitilakis and Anastasiadis 1998; Raptakis et al. 1994a, b; Raptakis 1995; Apostolidis et al. 2004). The dynamic properties of the main soil formations have been defined from extended laboratory testing including resonant column and cyclic triaxial tests (Pitilakis et al. 1992; Pitilakis and Anastasiadis 1998; Anastasiadis 1994). For the purpose of the present application the map shown in Fig. 7.2 is used, where three soil formations are defined according to the EC8 (CEN 2004) classification scheme (i.e., A, B, C soil classes).

The liquefaction susceptibility of the study area is defined based on the classification scheme introduced by Youd and Perkins (1978), which is also adopted in HAZUS (FEMA 2003). The assigned classes (Very High, High, Moderate, Low, Very Low and None) are categorised by Pitilakis et al. (2013) on the basis of deposit type, age and general distribution of cohesionless loose sediments (Fig. 7.3). The associated permanent ground deformations (PGD) are estimated at the vulnerable

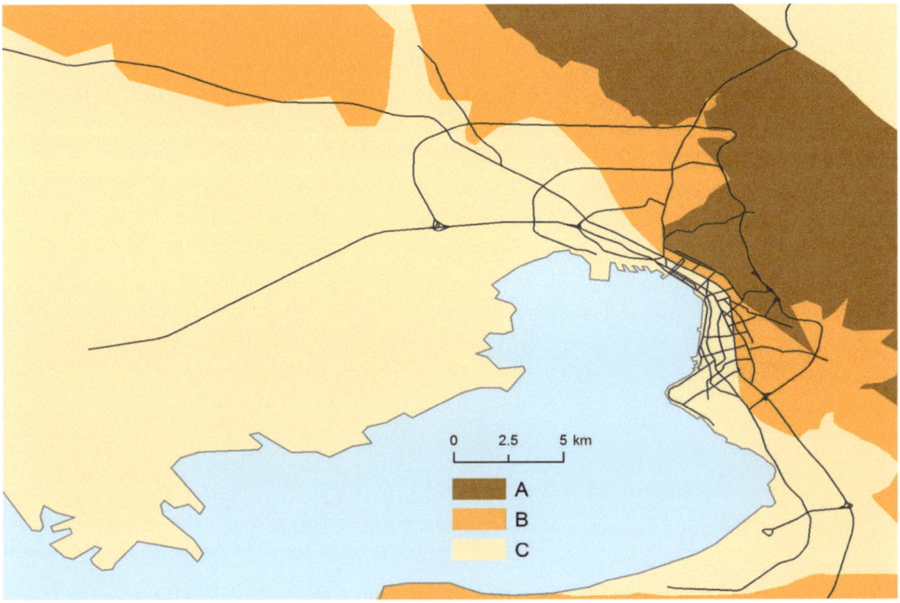

Fig. 7.2 Simplified geotechnical classification of the study area according to EC8 (Based on Pitilakis et al. 2013)

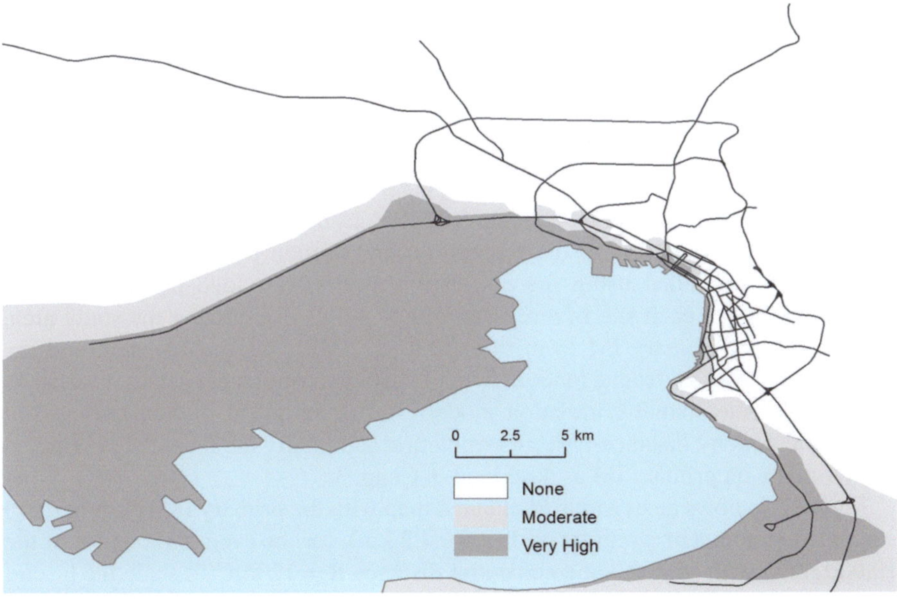

Fig. 7.3 Liquefaction susceptibility map of the study area according to HAZUS classification. The *solid black lines* represent the main edges of the Thessaloniki road network (Based on Pitilakis et al. 2013)

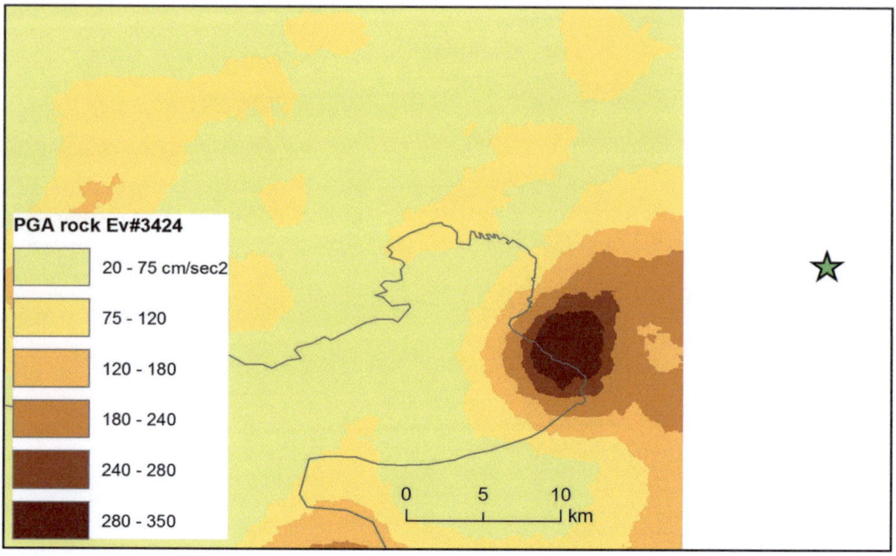

Fig. 7.4 Shake map in terms of PGA on rock at grid points, for a scenario M = 6.5, R = 15 km event on source #3424

sites based on the approach of HAZUS (FEMA 2003) and the modelling procedure described in Chap. 3. The landslide hazard is not considered in the present case study, because the landslide susceptibility is very low.

7.2.3 *Seismic Ground Motion*

The ground motion prediction equation (GMPE) by Akkar and Bommer (2010) is applied for the estimation of the outcrop ground motion parameters. The spatial variability is modelled using the correlation models provided by Jayaram and Baker (2009). For each site of a regular grid of points discretizing the study area, the averages of primary IM from the specified GMPE were calculated, and the residual was sampled from a random field of spatially correlated Gaussian variables according to the spatial correlation model. The primary IM is then retrieved at vulnerable sites by distance-based interpolation and finally the local IM is sampled conditionally on primary IM as described in Chap. 3.

Figure 7.4 shows an example Shakefield map with the primary IM (PGA at rock) computed at points of a regular grid (3.5 × 4.2 km), for the event #3566 (out of the 10,000 runs) with M = 6.0 and epicentral distance R = 12 km. To scale the hazard to the site condition, the amplification factors proposed in EC8 (2004) are used in accordance with the site classes that were defined in the study area (Fig. 7.2).

7.3 Application to the Electric Power Network

7.3.1 Methodological Aspects and Implementation Within the SYNER-G Model

The methodology of the analysis of electric power networks and its implementation within the SYNER-G model are described in Chaps. 2 and 5. A modern Electric Power Network (EPN) is a complex interconnected system that can be subdivided into four major parts: Generation, Transformation, Transmission and Distribution, and Loads. The electric power networks' components (that can be considered vulnerable or not) can be grouped on the basis of five different vulnerability analysis scales of the network: Network, Station, Distribution system, Substations' components (macro- and micro-components), and Line.

As in most of the systems, the analysis of an EPN in a seismically active environment can be carried out at two different levels (Chap. 5). The basic one focuses on connectivity only and can lead to a binary statement on whether any given node is connected with another node, specifically a source node, through the network (connectivity model). The other approach is a power flow analysis which follows the analysis of short-circuit propagation (capacity model). As noted, given the large size of this application, the analysis of EPN is kept as simple as possible and thus it is based on connectivity only. In addition, power flow analysis requires detailed data about the substations layout and their micro- and macro-components, which are not always available, and in this application would considerably complicate the analysis. Some of the network components are considered as non-vulnerable due to the lack of appropriate fragility curves (e.g. transmission lines), the extremely large number of system components (e.g. distribution substations) or due to the boundary conditions of the analysis (i.e. the generator should remain functional in order to perform a connectivity analysis). The set of subsystems connecting (1) generator (non-vulnerable) to transmission substations (vulnerable), (2) transmission substations to distribution substations (non-vulnerable), and (3) distribution substations and demand nodes, are analysed separately, in order to retrieve the functionality (isolated/non-isolated state) of each demand node. In this application distribution substations are not considered in the analysis, therefore transmission substations are directly connected to demand nodes.

During each simulation the nodes that are non-functional are removed from the system. In particular, the functionality of transmission substations is based on their physical damages (non-functional with damages greater or equal to moderate) and to their connectivity to the generator. When vulnerable, the functionality of distribution substations is based on their physical damages (non-functional with damages larger or equal to moderate) and to their connectivity to functional distribution substations. The damage states of substations are defined through the corresponding fragility curves (see Sect. 7.3.2). The adopted PI is the Electric power Connectivity Loss (ECL), described in Eq. 7.1 for a given seismic event i.

$$ECL_i = 1 - N_s^i/N_0 \tag{7.1}$$

where N_s^i and N_0 are the number of connected nodes, in seismic and non-seismic conditions respectively.

High voltage (150–20 kV) transmission substations, which are considered as vulnerable are connected with non-vulnerable transmission lines. An over-high voltage (400–150 kV) transmission/transformation station is considered as the system's non-vulnerable generator, providing constant power to the high voltage stations. Demand nodes (non-vulnerable) are located at the centroid of each sub-city district (SCD). The interaction with the WSS is simulated through the connection of WSS pumping stations (considered as demand node of EPN) with the reference EPN load bus (here substation).

The electric power network is made up of nodes and edges/lines connecting them. As a consequence, the EPN class is the composition of *EPNedge* and *EPNnode* classes, that are both abstract. In this connectivity analysis, the first one is the generalization of EPN Line (non-vulnerable), while the second one is the generalization of Generator (non-vulnerable), Transmission substations (vulnerable), distribution Substations (vulnerable) and EPN Demand nodes (non-vulnerable).

The general properties of the EPN class are summarized in Chap. 5. These include (i) several pointers, (ii) EPN global properties, (iii) subclass characteristics (including the main features of sides and nodes and structures), (iv) properties that record the state of EPN and each component for each event, and (v) properties required to assess the global performance of the EPN at the end of the simulation. The definition of EPN class includes functions to assess PIs (and relative statistics) for connectivity analysis, to assess damages and functionalities for each component, and to plot the system configuration and states (Chap. 5).

7.3.2 System Topology and Characteristics

Eight high voltage substations (150 kV/20 kV) supply the municipality of Thessaloniki. Medium voltage lines (20 kV) give power to medium-voltage customers and more than 1,500 distribution substations commonly located at the basement of residential buildings. Moreover, an over-high voltage substation connects the 400 kV grid with the 150 kV grid. The network includes about 120 transmission lines, in majority underground, with a total length of 600 km (80 km overhead lines). Distribution stations' voltages range between 630 and 1,630 kVA. They give power to low-voltage customers through an extremely extensive grid; underground at high-density load areas and overhead at low-density load areas.

The EPN is modelled as a directed graph, i.e., a graph in which all edges have a travelling direction, from node i to node j. The network for the case

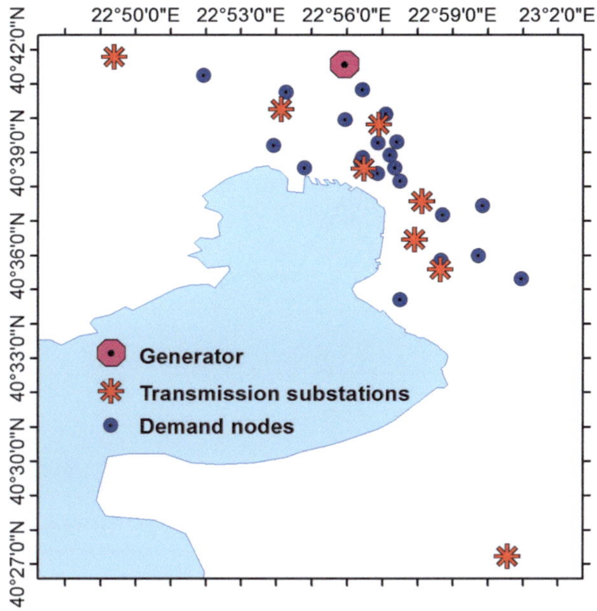

Fig. 7.5 Electric power network topology

study is composed of 30 nodes and 29 edges (Fig. 7.5). The nodes are subdivided into 1 generator, 8 transmission/transformation substations and 21 demand nodes (the locations of the WSS pumping stations). Only transmission substations are considered herein as the vulnerable components. Edges are non-vulnerable transmission lines (underground and overhead) connecting the generator with the transmission substations and the transmission substations with the demand points.

For the vulnerability analysis of the electric power transmission stations, the fragility curves proposed in the SRM-LIFE (2007) research project are used, which are provided in terms of peak ground acceleration (PGA). The fragility curves for transmission substations are classified in three classes (open, mixed and closed-type).

7.3.3 Results

Figure 7.6 (left) shows the moving average (μ) curve for ECL as well as the $\mu + \sigma$ and $\mu - \sigma$ curves. The jumps present in the plots are located in correspondence of simulation runs/samples in which at least one demand node is disconnected, leading ECL to yield values greater than 0. At the end of the analysis (10,000 runs) the moving average is stabilized. Figure 7.6 (right) shows the MAF of exceedance versus ECL (performance curve). The ECL with return period $T_R = 500$ years (i.e. with mean annual frequency of exceedance, λ, equal to 0.002) corresponds to 24 % loss, and the expected damages for the scenario with the maximum magnitude

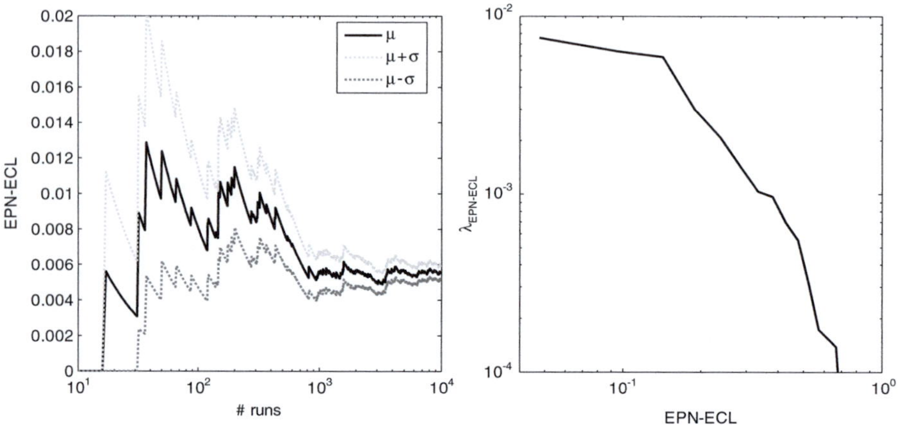

Fig. 7.6 Moving average μ, $\mu+\sigma$, $\mu-\sigma$ curves (*left*) and MAF curve (*right*) for ECL

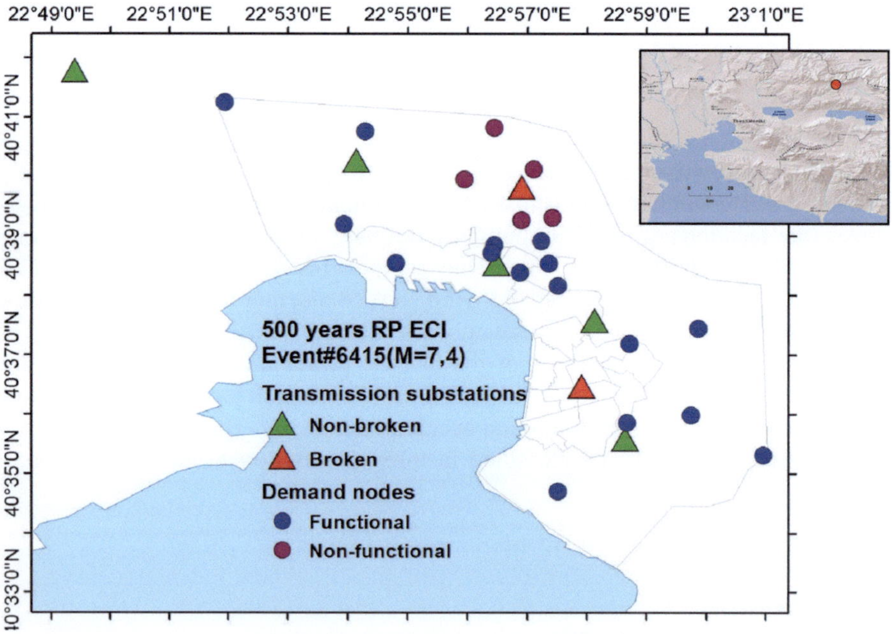

Fig. 7.7 Electric power network damages for an event (#6415, M = 7.4, R = 40 km) that corresponds to ECL with $T_R = 500$ years

corresponding to the specific return period of ECL (out of the 10,000) are shown in Fig. 7.7. Non-functional components (transmission substations and demand nodes – WSS pumping stations) for the specific event (i.e., #6415) are mostly concentrated to the North North-East part of the city.

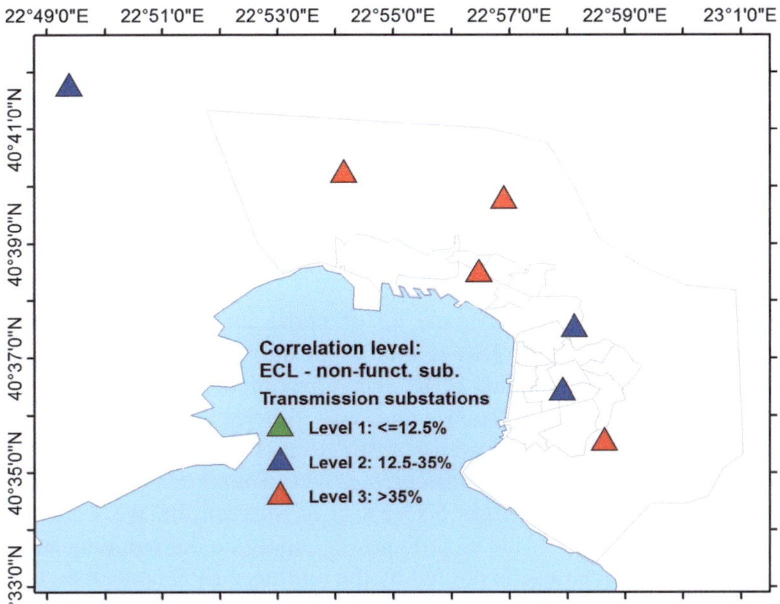

Fig. 7.8 Correlation of non-functional transmission substations to electric power network connectivity loss

Figure 7.8 shows the level of correlation between the ECL and non-functional transmission substations. In this way the most critical components of the network can be defined in relation with their contribution to the connectivity loss of the network. The majority of substations present high values of the correlation factor, close or higher than 35 %. This can be mostly attributed to the low level of redundancy and the low level of ramifications of the network, in combination to the substations vulnerability and distribution of PGA in average over all runs of the simulation.

7.4 Application to the Water Supply System

7.4.1 Methodological Aspects and Implementation Within the SYNER-G Model

The methods for systemic analysis of water supply systems and the implementation within the SYNER-G model are described in Chaps. 2 and 5. As for EPN, the analysis performed is based only on connectivity. The system connecting tanks to demand nodes is analysed in order to retrieve the functionality (isolated/non-isolated state) of each demand node. Demand nodes of the system are connected with edges (pipelines). Storage tanks and pumping stations comprise the system's water sources and they are considered as non-vulnerable. The only vulnerable elements of the system are the pipelines.

The considered PI is the Water Connectivity Loss (WCL), described in Eq. 7.2 for a given event i.

$$\mathrm{WCL_i} = 1 - \mathrm{N_s^i}/\mathrm{N_0} \tag{7.2}$$

where $\mathrm{N_s^i}$ and $\mathrm{N_0}$ are the number of connected nodes, in seismic and non-seismic conditions respectively.

The water supply system is made up of nodes and links connecting them. As a consequence, the WSS class is the composition of *WSSnode* and *WSSlink* abstract classes, of which the first is the generalization of the Pipe class, while the second is the generalization of the *DemandNode*, *WaterSource* and *PumpingStation* classes. In particular, the *WaterSource* abstract class is the generalization of the *VariableHeadWaterSource* and *ConstantHeadWaterSource* classes. An important interdependence considered within SYNER-G is between the WSS and EPN, in particular the dependency of the electric power supply on the pumping stations. If a pump serving a source node is not fed by the reference EPN node, then the pump itself is considered out of service and the relative WSS node removed from the system for the connectivity analysis.

The general properties of the WSS class are summarized in Chap. 5. These include (i) several pointers, (ii) WSS global properties, (iii) subclass characteristics (including the main features of sides and nodes and structures), (iv) properties that record the state of WSS and each component for each event, and (v) properties required to assess the global performance of the WSS at the end of the simulation. The definition of WSS class includes functions to assess PIs (and relative statistics), to assess damages and functionalities, to retrieve the EPN functionality states, and to plot the system configuration and states (Chap. 5).

7.4.2 System Topology and Characteristics

The city of Thessaloniki is supplied from various water sources (springs, wells, rivers). From the external aqueducts, the water is led to the main tanks and pumping stations. The treatment of raw water is confined only to its chlorination before the entrance to the distribution network. Water treatment units are placed at certain pumping stations, sedimentation tanks and wells.

The water supply system of the municipality of Thessaloniki includes 20 tanks with total capacity of 91,900 m^3. There is also a sedimentation tank of 8,000 m^3 capacity and a fire-fighting tank with total capacity of 2,100 m^3. The largest tank has a capacity of 10,000 m^3. Steel pipelines with a total length of 71 km comprise the main water transmission system. The distribution network has a 1,284.1 km approximate length and a supply capacity ranging between 240,000 and 280,000 m^3/day.

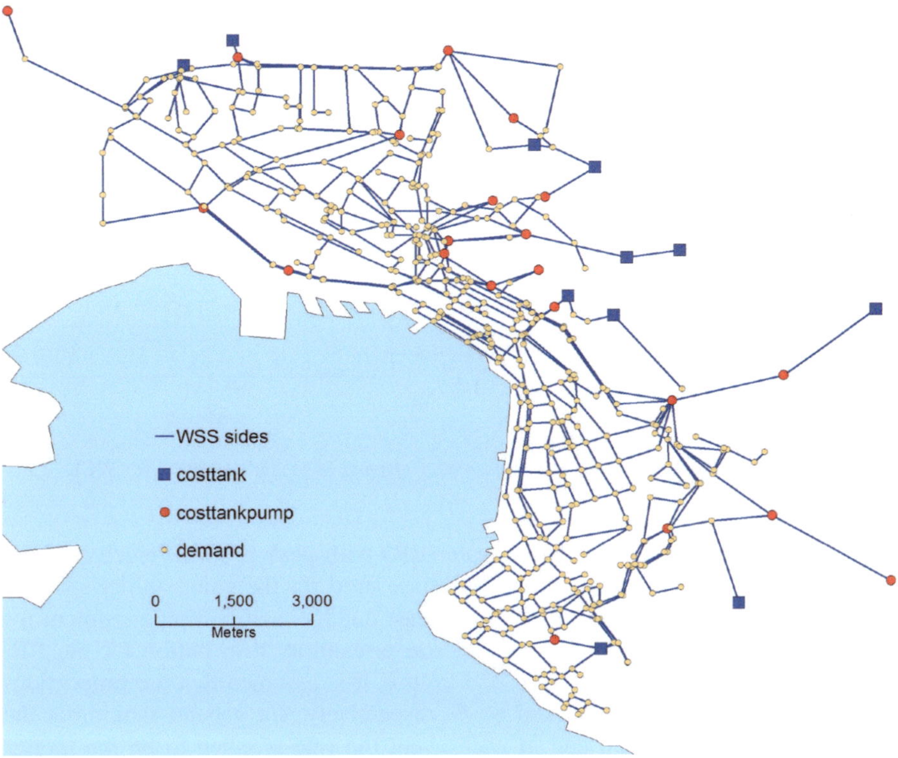

Fig. 7.9 Water supply system topology used for the application

The supplied customers are approximately 420,000 and the total supplied population about one million, 99 % of which are domestic users. The supplied area is approximately 55 km^2. The elevation is ranges from 0 to 380 m, and the water pressure between 2 and 5 bars. There are few areas in the municipality with independent water systems, supplied from local sources. Due to the complexity and oldness of the system, along with the fact that in some regions the detailed topology of the system is not perfectly known, a simplified (yet realistic) model for Thessaloniki's main WSS is used for the analysis. The WSS for the case study is comprised of 477 nodes and 601 edges with total length of about 280 km (Fig. 7.9). The nodes are subdivided in demand nodes, pumping stations (marked "costtankpump" in Fig. 7.9) and tanks (marked "costtank" in Fig. 7.9); the latter considered as water sources for the system. The simulated network includes 445 demand nodes, 21 pumping stations and 11 tanks.

The WSS is modelled as a directed graph. Pipelines have 24 different diameter values (ranging between 500 and 3,000 mm); their construction materials include asbestos cement, cast iron, PVC and welded steel. The site properties are specified according to the EC8 soil classification, depth to groundwater and liquefaction susceptibility classes (as described in Sect. 7.2.2). The fragility functions of

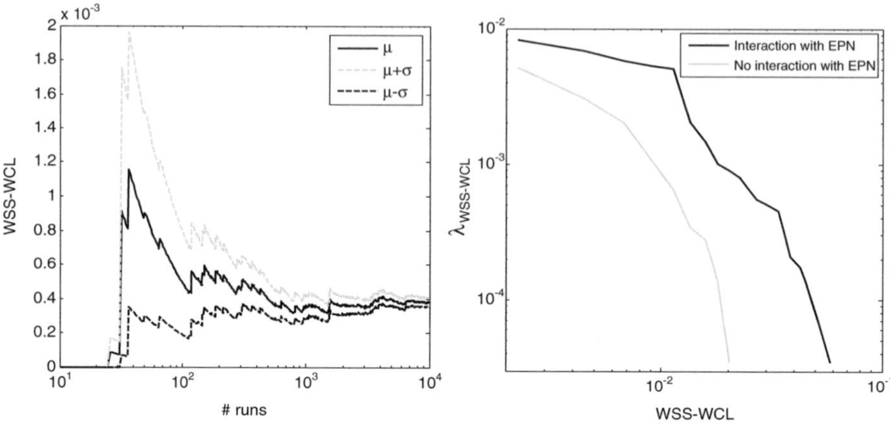

Fig. 7.10 Moving average μ, $\mu+\sigma$, $\mu-\sigma$ curves (*left*) and MAF curves (*right*) for WCL

ALA (2001) that correlate the repair rate (RR) with peak ground velocity (PGV) and permanent ground deformation (PGD) are used for the vulnerability analysis of pipelines. The probability of having at least one damage in a pipe segment for given intensity is estimated using a Poisson probability distribution for the RR. When a pipe is damaged due to ground failure, it is assumed that the proportions of leaks and breaks are 20 % and 80 % respectively; for ground shaking is the opposite. At each location, and in each event the rate is taken to be the largest between that obtained as a function of PGV and PGD, and multiplied by 0.2 and 0.8, respectively, since only breaks are considered in the connectivity analysis. It is noted that pipes are discretized between distant nodes into a number of smaller pipes to improve the description of the vulnerability (each pipe-section is considered as a separate vulnerable element, with the IM evaluated at the section centroid). Finally, the interaction with the EPN is simulated through the connection of WSS pumping stations with its specific reference EPN demand node.

7.4.3 Results

Figure 7.10 shows the moving average (μ) curve for Water Connectivity Loss (WCL) as well as the $\mu+\sigma$ and $\mu-\sigma$ curves. The jumps present in the plots are located in correspondence of simulation runs/samples in which at least one node is disconnected, leading WCL to yield values greater than 0. At the end of the analysis (10,000 runs) the moving average is stabilized. In the same figure the MAF of exceedance of WCL is shown with and without considering the interaction with electric power network in the analysis. This interaction can be important, as an example the connectivity loss is almost doubled, increased from 1 to 1.8 % for $\lambda=0.001$ ($T_R=1{,}000$ years) when the connections of water pumping stations to EPN are included in the analysis.

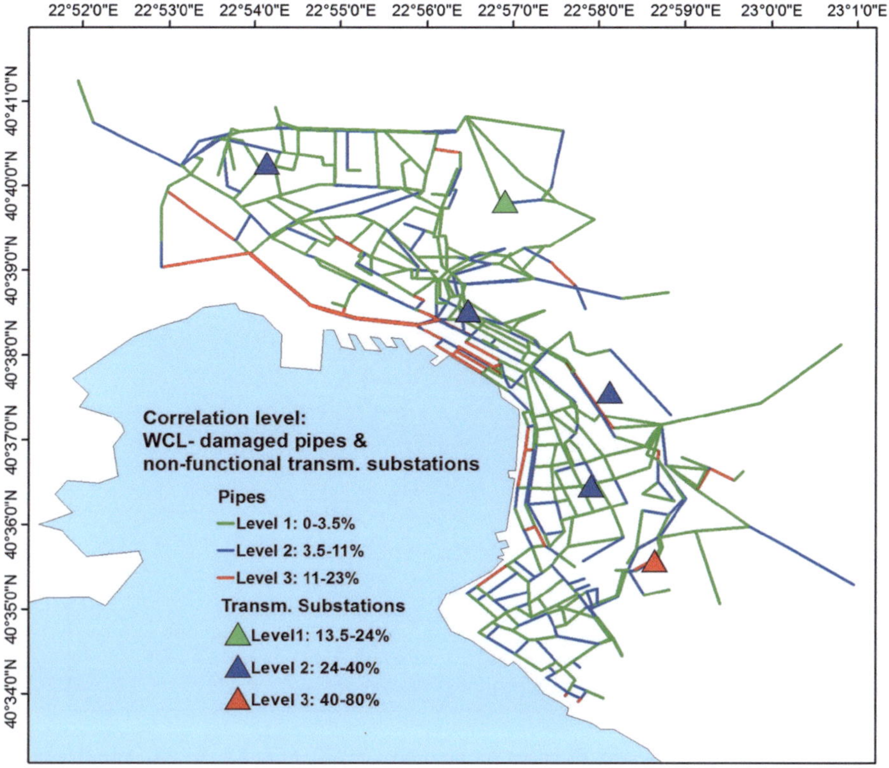

Fig. 7.11 Correlation of damaged pipes and non-functional EPN transmission stations to water network connectivity

Figure 7.11 shows the level of correlation between the WCL and damages in pipelines as well as the non-functionality in EPN substations supplying the water pumping stations. The most highly-correlated pipelines are concentrated along the coast where the ramification of the system is limited and the liquefaction susceptibility is high; therefore damage due to permanent ground displacement is more likely. Interestingly, a higher level of correlation is estimated for the EPN transmission substations, showing that the most critical components for the performance of one system (here WSS) may be external to the system (here EPN). The highest value of the correlation factor is 80 % and it is attributed to a component in the South-East part of the city, where several pumping stations, all connected to EPN, are located.

Figure 7.12 shows the expected distribution of damage for the event with the highest magnitude that corresponds to connectivity loss (WCL = 1.4 %) with return period $T_R = 500$ years (0.002 probability of exceedance). Only few broken pipes are observed, while the majority of non-functional pumping stations and not-connected demand nodes are accumulated at the South South-east part of the city.

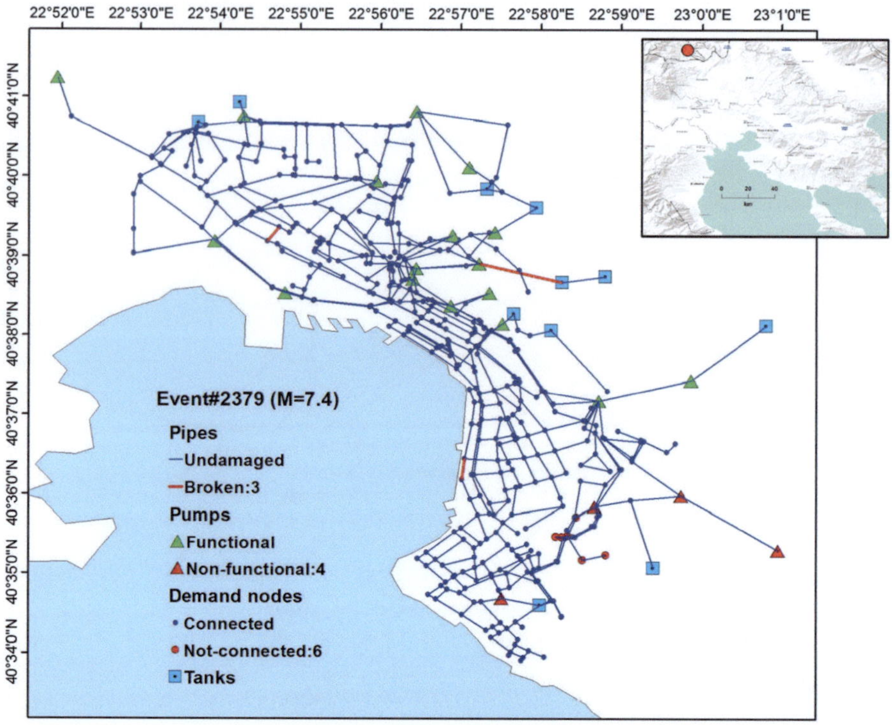

Fig. 7.12 Water supply system damages for an event (#2379, M = 7.4, R = 72 km) that corresponds to WCL with $T_R = 500$ years

7.5 Application to the Buildings

Within the object-oriented framework developed in SYNER-G (Chap. 2), each system is described as a class containing objects (i.e., instances) with similar features such as attributes and methods. Building-related attributes and methodologies are part of the cell and region classes (Chap. 5). Some attributes are initial parameters, which come from databases, and others are derived from computation of the states of the overall system (performance indicators).

The following Performance Indicators are calculated at the cell level:

- *Building damage*:
  - Number of buildings at yielding damage state;
  - Number of buildings at ultimate damage state.
- *Building usability* which is related to building damage states:
  - Number of occupants in non-usable buildings (N_{NU});
  - Number of occupants in partially usable buildings (N_{PU});
  - Number of occupants in fully usable buildings (N_{FU}).

Table 7.1 Empirically-derived usability ratios

UR	Damage state		
	None	Yielding	Ultimate
FU	0.87	0.22	0.00
PU	0.13	0.25	0.02
NU	0.00	0.53	0.98

Table 7.2 Utility loss tolerance thresholds (*ULT*) for Thessaloniki

Good weather		Bad weather	
Fully usable	Partially usable	Fully usable	Partially usable
0.9	0.8	0.4	0.3

$$\mathrm{N}_{FU\ or\ PU\ or\ NU} = \sum_{i=1}^{3} N_i \cdot NO_i \cdot UR_{i \cdot FU\ or\ PU\ or\ NU} \tag{7.3}$$

where:

i = damage level (i = 1, . . . ,3)
N_i = number of buildings having damage level i
NO_i = number of occupants (at the time of the event) in each building for each damage level i
UR_i = usability ratio (UR) for damage level i for each usability class (Table 7.1)

- *Building habitability* which is related to building usability, utility loss (UL) (water and electric power in this application) and a utility loss threshold (ULT) determined based on good and bad weather conditions:
  - Number of habitable buildings.

$$BH = (N_{FU} NH_{FU} + N_{PU} NH_{PU} + N_{NU} - N_d) \tag{7.4}$$

where:

N_{FU}, N_{PU}, N_{NU} = number of occupants in buildings that are fully, partially and non-usable
NH_{FU} = percentage of fully usable buildings that are non-habitable, where UL ≥ ULT
NH_{PU} = percentage of partially usable buildings that are non-habitable, where UL ≥ ULT
N_d = number of dead persons estimated in the casualty model described below

For Thessaloniki application the ULT was established on a context-specific basis by local experts and given in Table 7.2 (e.g., a utility loss tolerance threshold of 0.9 for fully usable buildings during good weather conditions means that buildings will be considered habitable for up to 90 % utility loss; conversely, during bad weather

Table 7.3 Casualty rates (%) by damage level and building type for Thessaloniki case study (Adapted by HAZUS, FEMA 2003)

	Yielding (RC)	Ultimate (RC)	Yielding (masonry)	Ultimate (masonry)
Injuries	0.03	20	0.04	20
Deaths	–	10[a]	–	10[a]

[a]It is assumed that a percentage of buildings in ultimate damage state actually collapses, therefore the number of buildings in ultimate damage is reduced by a factor of 0.05 for RC and 0.10 for masonry buildings, and is then multiplied with the casualty rates of the table

conditions a partially usable building will be considered habitable only as long as no more than 30 % of utility services is lost).

The building habitability results are used for the estimation of displaced persons in each cell, which is considered in the shelter needs model. The supply of EPN and WSS to building cells is estimated considering them as demand nodes in the connectivity analysis that is performed for these networks.

- *Casualties*:
  - Number of deaths;
  - Number of injuries.

The number of deaths and injuries is obtained by multiplying the casualty rates (i.e. proportion of occupants) of Table 7.3 with the number of buildings from each typology and the number of occupants within each building of each typology. The results for each typology are then aggregated at the cell level to give the number of deaths and injuries in each cell. Accordingly, the number of deaths (N_d) and injured (N_j) are determined using the following expressions.

$$N_d = \sum_{t=1}^{n} \sum_{i=1}^{3} N_{t,i} NO_t QD_{t,i} \quad N_j \sum_{t=1}^{n} \sum_{i=1}^{3} N_{t,i} NO_t QI_{t,i} \tag{7.5}$$

where:

t = building type ($t = 1, \ldots, n$)
i = damage level ($i = 1, \ldots, 3$)
$N_{t,i}$ = number of buildings of type t having damage level i
NO_t = number of occupants (at the time of the event) by building type
$QD_{t,i}$ = proportion of deaths by building type and damage level
$QI_{t,i}$ = proportion of injured by building type and damage level

7.5.1 System Topology and Characteristics

According to the 2011 census the municipality of Thessaloniki has a population of 322,240 people, while the Thessaloniki Urban Area has a population of 790,824.

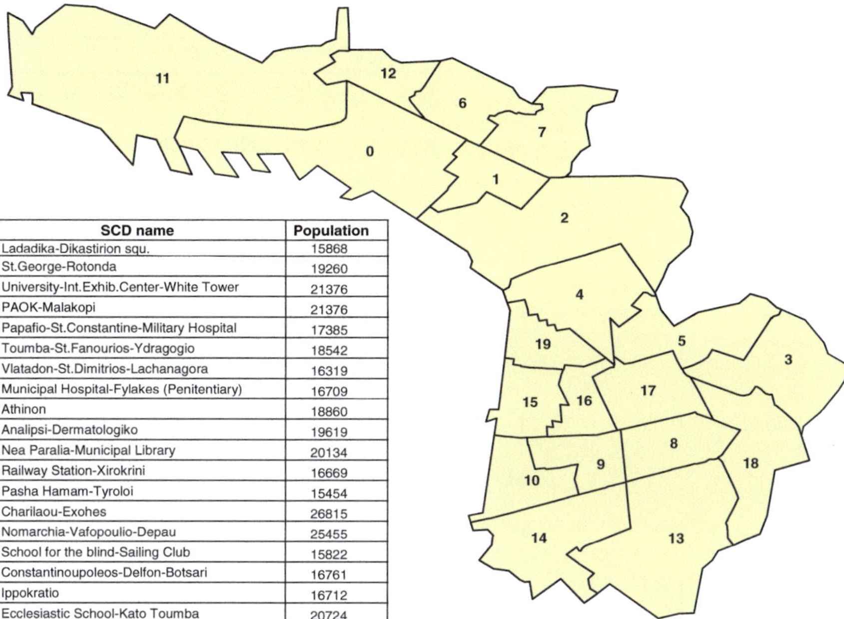

ID	SCD name	Population
0	Ladadika-Dikastirion squ.	15868
1	St.George-Rotonda	19260
2	University-Int.Exhib.Center-White Tower	21376
3	PAOK-Malakopi	21376
4	Papafio-St.Constantine-Military Hospital	17385
5	Toumba-St.Fanourios-Ydragogio	18542
6	Vlatadon-St.Dimitrios-Lachanagora	16319
7	Municipal Hospital-Fylakes (Penitentiary)	16709
8	Athinon	18860
9	Analipsi-Dermatologiko	19619
10	Nea Paralia-Municipal Library	20134
11	Railway Station-Xirokrini	16669
12	Pasha Hamam-Tyroloi	15454
13	Charilaou-Exohes	26815
14	Nomarchia-Vafopoulio-Depau	25455
15	School for the blind-Sailing Club	15822
16	Constantinoupoleos-Delfon-Botsari	16761
17	Ippokratio	16712
18	Ecclesiastic School-Kato Toumba	20724
19	Makedonia Palace-Paedagogiki Academia	16729

Fig. 7.13 Sub-city districts (*SCD*) of Thessaloniki study area as defined by Urban Audit

The Thessaloniki Metropolitan Area extends over an area of 1,455.62 km^2 and its population in 2011 reached a total of 1,006,730 inhabitants.

The study area comprises 20 Sub-City Districts (SCD) as they are defined by Eurostat through the European Urban Audit (EUA) approach (Fig. 7.13).

The total population in this area is 376,589 inhabitants. The same subdivision is used for the definition of the Land Use Plan (LUP) zones. The predominant use is residential (R), while some parts are characterized as commercial (C).

The building inventory in the study area is based on previous projects (Kappos et al. 2008); improvements and additions took place within SYNER-G project (Tenerelli and Crowley 2013). The reference unit of the inventory is the building block. The building inventory comprises 2,893 building blocks with 27,738 buildings, the majority of which (25,639) are reinforced concrete (RC) buildings, while the rest (2,099) are masonry buildings.

The classification of RC buildings is based on the following characteristics (Table 7.4):

- Structural system: frame buildings with masonry infills, frame buildings with open ground storey (pilotis) and wall-frame (dual) buildings;
- Level of seismic design: low (1959 seismic code), medium (1984 seismic code) and high (Eurocode 8);
- Height: low-rise (two storeys), medium-rise (four storeys) and high-rise (nine storeys) buildings.

Table 7.4 Building classes of Thessaloniki inventory

Structural type	Seismic code	Height	Category	SYNER-G class
RC – infilled frames	1959	L	RC1	MRF-EGB/C-RC/R/R/RI-X/ND/R-RC/F-X/L-X/LC
		M	RC2	MRF-EGB/C-RC/R/R/RI-X/ND/R-RC/F-X/M-X/LC
		H	RC3	MRF-EGB/C-RC/R/R/RI-X/ND/R-RC/F-X/H-X/LC
	1984	L	–	MRF-EGB/C-RC/R/R/RI-X/D/R-RC/F-X/L-X/MC
		M	–	MRF-EGB/C-RC/R/R/RI-X/D/R-RC/F-X/M-X/MC
		H	–	MRF-EGB/C-RC/R/R/RI-X/D/R-RC/F-X/H-X/MC
RC – pilotis	1959	L	RC4	MRF-EGB/C-RC/R/R/RI-P/ND/R-RC/F-X/L-X/LC
		M	RC5	MRF-EGB/C-RC/R/R/RI-P/ND/R-RC/F-X/M-X/LC
		H	RC6	MRF-EGB/C-RC/R/R/RI-P/ND/R-RC/F-X/H-X/LC
	1984	L	–	MRF-EGB/C-RC/R/R/RI-P/D/R-RC/F-X/L-X/MC
		M	–	MRF-EGB/C-RC/R/R/RI-P/D/R-RC/F-X/M-X/MC
		H	–	MRF-EGB/C-RC/R/R/RI-P/D/R-RC/F-X/H-X/MC
RC – dual	1959	L	RC7	W-EGB/C-RC/R/R/RI-X/ND/R-RC/F-X/L-X/LC
		M	RC8	W-EGB/C-RC/R/R/RI-X/ND/R-RC/F-X/M-X/LC
		H	RC9	W-EGB/C-RC/R/R/RI-X/ND/R-RC/F-X/H-X/LC
	1984	L	RC10	W-EGB/C-RC/R/R/RI-X/D/R-RC/F-X/L-X/MC
		M	RC11	W-EGB/C-RC/R/R/RI-X/D/R-RC/F-X/M-X/MC
		H	RC12	W-EGB/C-RC/R/R/RI-X/D/R-RC/F-X/H-X/MC
	EC8	L	RC13	W-EGB/C-RC/R/R/RI-X/D/R-RC/F-X/L-X/HC
		M	RC14	W-EGB/C-RC/R/R/RI-X/D/R-RC/F-X/M-X/HC
		H	RC15	W-EGB/C-RC/R/R/RI-X/D/R-RC/F-X/H-X/HC
Masonry-low-rise	Rigid floors		M1	SW-X/M-URM/R/R/X-X/ND/R-RC/F-X/L-2/NC
	Flexible floors		M2	SW-X/M-URM/R/R/X-X/ND/F-T/P-Ti/L-2/NC
Masonry-mid-rise	Rigid floors		M3	SW-X/M-URM/R/R/X-X/ND/R-RC/F-X/M-4/NC
	Flexible floors		M4	SW-X/M-URM/R/R/X-X/ND/F-T/P-Ti/M-4/NC

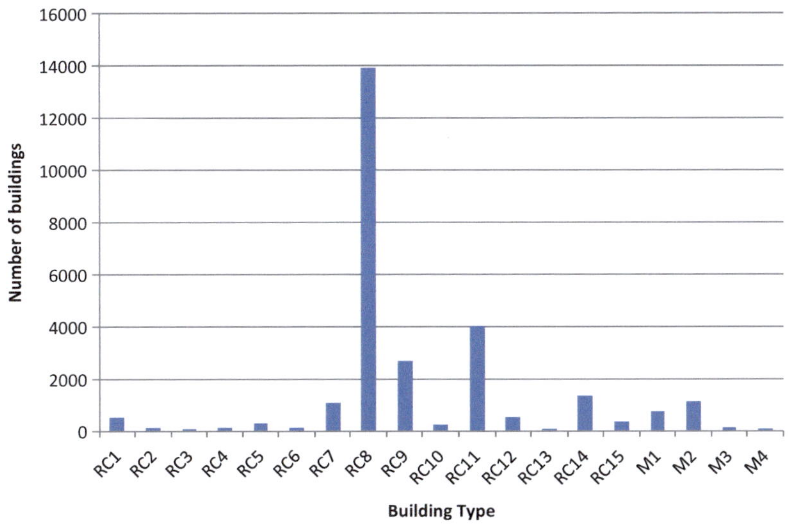

Fig. 7.14 Classification of the buildings of the study area

The classification of masonry buildings is described in Table 7.4 based on floor type (flexible or rigid) and building height (low or mid rise).

The classification of the buildings of the study area is illustrated in Fig. 7.14. Most of the buildings are either infilled dual RC system while the majority of RC buildings are pre-1980 constructions and thus have been designed with low level of seismic code. The description of the building classes is given in Table 7.4.

For the purpose of this study specific analytical fragility curves for RC (Fardis et al. 2012) and masonry buildings (Karantoni et al. 2012) have been developed and applied. Peak ground acceleration is adopted as the intensity measure and two damage grades, namely yielding and ultimate, are considered. An example for wall-frame buildings is shown in Fig. 7.15, where the important effect of design codes on the damage evaluation is evident (Pitilakis and Argyroudis 2013).

7.5.2 Results

Figures 7.16, 7.17 and 7.18 show the moving average (μ) curves for deaths, injuries and displaced persons as well as the $\mu + \sigma$ and $\mu - \sigma$ curves for these PIs. The values are given as percentages of the total population (790,824 inhabitants). At the end of the analysis (10,000 runs) the moving average is stabilized with an average value of 4 deaths, 11 injuries and 6,280 displaced people (in good weather conditions). This low fatality rate is reasonable in this case as the analysis averages the results over all possible magnitudes and epicentral distances, and the lower magnitude and longer distance events control the output. The MAF of

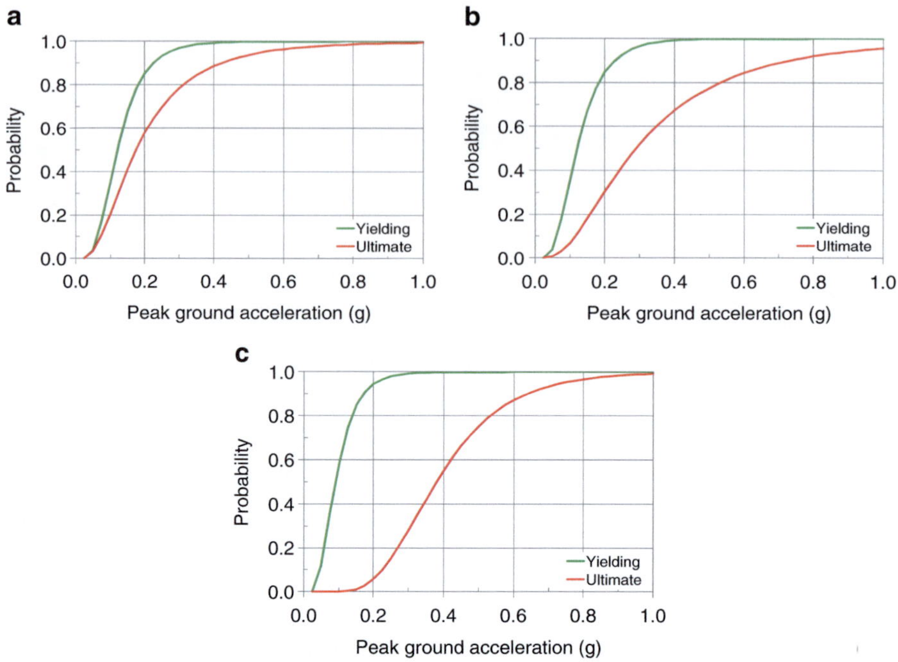

Fig. 7.15 Fragility curves for low-rise wall-frame buildings in Thessaloniki designed with low-level (**a**), medium-level (**b**) and high-level (**c**) seismic code

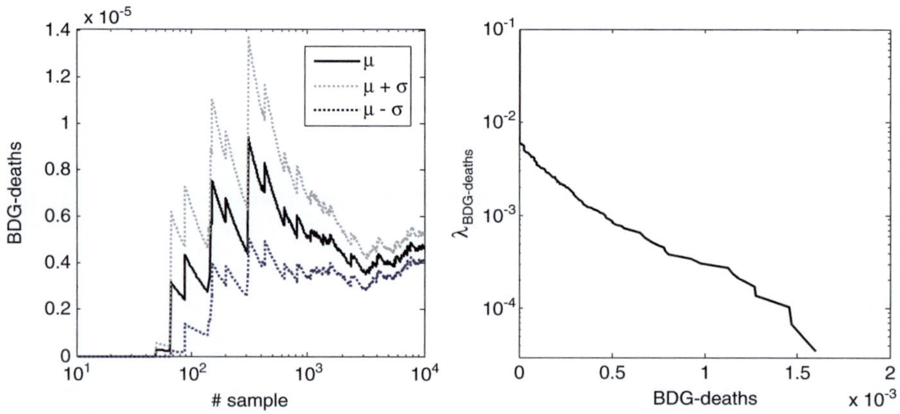

Fig. 7.16 Moving average μ, $\mu+\sigma$, $\mu-\sigma$ curves (*left*) and MAF curve (*right*) for deaths

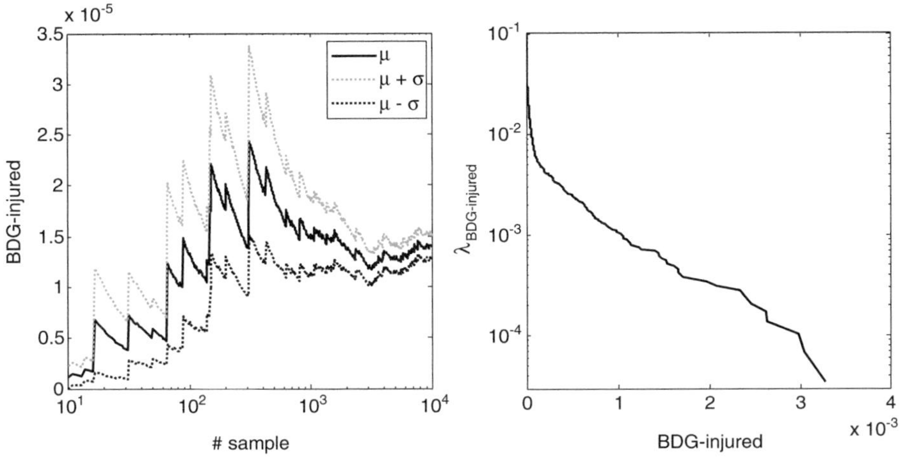

Fig. 7.17 Moving average μ, $\mu + \sigma$, $\mu - \sigma$ curves (*left*) and MAF curve (*right*) for injuries

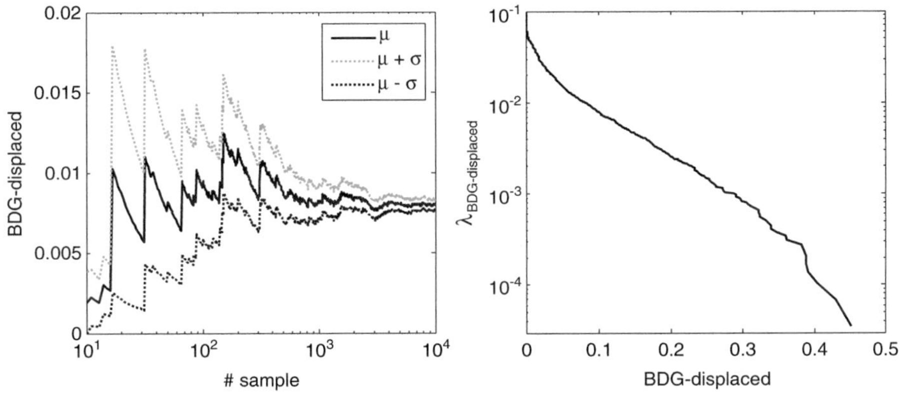

Fig. 7.18 Moving average μ, $\mu + \sigma$, $\mu - \sigma$ curves (*left*) and MAF curve (*right*) for displaced people

exceedance curves for deaths, injuries and displaced persons (as percentages of the total population) are shown in the same figures. The expected deaths for $\lambda = 0.002$ (return period $T_R = 500$ years) are 218.

The distribution of losses for an event that corresponds to this return period of deaths is shown in Figs. 7.19, 7.20, and 7.21. The estimated losses are: 1,578 buildings at the ultimate damage state and 13,113 buildings at the yielding damage state, 218 deaths, 503 injuries and about 145,000 potentially displaced people. It is noted that in this specific simulation the utility losses (EPN and WWS) are minor and therefore the habitability of buildings as well as the number of displaced

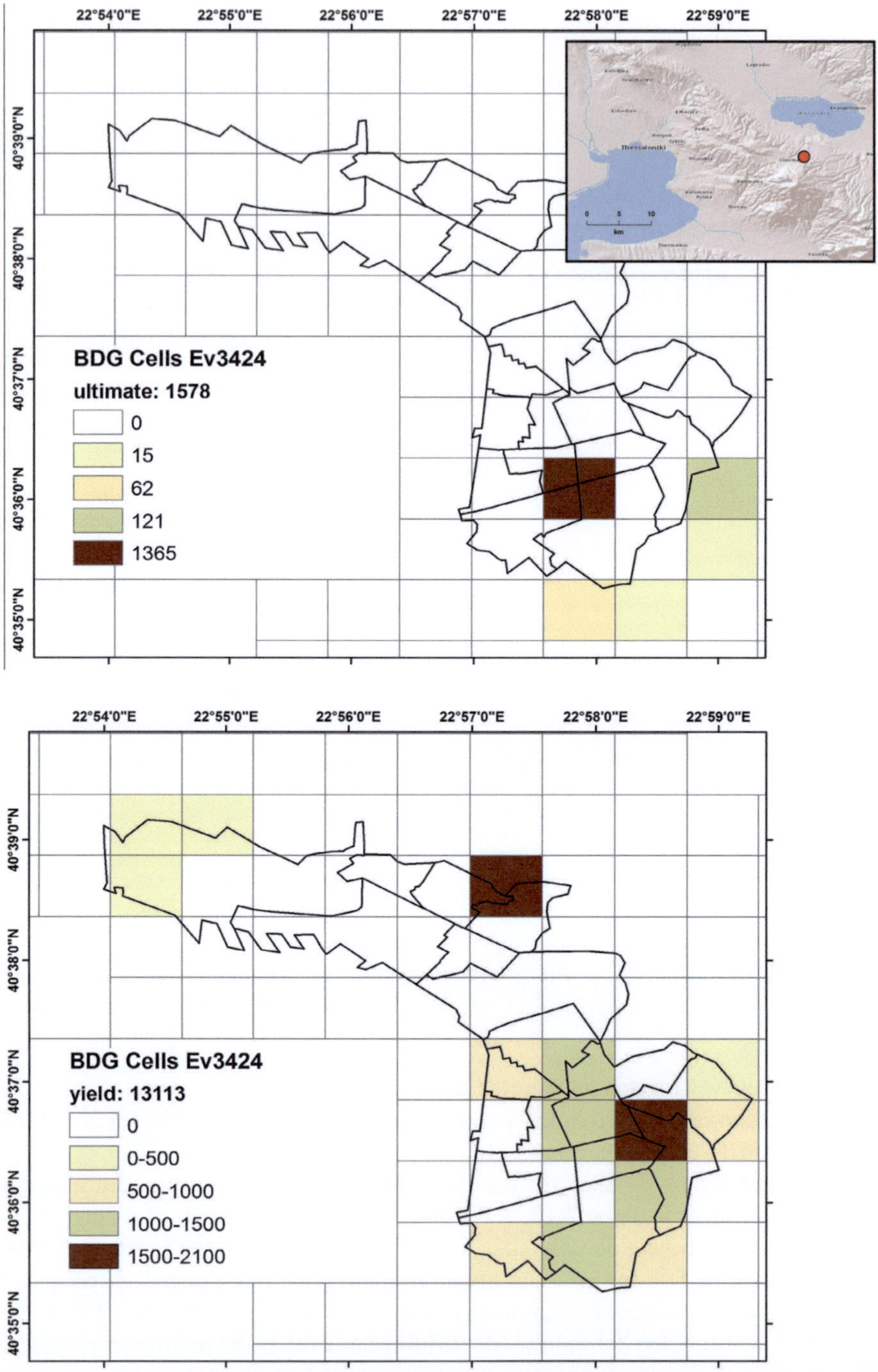

Fig. 7.19 Distribution of estimated building damages into cells of the study area for an event (#3424, $M = 6.5$, $R = 15$ km) that corresponds to death rate with $T_R = 500$ years

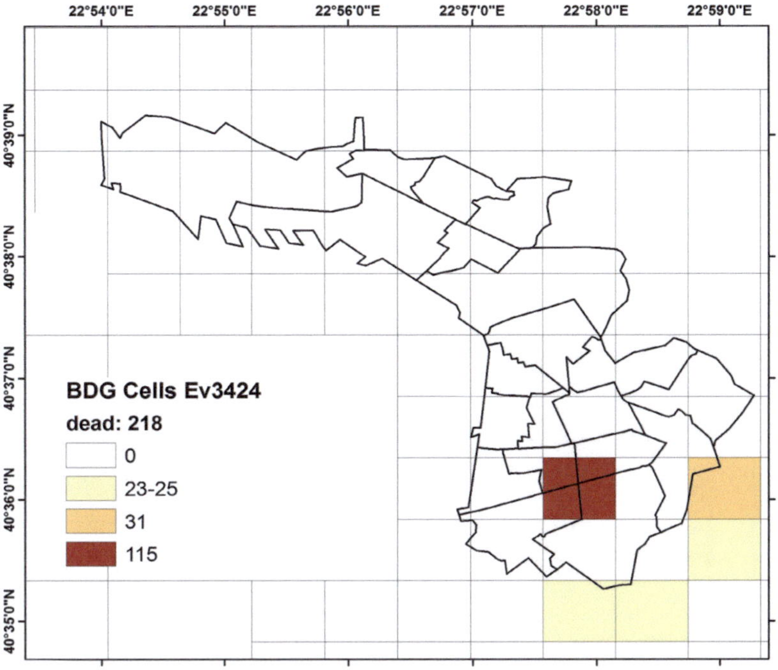

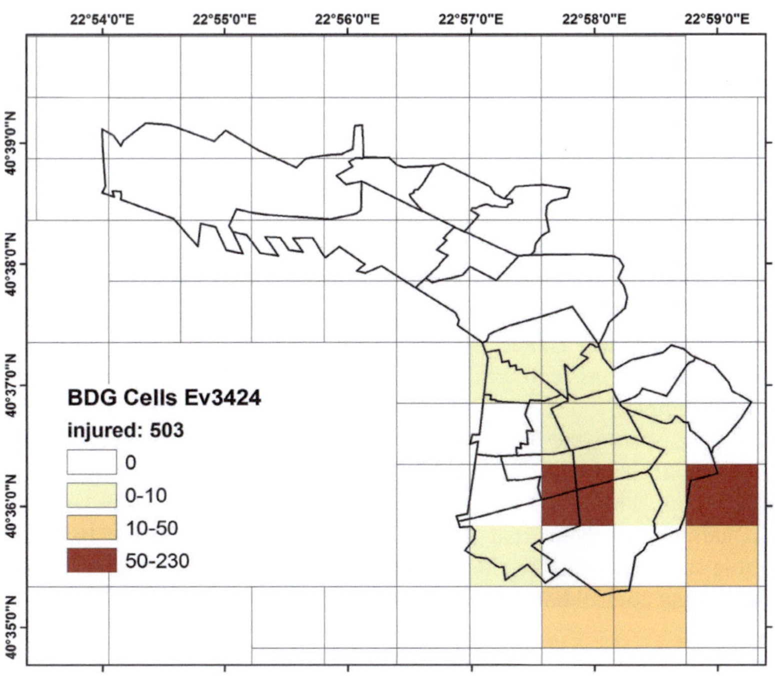

Fig. 7.20 Distribution of estimated casualties (deaths, injuries) into cells of the study area for an event (#3424, M = 6.5, R = 15 km) that corresponds to death rate with $T_R = 500$ years

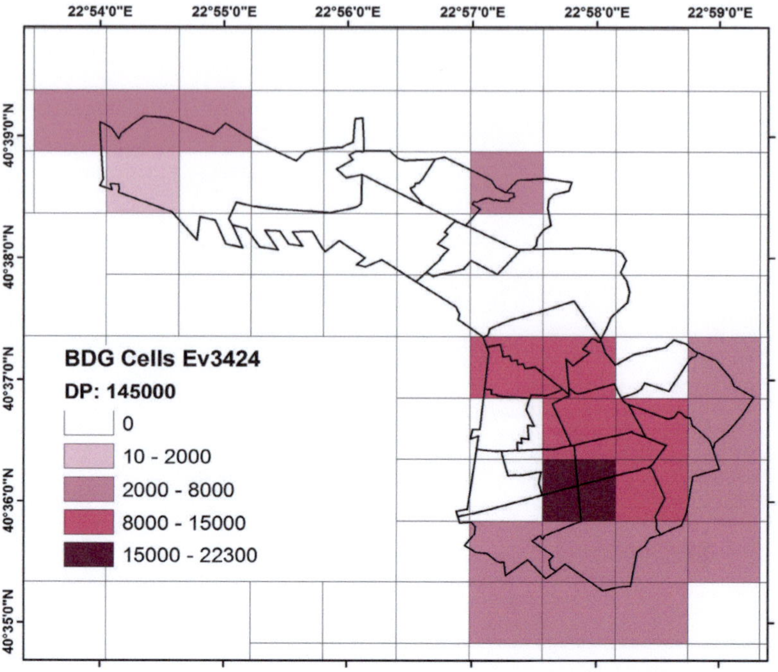

Fig. 7.21 Distribution of estimated displaced persons into cells of the study area for an event (#3424, M = 6.5, R = 15 km) that corresponds to death rate with $T_R = 500$ years

people are not affected for different weather conditions (see Chap. 4). The estimated numbers correspond to a worst-case scenario and they can be reduced if optimal thresholds of building usability are considered. In this specific case it is evident that the distribution of damages and losses depends very strongly on where the positive residuals cluster with respect to the density of buildings and population. The shake map for this particular simulation is shown in Fig. 7.4 (PGA on rock), where it is clear that high residuals are observed in the area where the building damage is concentrated. Hence, this sample event is representative for the specific epicenter, magnitude as well as the specific realization of the spatially correlated ground motion field.

Figure 7.22 shows the level of correlation between the damaged WSS and EPN components and the displaced people. The correlation is higher with the EPN substations, which highlights the importance of the interaction between EPN loss and habitability. Note that the impact of EPN and WSS on habitability is comparable, and thus this result is also probably due to the fact that, as noted above, EPN may control habitability both directly and indirectly, through WSS (see Fig. 7.12).

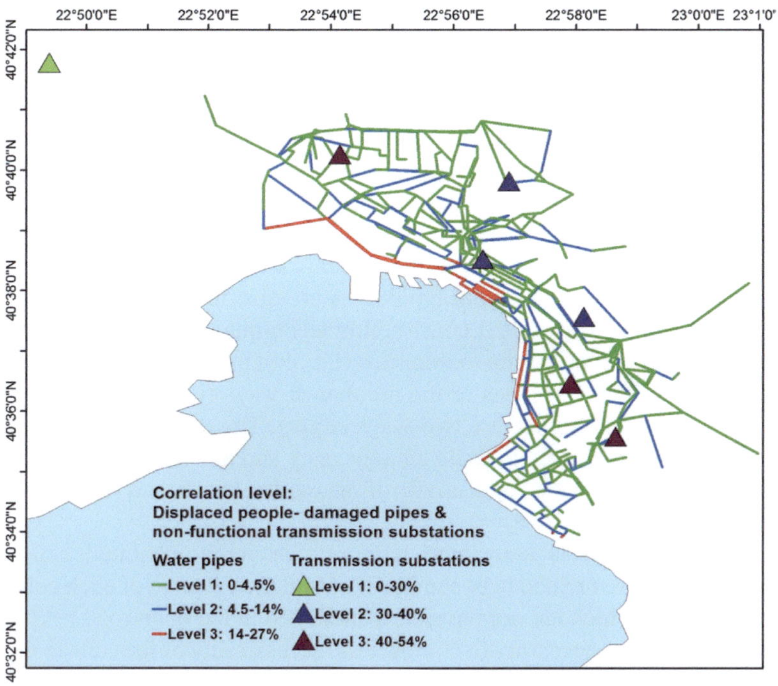

Fig. 7.22 Correlation of damaged EPN and WSS components to displaced people

7.6 Application to the Roadway Network

7.6.1 Methodological Aspects and Implementation Within the SYNER-G Model

The methods of systemic analysis for transportation networks are briefly described in Chap. 5. In the present application the analysis is focused on the functioning of the network in terms of pure connectivity (Argyroudis et al. 2014). The network is represented as a directed graph. It is made up of nodes and links/edges connecting them.

Among the types of RDN nodes, *Intersection* nodes simply represent the summits of the graph that are used to define the edges that can link them: these nodes have no specific properties, except general information such as coordinates, altitude and soil type. *TAZ nodes* (i.e., Traffic Analysis Zones) are nodes that are defined around inhabited areas and they are used to evaluate the connectivity of a given neighbourhood to others TAZs (i.e., they are used to build the origin –destination matrix): they have additional properties (such as number of households or the pointer to the reference cell) that can be used to evaluate traffic demand and connectivity loss for the associated cells. Finally, *ExternalStation* nodes are a type of TAZs that are not associated with the inhabited cells, but they are used to link the studied portion of road network to the 'outside' (i.e., definition of inward/outward traffic demand in the case of an open system).

The types of RDN edges are defined with respect to the physical properties of the road segments (i.e., bridges, tunnels, simple road segments, and roads in cuts, on embankment or on slope) and the different vulnerability models that may be used for each one of them (i.e., different damage mechanisms or intensity measures have to be considered for bridges or for road segments on unstable slope). Within each of these edge sub-classes, different typologies are also defined, depending on the material used, the soil type or the construction technique. Some other properties of edges include the pointers of the extremities (i.e., end and start nodes, as the graph is directed), the number of lanes or the number of ways (i.e., in order to generate two directed paths when there are two-ways edges). The definition of edges along with their extremities is used to build an adjacency and an incidence matrix, which are then used to describe the connectivity of the road network, and subsequently the accessibility of TAZ nodes.

When a TAZ node is associated with one or more inhabited cells, some dependency edges are created between the TAZ and the centroid of each cell. These lower order roads, which are not physically modelled in the main road network, are abstract edges that are necessary to compute the accessibility of the cells to the TAZ, and finally to the main road network.

The various performance indicators used to measure the RDN seismic performance have been introduced in Chap. 5 together with further details on the systemic vulnerability methodology. In the present application two performance indicators are used at the system level: the Simple Connectivity Loss (SCL) and the Weighted Connectivity Loss (WCL). SCL measures the average reduction in the ability of sinks to receive flow from sources. WCL upgrades the simple connectivity loss by weighting the number of sources connected to each sink, in the seismically damaged network and in non-seismic conditions.

Within the object-oriented framework developed in SYNER-G each system is described as a class containing objects (i.e., instances) with similar features such as attributes and methods. A list of main properties of the RDN class is given in Chap. 5. Direct (physical) and indirect failures are considered in the application namely (i) bridge damage due to ground shaking (peak ground acceleration) or road damage due to liquefaction (permanent ground deformation), (ii) road blockage due to collapsed buildings and (iii) road blockage due to collapsed bridges (overpasses).

In particular, during the simulation, a RDN edge can be in the following states:

- *Broken*: 0 or 1 (direct physical failure);
- *BlockedbyBuilding*: 'open', 'open for emergency' or 'closed';
- *BlockedbyBridge*: 0 or 1.

These state variables appear in the output attributes of the simulations and they are used to update the adjacency matrix of the RDN class. This matrix represents all working edges that link two nodes by a 0, and the values are 1 otherwise. For each

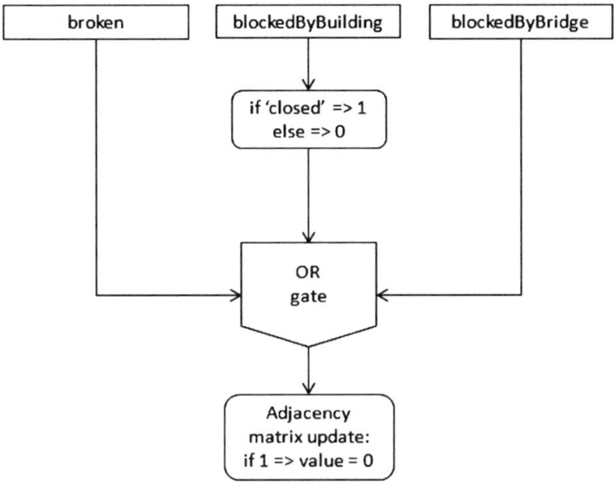

Fig. 7.23 Update procedure of the adjacency matrix

simulation, the values in the adjacency matrix are updated to account for the loss of functionality of some edges (i.e., connectivity analysis). Since here the edge can be disrupted by several causes, the logical tree presented in Fig. 7.23 is adopted to update the adjacency matrix (i.e., use of an OR gate).

7.6.2 System Topology and Characteristics

Thessaloniki is extended along the seaside and consequently the urban road system is parallel and perpendicular to the sea shore. The roadway network of the urban area is rather insufficient, especially in the centre districts, where the densely built up area creates a complex network, with narrow streets and inadequate parking areas. In the present application the main network of the urban area is considered together with the ring road and the main exits of the city where the majority of bridges and overpasses are located.

The road network for the case study is composed of 594 nodes and 674 edges (Fig. 7.24). The nodes are subdivided into 15 external nodes, 127 Traffic Analysis Zone (TAZ) centroids and 452 simple intersections. The RDN is modelled as a directed graph. For this particular network, 495 edges are two-ways roads and 179 are one-way roads. Further information for roadway edges, which are the only vulnerable components, includes the road width, road class (minor, principal or highway), distance from buildings, existence of buildings in one or two sides of the edge, capacity, number of ways and free flow speed. The distance from buildings is based on data extracted through remote sensing techniques (Tenerelli and Crowley 2013) and on information from previous studies (Argyroudis and Pitilakis 2011).

In the framework of SYNER-G specific fragility curves were constructed for 22 bridges, out of the 60 that are located in the study area, for which detailed

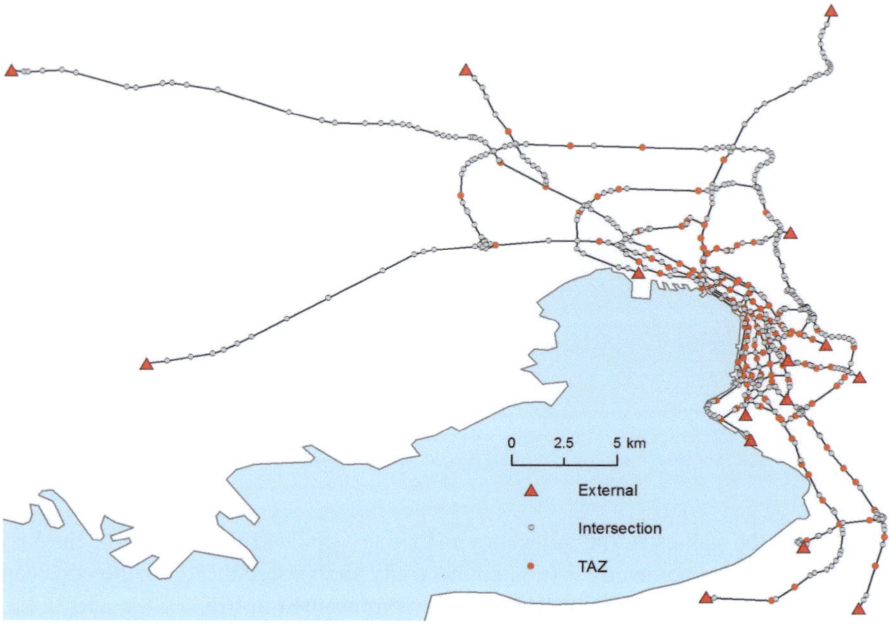

Fig. 7.24 Road network of Thessaloniki case study

construction drawings were made available by the competent authorities. The most appropriate among these fragility curves are assigned to the remaining bridges on the basis of their structural characteristics. They are classified based on the number of spans, deck continuity, deck-pier connection, transverse translation at the abutments, year of construction and pier type (Tsionis and Fardis 2012; Pitilakis and Argyroudis 2013). Peak ground acceleration is adopted as intensity measure and two damage grades, namely yielding and ultimate, are considered. The obtained fragility curves show that older bridges, without or low seismic design, are likely to experience damage for low to medium levels of earthquake excitation (e.g. Fig. 7.25a). On the other hand, modern bridges are markedly less vulnerable (e.g. Fig. 7.25b).

Edges are assumed to be the only vulnerable components in the network. In this particular application they are classified into road pavements and bridges, with fragility models expressed in terms of permanent ground deformation (PGD) due to liquefaction and peak ground acceleration (PGA) for ground shaking, respectively. The fragility curves provided in HAZUS (FEMA 2003) are used for the road pavements, which are classified to urban (two traffic lanes) or major (four or more lanes) roads. The effect of collapsed buildings to the road functionality is also considered based on the results of the fragility analysis of buildings and other models for the estimation of the induced debris and road blockages (Argyroudis et al. 2014). It is noted that road lines are discretized between distant nodes into a

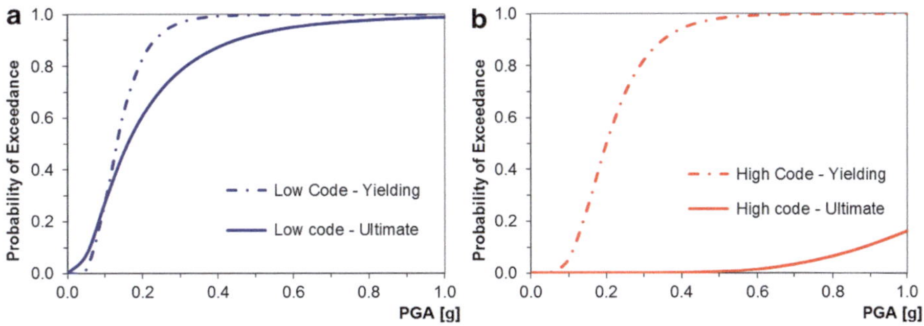

Fig. 7.25 Fragility curves for (**a**) a bridge with the deck supported on bearings, constructed in 1985 (low seismic code) and (**b**) an overpass with monolithic deck-pier connection, constructed in 2003 (high seicmic code)

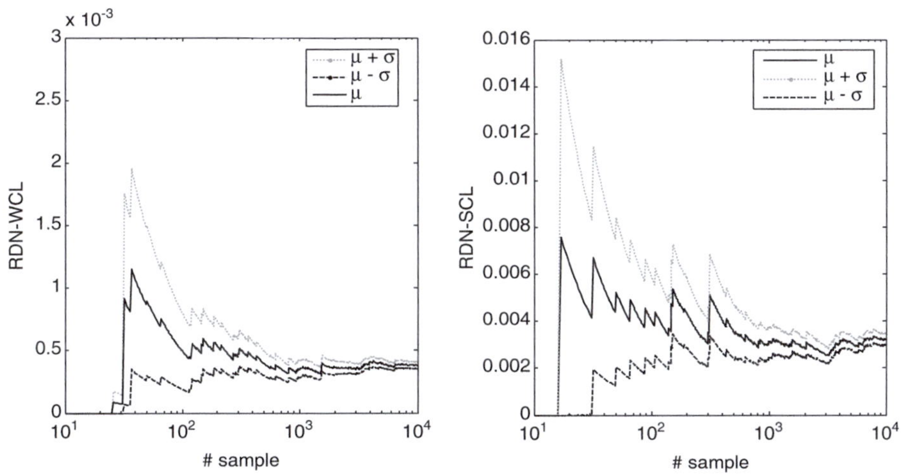

Fig. 7.26 Moving average μ, $\mu + \sigma$, $\mu - \sigma$ curves for WCL (*left*) and SCL (*right*)

number of smaller edges to improve the description of the vulnerability (each road-section is considered as a separate vulnerable element, with the IM evaluated at the section centroid).

7.6.3 Results

The analysis results as obtained from the plain MCS of 10,000 runs are presented in the following. Figure 7.26 shows the moving average (μ) curves for SCL and WCL,

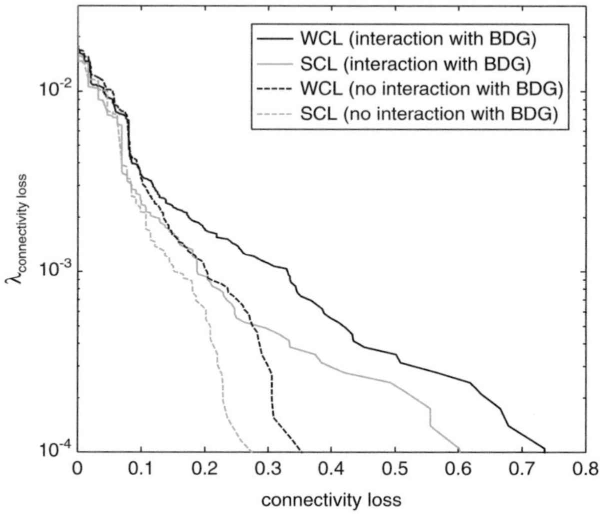

Fig. 7.27 MAF curves for simple (*SCL*) and weighted (*WCL*) connectivity loss with and without interaction with building collapses

as well as the $\mu + \sigma$ and $\mu - \sigma$ curves for the two PIs. The figures indicate that the expected value of connectivity loss given the occurrence of an earthquake is higher for WCL than for SCL, as expected. This is because WCL takes into account not only the existence of a path between two TAZs, but also the increase in travel time due to the seismically induced damage suffered by the RDN. The jumps present in the plots are located in correspondence of simulation runs/samples in which at least one TAZ node is disconnected, leading SCL and WCL to yield values greater than 0. At the end of the analysis the moving average is stabilized.

Figure 7.27 shows the MAF of exceedance curves for SCL and WCL i.e., the performance curves. As expected, weighting the computation of connectivity loss with the path travel times yields higher values of exceedance frequency. The estimated MAF of exceedance curves when the road blockage due to collapsed building is not considered in the analysis are presented in the same figure. The interaction with building collapses can be important especially for return periods higher than 500 years ($\lambda = 0.002$). As an example the connectivity loss is increased from 20 to 33 % for $\lambda = 0.001$ ($T_R = 1{,}000$ years) when the building collapses are included in the analysis.

Figures 7.28 and 7.29 show the level of correlation between the WCL and the distribution of damages in bridges and road blockages respectively. Relatively higher correlation factors are found for RDN edges blocked by building collapse, demonstrating the importance of this failure mechanism for RDN analysis. In particular, the roads with highest correlation to building collapse are mainly in

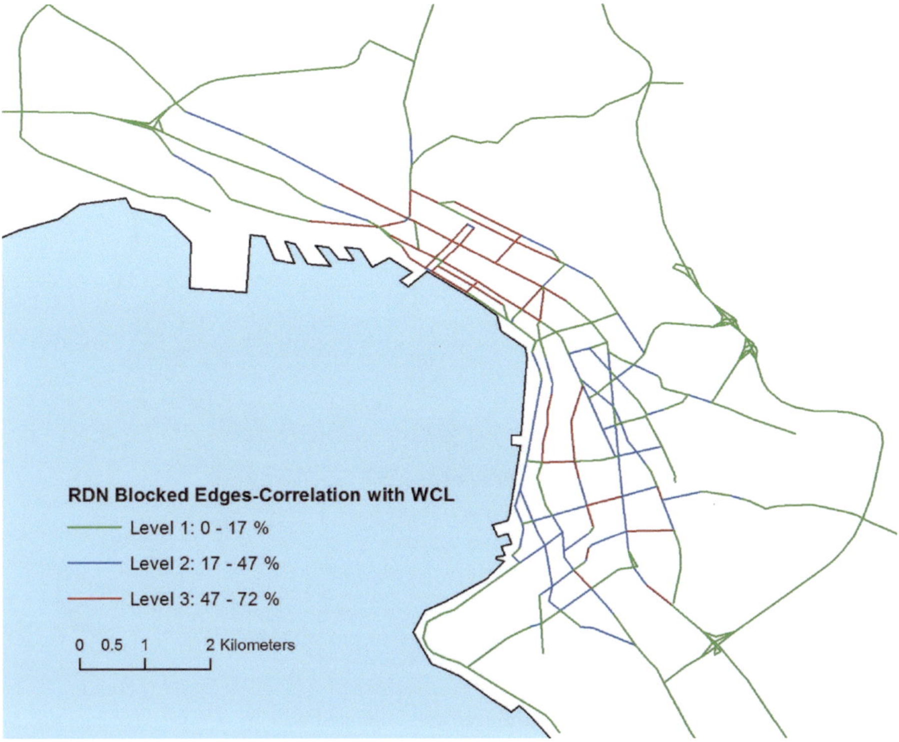

Fig. 7.28 Correlation of edges blocked by buildings to road network connectivity (PI = WCL)

the historical centre of the city, where the vulnerability of buildings (mostly built with the oldest seismic code in Greece) is higher, and the road-to-building distances are shorter. Several road segments in the city centre and the southeast part of the study area present a medium correlation to building collapses. A few roads near the coastline which are subjected to ground failure due to liquefaction are also highly correlated to the network connectivity. The high risk of failure for bridges is attributed to their typology characteristics (old, simple span bridges) and the high values of PGA, which increase their vulnerability.

Figure 7.30 shows the expected distribution of damages for the event with the highest magnitude that corresponds to connectivity loss (WCL = 18 %) with return period $T_R = 500$ years (0.002 probability of exceedance). For this event no blockages are expected due to building collapses. However, seven bridges will be severely damaged and few road blockages due to damage of overpass bridges are expected in the main road network. Damages due to liquefaction are concentrated in the airport access roads.

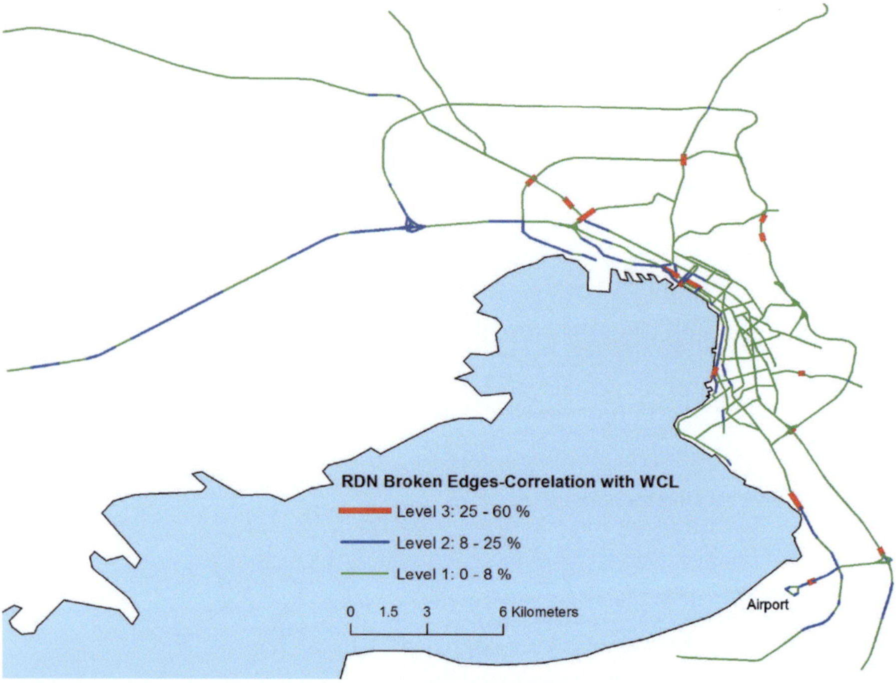

Fig. 7.29 Correlation of broken edges (bridges due to ground shaking or road segments due to liquefaction) to road network connectivity (PI = WCL)

7.7 Shelter Demand Analysis[1]

To demonstrate the application of the socio-economic methodology developed in SYNER-G, a shelter needs analysis has been applied in the Thessaloniki study area. The results of the simulations for building damage, displaced people and utility losses as described in the previous sections are used in the analysis. The goal here is to demonstrate how such a framework can be used as a communication tool for decision makers in disaster risk management through the interactive modelling of parameters influencing the displacement of populations and whether or not they will seek shelter via a system of socio-economic indicators and their respective importance weights. The focus in the shelter needs model is to obtain shelter demand as a consequence of building usability, building habitability and social vulnerability of the affected population rather than building damage alone (see Chap. 4). The shelter model simulates households' decision-making and considers physical, socio-economic, climatic, spatial and temporal factors in addition to modelled building

[1]Bijan Khazai from Karlsruhe Institute of Technology, Germany, is co-author of this subsection, e-mail: khazai@gmail.com.

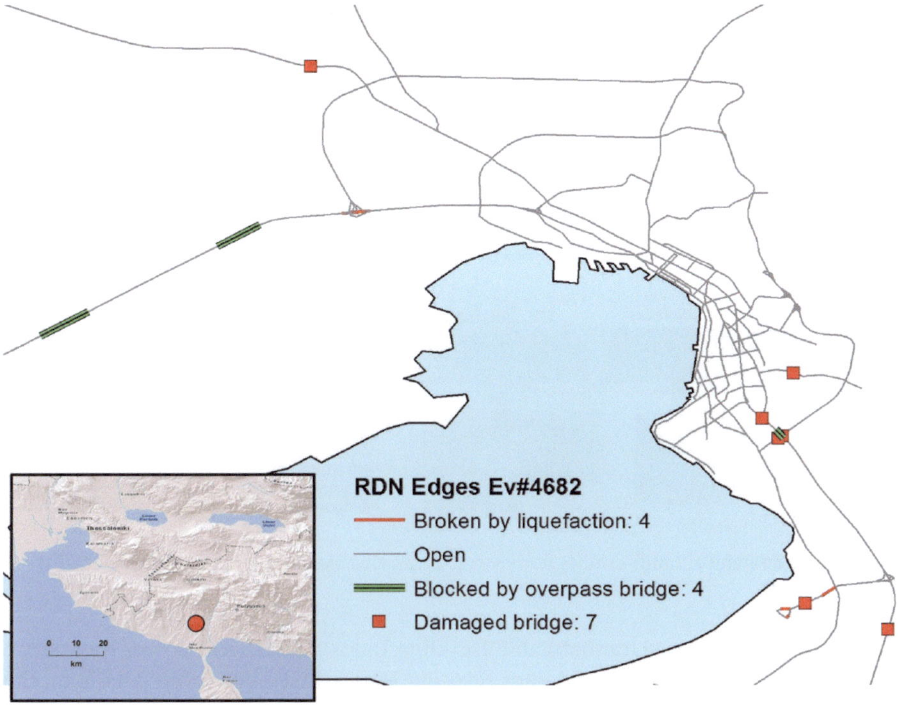

Fig. 7.30 Road network damages for an event (#4682, WCL = 18 %, M = 7.4, R = 40 km) that corresponds to WCL with $T_R = 500$ years

damage states. A group of proposed socio-economic indicators were harmonized for Thessaloniki based on data available for Europe from the EUROSTAT Urban Audit Database.

In particular, a new advancement to shelter estimation methodology was explored through three types of key inputs: (1) the "habitability" of buildings which combines inputs from the physical models (building usability, utility loss and climate factors) considering interactions between building stock (BDG), water supply system (WSS) and electric power network (EPN) to provide information on the habitability of a building and can be used as a better determinant in influencing the decision to evacuate than building damage alone; (2) GIS-based shelter accessibility analysis considering connectivity of the Roadway Network (RDN) as an input to the shelter-seeking model; and (3) a multi-criteria decision model for implementing a shelter-seeking logic model based on complex socio-economic factors which ultimately lead to the decision to evacuate and seek public shelter. These three inputs are combined into a dynamic shelter model and software tool developed within the SYNER-G platform to provide stakeholders an interactive framework in decision-making process for shelter planning and preparedness as well as resource allocation. The multi-criteria framework is described schematically in Fig. 7.31 as composed of

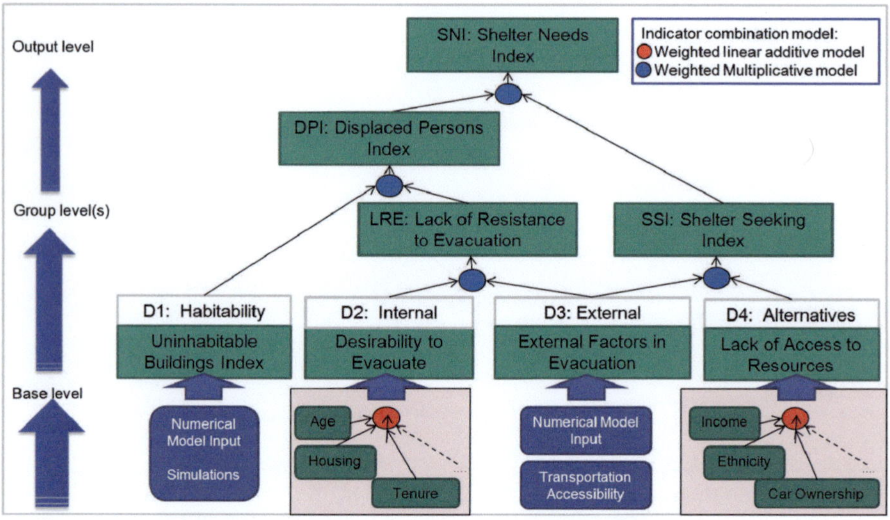

Fig. 7.31 Hierarchical multi-criteria framework to describe shelter needs (Khazai et al. 2012)

the three measures: (a) Uninhabitable Building Index (UBI), (b) Lack of Resistance to Evacuation (LRE) and (c) Shelter Seeking Index (SSI). For more details on the shelter demand analysis the reader is referred to Chap. 4.

The first step in the decision to evacuate after an earthquake is based on the structural stability of a building and functional lifeline structures, such as access to water and electric power services. Weather conditions can further influence the decision to evacuate buildings with disrupted lifeline services. The "displaced persons" model provides an estimate of proportion of persons in habitable and uninhabitable buildings using the following inputs for each simulation (Fig. 7.32a):

- Building Usability as estimated in Sect. 7.5 (building structural damage which leaves the building unusable, partially or fully usable depending on the level of damage and possibility of repairs);
- Utility Loss in water supply and electric power as estimated in Sects. 7.3 and 7.4;
- Weather conditions (which determine the tolerance to utility loss).

The decision to evacuate one's home after an earthquake and to utilize public shelter is correlated with a variety of social, economic and demographic factors. The EUROSTAT Urban Audit data has been analyzed for 34 indicators collected for 20 SCD of Thessaloniki. The desirability to evacuate was ranked for each SCD (Fig. 7.32b) based on specific indicators (i.e., proportion of dwellings lacking basic amenities; proportion of non-conventional dwellings; lone-parent households with children aged 18 or under; lone pensioners; proportion of total population aged 0–4; proportion of total population aged 75 and over).

Not all the displaced population will seek public shelter, and some may find alternative shelter accommodations (rent motel rooms or apartments), stay with

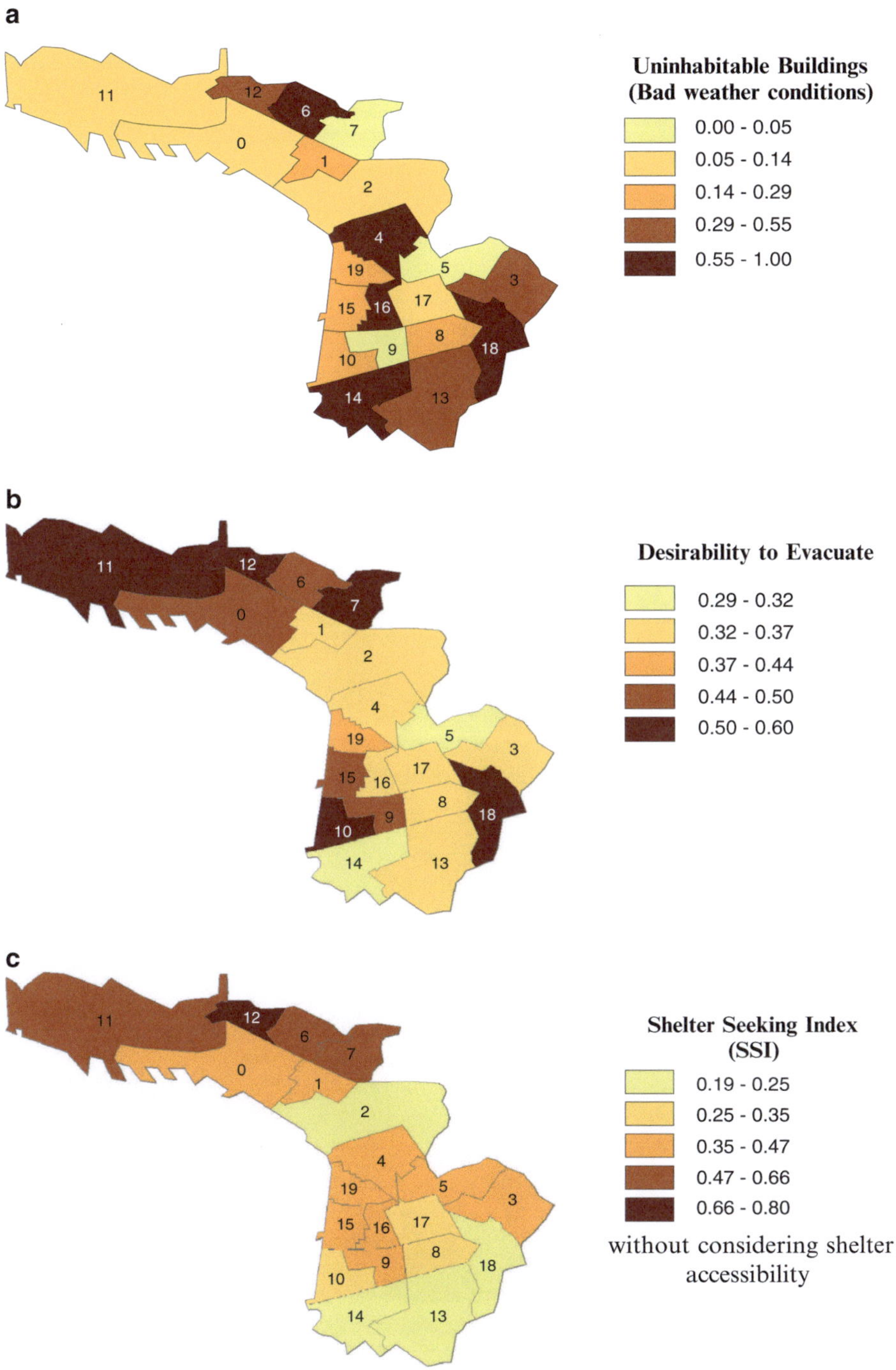

Fig. 7.32 Ranking of different indexes for shelter demand analysis in Thessaloniki

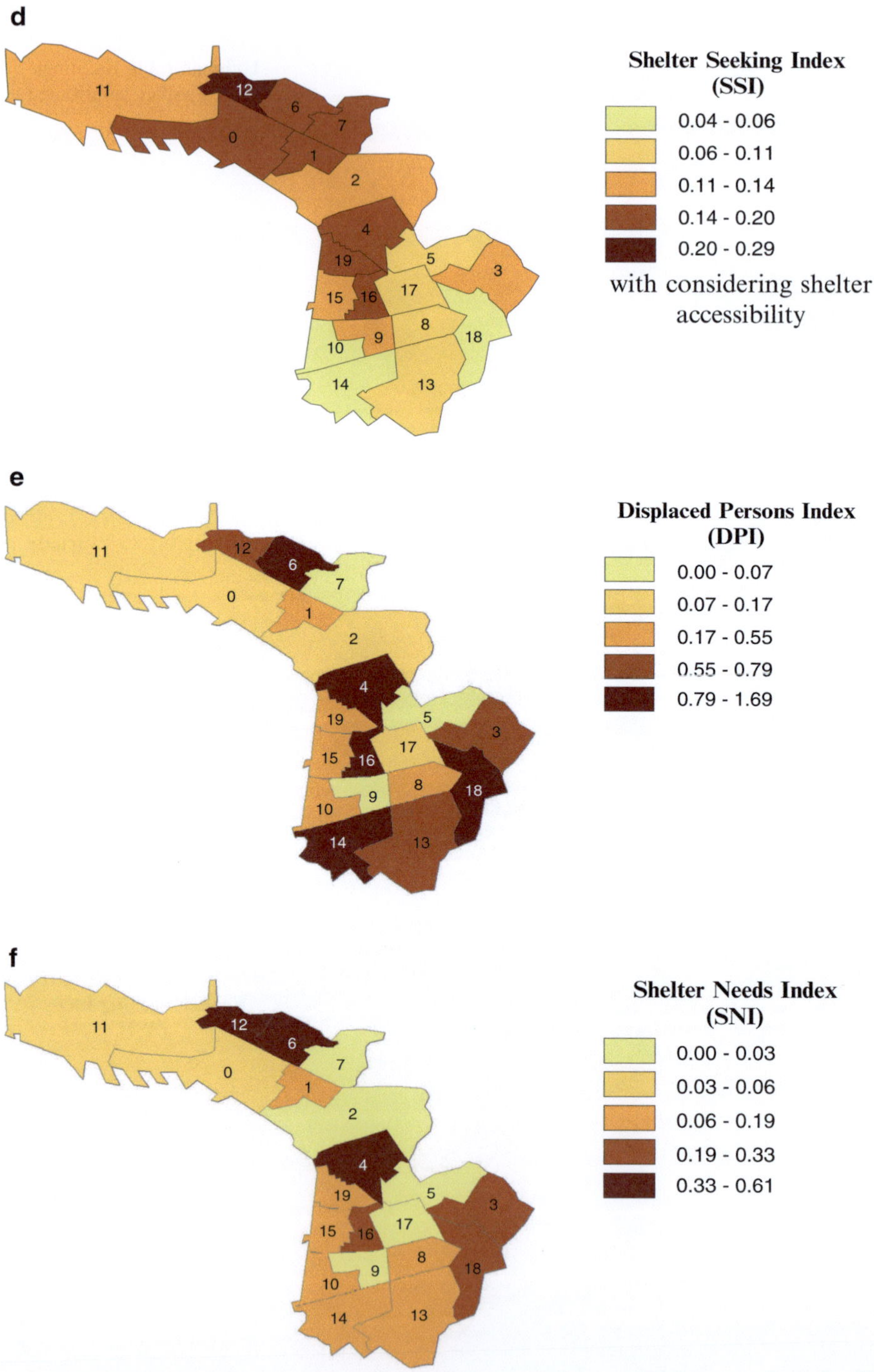

Fig. 7.32 (continued)

family and friends, or leave the affected area. The desirability to seek public shelter was ranked for each SCD (Fig. 7.32c) based on specific indicators (e.g. unemployment, low income migrants, and education). The proximity and ease of access of shelter locations might be key criteria for these households whose decision of leaving is not founded on aspects of vulnerability but on individual preferences (Fig. 7.32d).

The integrated shelter needs model is based on a Multi-Criteria Decision theory (MCDA) framework which combines parameters influencing the physical inhabitability of buildings, with social vulnerability (and coping capacity) factors of the at-risk population as well as external factors (like building damage, loss of utilities, and weather conditions) to determine the desirability to evacuate and seek public shelter. The multi-criteria framework is composed of two main criteria: overall population at risk of being displaced after an earthquake (DPI) and the proportion of this population likely to seek public shelter (SSI). The Displaced Persons Index (DPI) is given as occupants in uninhabitable buildings (UBI) amplified by external and internal socio-economic factors related to desirability to evacuate (Fig. 7.32e).

The total demand for public shelter for a particular location (i.e., city district) can be described as a product of the population at risk of being displaced (D1, D2 and D3) to the population likely to seek public shelter (D4) (Fig. 7.32f). The results of the analysis can provide valuable information for an efficient planning of shelter capacities and allocation.

7.8 Conclusions

The SYNER-G methodology has been applied to the city of Thessaloniki, the second largest city of Greece, an area characterized by intense seismic activity. The main study area is the municipality of Thessaloniki, which is divided in 20 Sub City Districts as defined by Eurostat and Urban Audit approach. The case study includes the following elements: building stock (BDG), road network (RDN), water supply system (WSS) and electric power network (EPN). New analytical fragility curves are developed for buildings (RC/masonry) and bridges based on the inventory of Thessaloniki area. Yielding and ultimate damage states are considered with PGA as intensity measure.

Buildings and networks as well as the seismic hazard acting upon them are modelled in the SYNER-G prototype software through the object-oriented paradigm. Five seismic zones have been selected based on the SHARE European research project. A plain Monte Carlo simulation (MCS) has been carried out sampling earthquake events for these zones and computing selected performance indicators (PIs) for each system. A connectivity analysis has been performed for EPN, WSS and RDN considering specific interdependencies between systems according to the SYNER-G methodology: EPN with WSS (electric power supply to pumping stations), RDN with BDG (road blockage due to building collapses), BDG with EPN and WSS (displaced people due to utility loss). Results coming from the analyses carried out indicate the important role of interdependencies in the overall performance of the networks. The estimated losses are much higher when interactions are considered.

The overall performance of each network is expressed through the moving average μ and moving standard deviation σ (averaged over simulations), as well as the Mean Annual Frequency (MAF) of exceedance of different levels of decrease of PIs (performance loss curve). The average loss is defined based on the moving average graph. Through the MAF graphs the annual probability of exceeding specific levels of loss can be defined and the loss for specific return period of the particular PI can be estimated. The earthquake event(s) that correspond to a particular return period (e.g., 500 years) are identified and maps with the distribution of damages are produced for this event(s). The spatial distribution of damages and losses for such sample events are strongly related not only to the location and the size of the earthquake, but also to the (randomly sampled) spatial distribution of intensities. Correlation factors are estimated which relate the estimated damages of specific components (either internal or external to the selected system) with the system's functionality. This type of analysis is based on the results of each single event, and thus it preserves the information about systems' topology and its behaviour in case of spatial correlated damages (related to single earthquakes). Thus, it allows identifying the most critical elements for the functionality of each system (i.e., the damaged components that more closely control the performance of the network).

To demonstrate the application of the socio-economic methodology in SYNER-G a shelter needs analysis has been applied. The shelter model simulates households' decision-making and considers physical, socio-economic, climatic, spatial and temporal factors in addition to modelled building damage states and utility loss. From the analysis, different SCDs of Thessaloniki are identified, as "Hot Spots" for shelter needs. These results are supporting an efficient planning of shelter allocation.

Acknowledgments The authors would like to acknowledge Francesco Cavalieri and Paolo Franchin from Sapienza University of Rome and Pierre Gehl from BRGM for their contribution and support to the implementation of the prototype software. Also, Michael Fardis and George Tsionis from University of Patras for their efforts in the fragility assessment of buildings and bridges for Thessaloniki.

References

Akkar S, Bommer JJ (2010) Empirical equations for the prediction of PGA, PGV and spectral accelerations in Europe, the Mediterranean region and the Middle East. Seismol Res Lett 81(2):195–206

ALA (2001) Seismic fragility formulations for water systems. American Lifeline Alliance, ASCE, Washington, DC

Anastasiadis A (1994) Contribution to the determination of the dynamic properties of natural Greek soils. PhD thesis, Department of Civil Engineering, Aristotle University of Thessaloniki (in Greek)

Anastasiadis A, Raptakis D, Pitilakis K (2001) Thessaloniki's detailed microzoning: subsurface structure as basis for site response analysis. Pure Appl Geophys 158(12):2597–2633
Apostolidis P, Raptakis D, Roumelioti Z, Pitilakis K (2004) Determination of S-wave velocity structure using microtremor and SPAC method applied in Thessaloniki (Greece). J Soil Dyn Earthq Eng 24:49–67
Argyroudis S, Pitilakis K (2011) Seismic risk performance of urban transportation systems considering site effects and interaction with the built environment. In: Proceedings of 8th international conference on urban earthquake engineering. Tokyo Institute of Technology, Tokyo, Japan
Argyroudis S, Selva J, Gehl P, Pitilakis K (2014) Systemic seismic vulnerability assessment of road networks. Computer-Aided Civil and Infrastructure Engineering (in review)
CEN (2004) EN 1998-1 Eurocode 8: design of structures for earthquake resistance – Part 1: General rules, seismic actions and rules for buildings. European Committee for Standardization, Brussels
Fardis MN, Papailia A, Tsionis G (2012) Seismic fragility of RC framed and wall-frame buildings designed to the EN-Eurocodes. Bull Earthq Eng 10(6):1767–1793
FEMA (2003) HAZUS-MH technical manual. Federal Emergency Management Agency, Washington, DC
Giardini D, Woessner J, Danciu L, Crowley H, Cotton F, Grünthal G, Pinho R, Valensise G, Akkar S, Arvidsson R, Basili R, Cameelbeeck T, Campos-Costa A, Douglas J, Demircioglu MB, Erdik M, Fonseca J, Glavatovic B, Lindholm C, Makropoulos K, Meletti C, Musson R, Pitilakis K, Sesetyan K, Stromeyer D, Stucchi M, Rovida A (2013) Seismic Hazard Harmonization in Europe (SHARE). Online data resource, http://portal.share-eu.org:8080/jetspeed/portal/. doi:10.12686/SED-00000001-SHARE
Jayaram N, Baker JW (2009) Correlation model of spatially distributed ground motion intensities. J Earthq Eng Struct Dyn 38(15):1687–1708
Kappos AJ, Panagopoulos G, Penelis G (2008) Development of a seismic damage and loss scenario for contemporary and historical buildings in Thessaloniki, Greece. Soil Dyn Earthq Eng 28:836–850
Karantoni F, Lyrantzaki F, Tsionis G, Fardis MN (2012) Seismic fragility functions of stone masonry buildings. In: Proceedings of the 15th world conference on earthquake engineering, 24–28 Sept, Lisbon, Portugal
Khazai B, Daniell JE, Franchin P, Cavalieri F, Vangelsten B, Iervolino I, Esposito S (2012) A new approach to modeling post-earthquake shelter demand in the aftermath of earthquakes: integrating social vulnerability in systemic seismic vulnerability analysis, 15 WCEE. Paper no. 2105, conference proceedings, Lisbon, Portugal
Papazachos BC, Papazachou C (1997) The earthquakes of Greece. Ziti Publications, Thessaloniki (in Greek)
Pitilakis K, Anastasiadis A (1998) Soil and site characterization for seismic response analysis. In: Invited lecture. Proceedings of the XI ECEE, Paris, pp 65–90, 6–11 Sept 1998
Pitilakis K, Argyroudis S (eds) (2013) Systemic seismic vulnerability and loss assessment: validation studies. SYNER-G reference report 6, Publications Office of the European Union. Luxembourg, ISBN 978-92-79-30840-6
Pitilakis K, Anastasiadis A, Raptakis D (1992) Field and laboratory determination of dynamic properties of natural soil deposit. In: Proceedings of the 10th world conference on earthquake engineering, vol 5, Madrid, pp 1275–1280
Pitilakis K, Riga E, Anastasiadis A (2013) New design spectra in Eurocode 8 and application to the seismic risk of Thessaloniki, Greece. Invited lecture in international conference on earthquake geotechnical engineering: from case history to practice, Istanbul, Turkey, 17–19 June 2013
Raptakis D (1995) Contribution to the determination of the geometry and the dynamic characteristics of soil formations and their seismic response. PhD thesis, Department of Civil Engineering, Aristotle University of Thessaloniki (in Greek)
Raptakis D, Anastasiadis A, Pitilakis K, Lontzetidis K (1994a) Shear wave velocities and damping of Greek natural soils. In: Proceedings of the 10th European conference on earthquake engineering, vol 1, Vienna, Austria, pp 477–482

Raptakis D, Karaolani E, Pitilakis K, Theodulidis N (1994b) Horizontal to vertical spectral ratio and site effects: the case of a down-hole array in Thessaloniki (Greece). In: Proceedings of the XXIV general assembly, vol III. Athens, pp 1570–1578

SRMLIFE (2007) Development of a global methodology for the vulnerability assessment and risk management of lifelines, infrastructures and critical facilities. Application to the metropolitan area of Thessaloniki. Research project, General Secretariat for Research and Technology, Greece

Tenerelli P, Crowley H (eds) (2013) Development of inventory datasets through remote sensing and direct observation data for earthquake loss estimation. SYNER-G reference report 3. Publications Office of the European Union. Luxembourg, pp 82, doi: 10.2788/86322

Tsionis G, Fardis MN (2012) Seismic fragility of concrete bridges with deck monolithically connected to the piers or supported on elastomeric bearings. In: Proceedings of the 15th world conference on earthquake engineering. Lisbon, 24–28 Sept

Youd TL, Perkins DM (1978) Mapping of liquefaction induced ground failure potential. J Geotech Eng Div ASCE 1044:433–466

Chapter 8
Application to the City of Vienna

Helmut Wenzel, David Schäfer, and Anna Bosi

Abstract The Vienna use case is an attempt to apply the methods developed in SYNER-G to a small area of the city with very detailed input data. This introduces some major difficulties for both the software as well as the need for systematic high resolution data collection, which are addressed here. The Vienna test case applies a deterministic methodology implemented in EQvis using high resolution building level data with a probabilistic analysis using the SYNER-G methodology and prototype software by accounting for system interdependencies. EQvis, an advanced seismic loss assessment and risk management software based on the Mid-America Earthquake Center software tool MAEviz (MAEviz, MAEviz software tool. http://mae.cee.illinois.edu/software_and_tools/maeviz.html. Accessed Sept 2010, 2010), was further developed and adapted in SYNER-G as a platform for deterministic risk analysis on the above mentioned area. The EQvis case highlights the importance of a user friendly and easy to handle software package. In addition, a powerful visualisation tool for the results plays a major role when dealing with stakeholders. EQvis has brought together all these components in one software solution which is easy to handle.

8.1 Introduction

8.1.1 Test Case

The city of Vienna is located in the North-Eastern part of Austria. It is the capital of Austria with a population of about 1.7 million people. The city of Vienna is

H. Wenzel (✉) • D. Schäfer • A. Bosi
Department of Natural Hazards, Vienna Consulting Engineers (VCE), Vienna, Austria
e-mail: wenzel@vce.at; schaefer@vce.at; bosi@vce.at

K. Pitilakis et al. (eds.), *SYNER-G: Systemic Seismic Vulnerability and Risk Assessment of Complex Urban, Utility, Lifeline Systems and Critical Facilities*, Geotechnical, Geological and Earthquake Engineering 31, DOI 10.1007/978-94-017-8835-9_8,

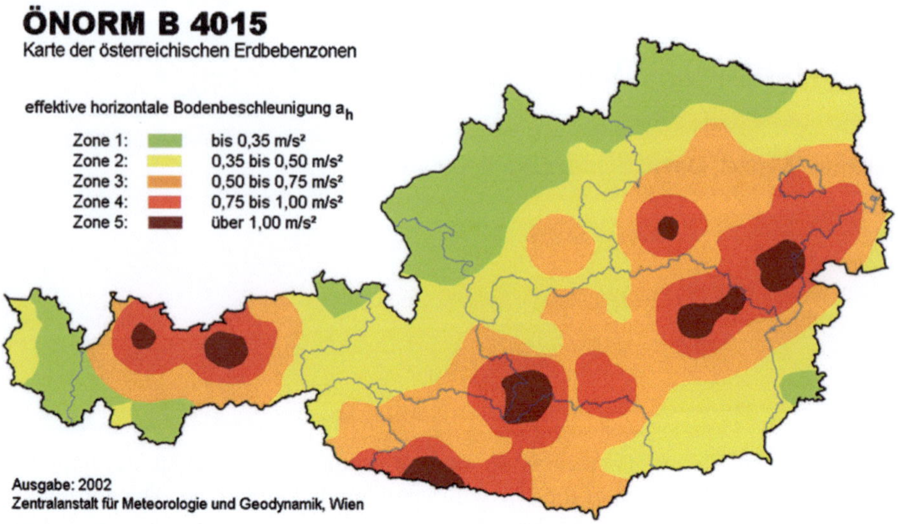

Fig. 8.1 Historic seismicity of Austria

placed east of the Alps, at the west end of the tertiary Wiener Beckens. Three main geological formations can be identified:

- Brash and sand from the river Danube
- Loose rock – tertiary loose rock from the Vienna basin
- Solid rock from the flysch zone and the limestone alps

There is a system of north-south aligned faults and cracks that goes through the city of Vienna. The majority of seismic risk in Austria is associated with the Vienna transform fault zone (Fig. 8.1), which runs through the eastern part of Austria beneath the city of Vienna and surrounding areas (Achs et al. 2010).

The region of interest selected in the city for the case study is the Brigittenau district, which is the 20th district of Vienna. It is located north of the central district, north of Leopoldstadt on the same island area between the Danube and the Danube Canal. Brigittenau is a heavily populated urban area with many residential buildings.

The reasons for the choice of this particular area can be summarized as follows:

- The district consists of various types of buildings, with construction practices that start from 1848 until recently (Flesch 1993).
- The topic of transportation is covered even in this relatively small area as there are railroads/railway stations, underground and tramway lines as well as bus lines and numerous very frequently used bridges across the Danube.
- There is a huge amount of data available for the whole district (lifelines, essential facilities, etc.) (Fig. 8.2).

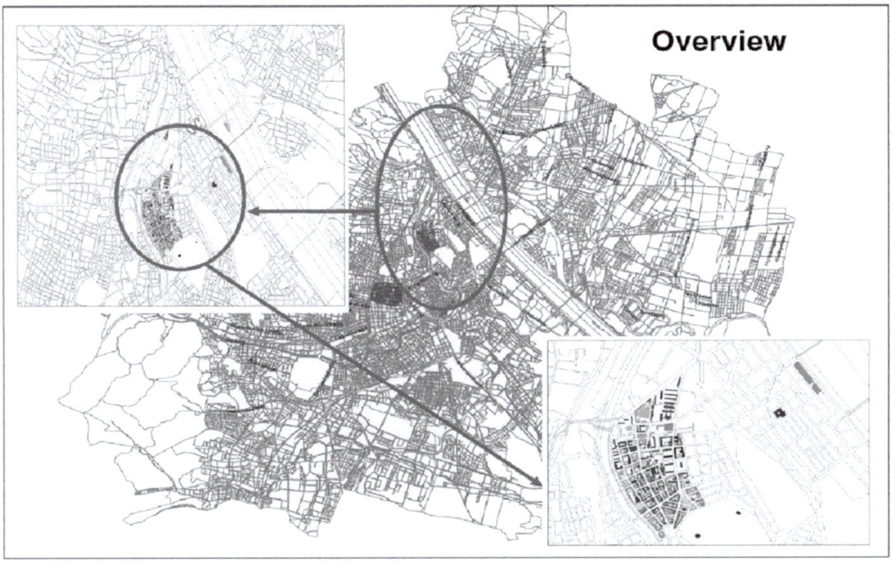

Fig. 8.2 Brigittenau district in the city of Vienna

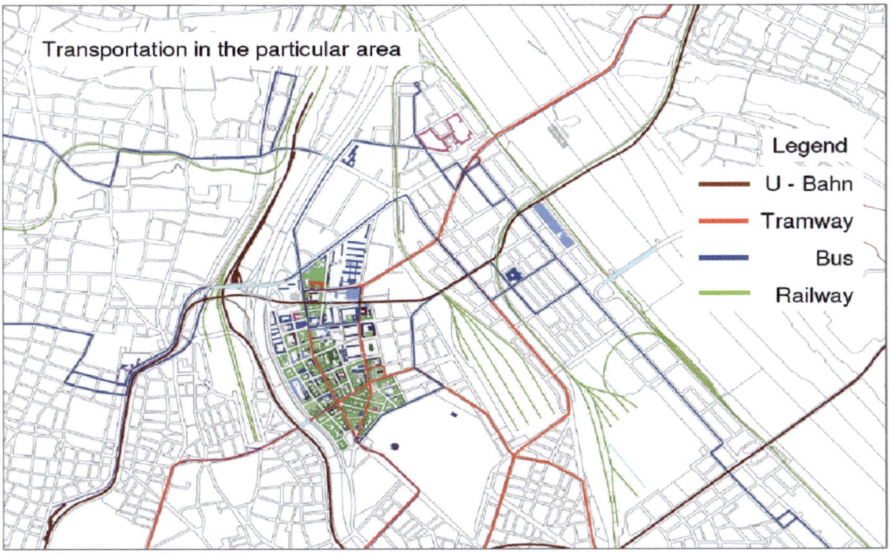

Fig. 8.3 Overview of the transportation networks in the considered area of interest

- There are numerous essential facilities like fire stations, police stations, schools, ambulance stations, an important hospital, the Millennium Tower (one of the tallest buildings in Vienna), etc. (Fig. 8.3)

The Vienna test case is an attempt to look at SYNER-G methods at the building level using a very high resolution data. Usually, this data is not available. Therefore, in order to collect and to store this vast amount of data in a systematic way, a methodology that enabled performing the task in a standardized way was established. The methodology is described in Sect. 8.2.

8.2 The Building Identification Procedure (BIP)

The building identification procedure was formulated to identify and inventory buildings that will be considered in the present case study (FEMA 2002). The procedure can be implemented relatively quickly and inexpensively to develop a list of potentially vulnerable buildings without the high cost of a detailed seismic analysis of individual buildings. The inspection, data collection, and decision-making process typically occur at the building site, taking an average of 15–30 min per building (30 min to 1 h if access to the interior is available). The main purpose of this procedure is to identify and categorize buildings in a relatively big area. The output of this procedure is a fact sheet for every building, which contains all the essential information with respect to earthquakes and the overall condition of the building. The Data Collection Form includes space for documenting building identification information, including its use and size, a photograph of the building, and documentation of pertinent data related to seismic performance.

Buildings may be reviewed from the sidewalk without the benefit of building entry, structural drawings, or structural calculations. Reliability and confidence in building attribute determination are increased, however, if the structural framing system can be verified during interior inspection, or on the basis of a review of construction documents. The BIP procedure is intended to be applicable nationwide, for all conventional building types. Bridges, large towers, and other non-building structure types, however, are not covered by the procedure.

8.2.1 Completing the Building Identification Protocol

The purpose of the chapter is to give instructions how to complete the Building Identification Protocol for each building screened, through execution of the following steps:

- Verifying and updating the building identification information.
- Walking around the building to identify its size and shape and looking for signs that identify the construction year.
- Determining and documenting occupancy.
- Determining the construction type.
- Identifying the number of persons living/working in the building.

Protocol for building identification procedure		Name.:	No.
		Date:	
		Time:	
Address:	Street / No.		
	PLZ		

Fig. 8.4 Verifying and updating the building identification information

- Characterizing the building through the plan view and determining the distance to traffic area.
- Characterizing the building elevation; using the laser telemeter to define building height; identifying soft stories or added attic space.
- Identifying façade elements inclusively number of windows and doors.
- Determining non-structural members.
- Determining the overall condition of the building.
- Noting any irregularities/anomalies.
- Taking pictures with the digital camera.

All these steps have to be done carefully. Each step is now explained in detail.

(a) Verifying and updating the building identification information (Fig. 8.4).

This is the first step in the whole procedure. Arriving at the site, the "identity" of the building must be first checked. Afterwards, the data collection can start: date, name and time are registered.

(b) Walking around the building to identify its size and shape and looking for signs that identify the Construction year.

At first the building should be looked at to identify its size and shape and to get a first impression of the building. The construction year of a building can be determined if there is a sign at the facade of the building. If there is not such a sign, and the construction year cannot be determined, the field Construction Year is left empty (Fig. 8.5).

(c) Determining and documenting occupancy.

This field describes the general usage of the building, like apartment building, school, kindergarten, hospital, office building, etc. If the building usage is not limited to one category the percentage of the usage categories are identified. Example: Apartments (70 %), Offices (30 %) (Fig. 8.5).

(d) Determining the construction type.

The construction type can be difficult to identify in the field and without appropriate additional knowledge. However what can be determined easily is the construction material. Predominately this can be identified by visual inspection however the construction year can provide a good indication as construction materials (and

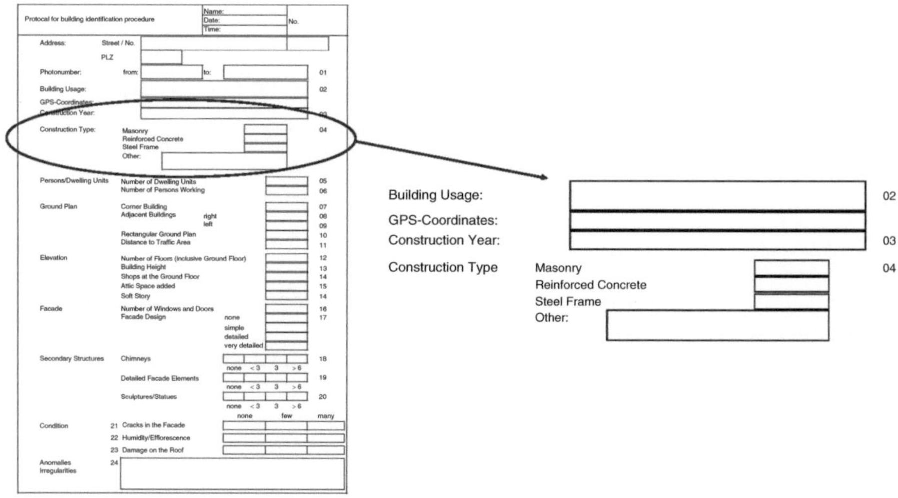

Fig. 8.5 Steps b, c and d

construction types) have been largely used in specific historical periods. It can also be helpful to have a look into the building, if possible. Often the interior walls can give clues as to what building type is present. Sometimes it also helps to get into the basement, because the walls are not always covered in basements (Fig. 8.5).

(e) Identifying the number of persons living/working in the building.

This step in the whole procedure allows estimating future casualties in case of an earthquake. To this aim, the number of occupants (living/working) of the building must be identified. The number of dwelling units can easily be determined in the entrance area of a building by looking at the number of door bells or the number of mail boxes. All dwelling units, even not used ones, should be counted. The next field addresses the number of people living or working in areas not depicted as dwelling units like shops at the basements, cafes, etc. The number can only be approximated, but this number should depict the maximum number of persons that can stay/work in the building. A practical example is the case of a building with a café on the ground floor and several dwelling units on the other floors. The maximum number of persons is given by the sum of the people leaving in the dwelling units (assessed as explained before) plus the maximum number of people that can occupy the café taking into account customers and employees (Fig. 8.7).

(f) Characterizing the building through the ground plan and determining the distance to traffic area.

The characterization of the building through the ground plan can mostly be made with a plan of the city (Fig. 8.6). There are three questions to be addressed: Is the building a Corner Building? Are there any adjacent buildings? Does the building have a rectangular ground plan?

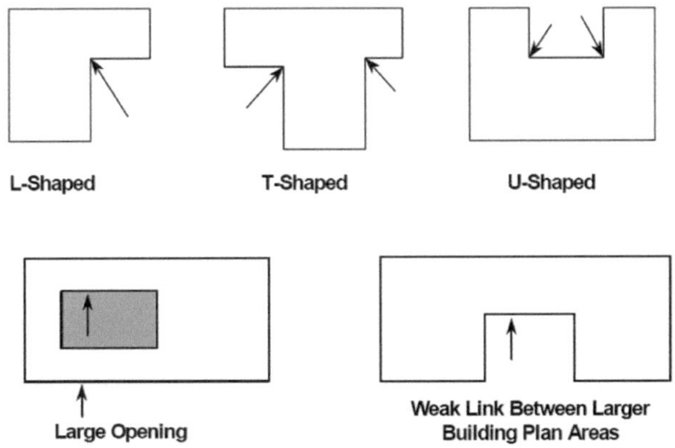

Fig. 8.6 Plan views of various building configurations showing plan irregularities; *arrows* indicate possible area of damage (FEMA 2002)

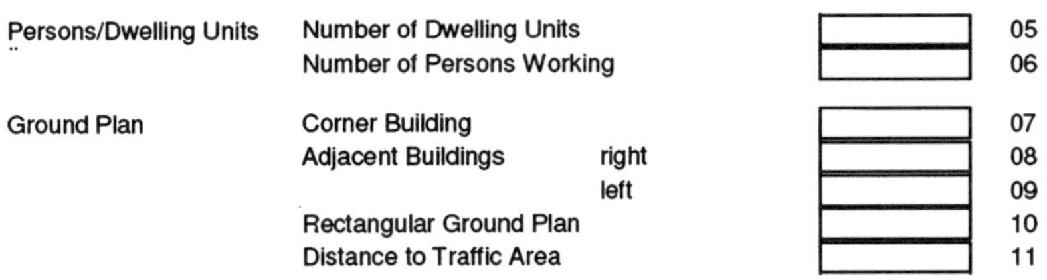

Fig. 8.7 Steps e and f

The distance to traffic area means the lowest distance between building and street. Parking areas and sidewalks do not count as traffic areas and should not be considered. The purpose of this distance is to know whether street can be blocked by building debris or not (Fig. 8.7).

(g) Characterizing the building elevation; Using the laser telemeter to define building height; identifying soft stories or added attic space.

Number of floors, including the ground floor has to be registered considering carefully also the additional attic space. Attic space counts only if the housing area is more than 50 % of the ground floor area. Also building height (defined as the height from the top edge of the sidewalk to the beginning of the cornice) must be measured. The easiest way is by means of a laser – telemeter. A remark is added in case that the building height can be only approximately measured. If the building height cannot be directly measured; an "*a posteriori*" assessment can be performed: a measuring tape is put it to the wall of the building and a photo is taken. The building height can then be determined afterwards. The same procedure can also be done with balconies, etc.

The presence of shops or cafes at the ground floor, that might represent a soft story, is also evaluated. A soft story is a floor (does not have to be the ground floor)

Fig. 8.8 Identifying soft stories and additional attic spaces

where most of the interior walls are missing due to the space needed. Soft stories perform poorly under seismic excitation (Fig. 8.8).

Additional attic space can often be determined by looking at the windows at the attic or due to the existence of balconies. Also added attic spaces, representing a vertical discontinuity, have poor seismic performances.

(h) Identifying façade elements inclusively number of windows and doors.

Number of windows and doors at the façade facing the street and when possible also for the sides facing the courtyard are identified.

Evaluation of façade elements alias how detailed the façade design is, is also registered. Examples are given in the figures below (Fig. 8.9).

(i) Determining non-structural members.

Chimneys represent another vertical discontinuity with poor seismic performances. When possible, they must be counted, otherwise this field is left open (Fig. 8.10).

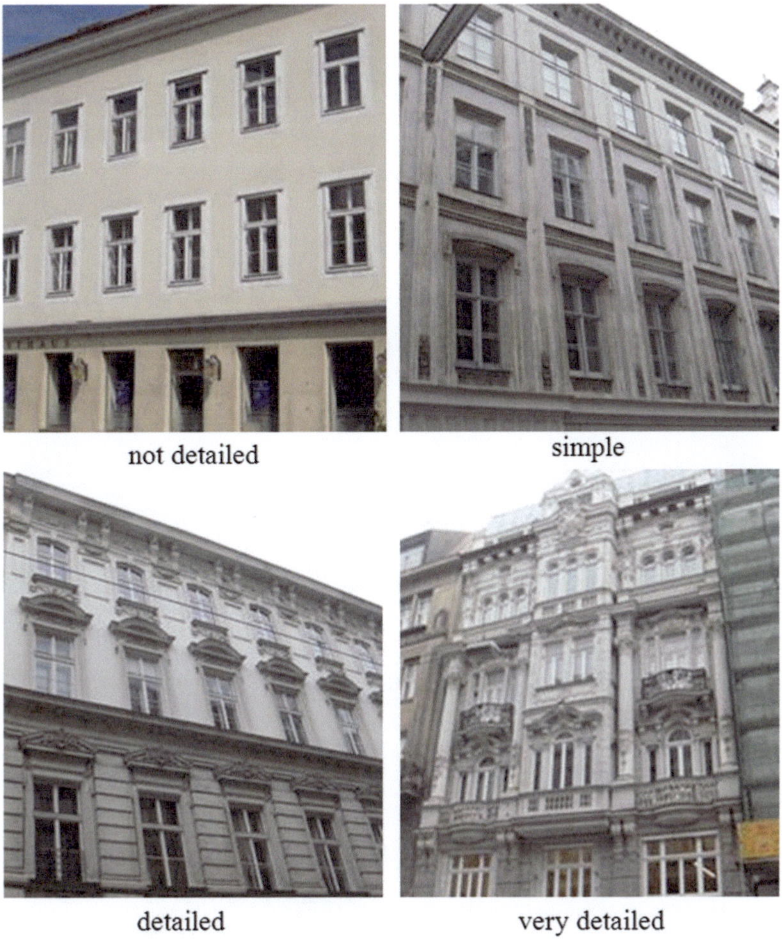

Fig. 8.9 Identifying façade elements

Also all those façade elements that can fall off the building and on the street are registered. This includes sculptures, balconies, statues, etc. It is important to count all potentially hazardous elements on the façade: shop signs are also considered here (Fig. 8.11).

(j) Determining the overall condition of the building.

This part focuses on the overall condition of the building. The main attributes are the presence of water leakage, damages to the roof and cracks in the walls. This mainly means the cracks in the walls. It is, when possible, distinguished between cracks on the outside layer of façade (that do not represent a structural problem and therefore are not counted) and cracks in the walls. If the crack is going diagonal it should be counted anyways (Fig. 8.12).

Fig. 8.10 Example for a building with eight chimneys

Fig. 8.11 Examples for non-structural members

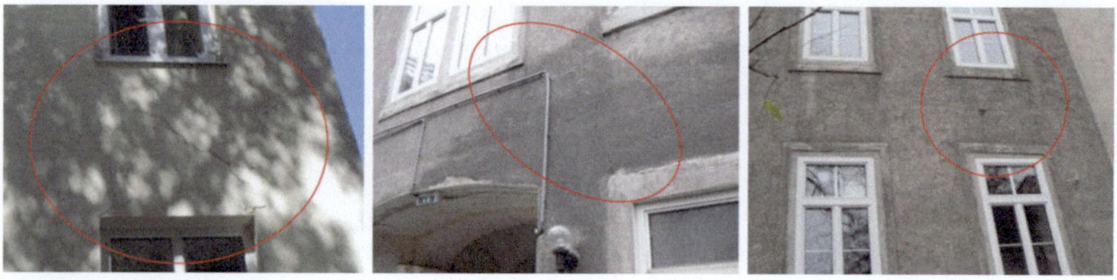

Fig. 8.12 Examples of cracks

Fig. 8.13 Condition assessment

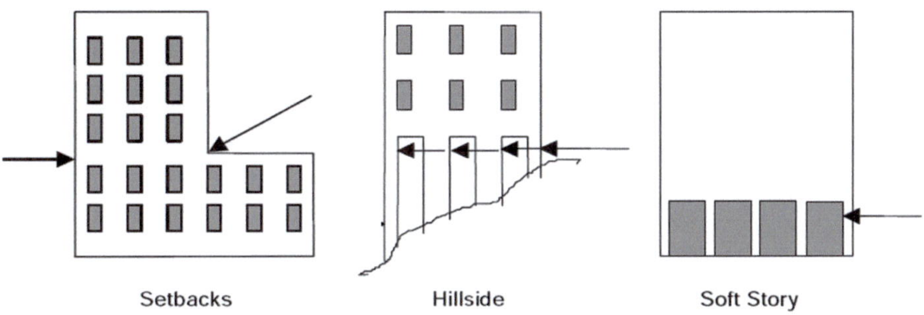

Fig. 8.14 Irregularities/anomalies and soft story

Also the presence of humidity and efflorescence (revealed by a change in the façade color) is registered (Fig. 8.13).

The estimation of damage at the roof level can be rather difficult. Nevertheless, the presence of humidity on the façade at the conjunction roof-upper floor is synonymous of possible roof damage. When the access to the building is possible, photos can document the degrade state. Damages at the roof can be very dangerous if not properly treated.

(k) Noting irregularities/anomalies.

If there is anything out of the ordinary that is not explicitly in the checklist this is the place to write it down. If anything is written down here, it should always be documented with a photo if possible (Fig. 8.14).

(l) Taking pictures with the digital camera.

A software program can modify pictures and combine them. The software is designed to reconstruct a coherent building out of your photographs. Note that since it is an automatic process, it can always lead to unexpected results. In order to avoid these degenerated cases, photos must be taken with care and following some guidelines. For each new reconstruction project focus must be put on only one building, or even only one façade (Fig. 8.15).

Fig. 8.15 Focus on one building for each reconstruction project

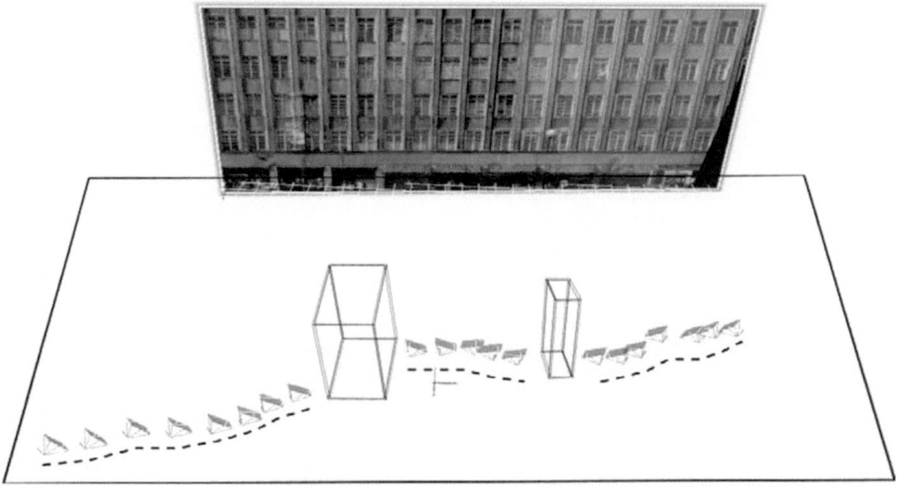

Fig. 8.16 Plan the path in front of the building. Move around occludes

Having chosen the façade, a careful track must be planed (Fig. 8.16): pictures should be shut moving in approximately a half-circle around the façade Note that those coherent paths, with distance of about 1–3 steps between the shots, deliver best results.

In order to obtain a coherent point-cloud of the object, the biggest possible part of the object (façade) should be kept in each photograph (Fig. 8.17).

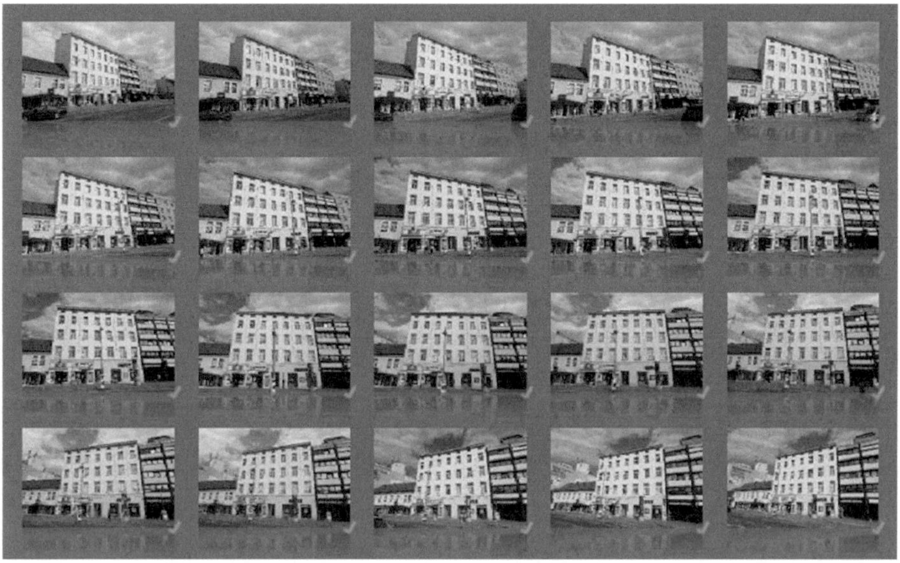

Fig. 8.17 Always try to keep the whole façade in the view port of the camera

It is important to avoid delivering images difficult to distinguish: highly repetitive façade can confuse the software and produce mismatches. In general it is better to supply fewer images with good quality, than too many poor photos (Figs. 8.18 and 8.19).

8.2.2 Building Identification Process – An Example

This section provides an example for the building identification process (Fig. 8.20). The following example describes a part of the process for the city of Vienna. The first thing to do is choosing an area of interest and collecting all information about the area that does not need field work: street plans, building plans, geology maps, etc. Once this information has been gathered the route of the screeners can be identified. If the buildings to be identified are selected, the screeners can begin to investigate the area. It has been shown that the best way to begin the process is to have a very detailed route and detailed plans for the field observations. The last step is transferring the information on the BIP Data Protocols into the relational electronic Building BIP Database. This requires that all photos are numbered (for reference purposes), and that additional fields (and tables) be added to the database for those attributes not originally included in the database. After arriving at the site the screeners observe the building as a whole and begin the process of gathering the information in the building identification protocol, starting with name, date, time, protocol number and the street address. The next step is to take photos of

Fig. 8.18 Façade with highly repetitive content. Making close-up photographs from two ends of such building will produce ambiguous result. This is the type of input to be avoided

Fig. 8.19 Unclear data: difficulties in distinguishing among the sides of the building

the building. This step can also be performed at the end of the screening process, after filling all the fields of the protocol. After determining the building usage, the construction year and the construction type are being determined. These two fields can also be left empty, in case that the construction year or type cannot be deter-

Protocal for building identification procedure	Name: David Vukovic Date: 29.4.10 Time: 09:20 am	No.	16

Section	Item	Entry	Code
Address:	Street / No.	Brigittaplatz 20	
	PLZ	1200	
Photonumber:	from: 3441	to: 3477	01
Building Usage:	Residential / Commercial / % R / % C	X / X / 25 / 75	02
GPS-Coordinates:			
Construction Year:		1912	03
Construction Type:	Masonry	X	04
	Reinforced Concrete		
	Steel Frame		
	Other:		
Persons/Dwelling Units	Number of Dwelling Units	18	05
	Number of Persons Working	5	06
Ground Plan	Corner Building	X	07
	Adjacent Buildings right	X	08
	Adjacent Buildings left	X	09
	Rectangular Ground Plan		10
	Distance to Traffic Area	2 m	11
Elevation	Number of Floors (inclusive Ground Floor)	4	12
	Building Height	16	13
Facade Width: 25 (1), 17 (2)	Shops at the Ground Floor	1	14
	Attic Space added		15
	Soft Story		15
Facade	Number of Windows and Doors	24 / 37	16
	Facade Design: none		17
	simple		
	detailed	X	
	very detailed		
Non-Structural Members	Chimneys (none / < 3 / 3 / > 6)	> 6: X	18
	Detailed Facade Elements (none / < 3 / 3 / > 6)	> 6: X	19
	Sculptures/Statues (none / < 3 / 3 / > 6)	none: X, < 3: X	20
Condition (none / few / many)	21 Cracks in the Facade	none: X	
	22 Humidity/Efflorescence	many: X	
	23 Damage on the Roof	none: X	
Anomalies Irregularities	24	Humidity, currently repaired	

Fig. 8.20 Building identification protocol

mined for sure. The next big block of fields is pretty easy to determine, number of persons/dwelling units, ground plan, elevation and façade. Non-structural members cannot always be determined properly like number of chimneys. The procedure is the same as for the construction year, if the number cannot be determined for sure, the field should be left empty. After determining the overall condition of the building there is a big field for irregularities. In each example there is an oriel starting at the first floor. This is written in this field and a photo is taken.

8.3 Deterministic and Probabilistic Analysis, Inputs and Outputs

A deterministic and a probabilistic analysis were performed in the area of interest. The EQvis software is based on Mid-America Earthquake Center software tool MAEviz (MAE 2013) and adapted in SYNER-G as a platform for performing deterministic earthquake simulations as well as various other tools for pre and post processing of the input and output data. The SYNER-G prototype software for probabilistic analysis was integrated into the EQvis platform. In this way we could perform the probabilistic analysis and at the same time use the tools for pre and post processing of the input and output data available into EQvis. Hence, EQvis platform was improved adding the probabilistic analysis to the original deterministic one.

The case study here described represents an application and a validation of the functions of the EQvis platform.

It is noteworthy to clarify that the deterministic (performed with EQvis) and the probabilistic analyses (as developed in the SYNER-G project and implemented in the SYNER-G prototype software) differ on the resolution level of input and output data. In EQvis input and output data refer to every individual building: for every single building, the analysis requires data on the structural characteristics, number of occupants, proximity to the streets, etc. Accordingly, the software provides the assessment of the damage and of the casualties related to every single building.

In the SYNER-G prototype software instead, the buildings are grouped into zones (census tracts) and for each of those zones, the structural features of the buildings are statistically classified (Wen et al. 2003). Also the results of the calculation represent a statistical distribution inside the census tracts. Due to this difference in the resolution, although referring to the same area, input data of the deterministic and probabilistic analysis are not coincident. The following paragraphs present first the deterministic analysis and subsequently the probabilistic analysis.

The reason for choosing both of the analysis types is that decision makers can benefit from both of them. The probabilistic case gives them an overview of the general situation and a full consideration of system of systems and the “risk” of having a damaging earthquake together with the probabilistic values for the consequences.

The deterministic case can help in getting some concrete values on the damage expected for a given earthquake. The decision maker can for example choose a worst case scenario and see directly what the consequences will be. Another important usage of the deterministic case has been shown in a demonstration case in Hungary where immediately after an earthquake EQvis has been used as a management platform to steer the rescue process (Schäfer et al. 2013).

8.4 EQvis Deterministic Analysis: Input Data

EQvis is an advanced seismic loss assessment, and risk management software which is based on the Consequence-based Risk Management (CRM) methodology. CRM provides the philosophical and practical bond between the cause and effect of the disastrous event and mitigation options. It enables policy-makers and decision-makers to ultimately develop risk reduction strategies and implement mitigation actions (Schäfer et al. 2013; Mid-America Earthquake Center 2009). In EQvis, a wide range of user-defined parameters are introduced. The breadth of user-defined parameters enables emergency planners to model a virtually unlimited number of scenarios.

It has an open-source framework which employs the advanced workflow tools to provide a flexible and modular path (Clayberg and Rubel 2008). It can run over 50 analyses ranging from direct seismic impact assessment to the modeling of socioeconomic implications. It provides 2D and 3D mapped visualizations of source and result data and it provides tables, charts, graphs and printable reports for result data. It is designed to be quickly and easily extensible (McAffer et al. 2010).

8.4.1 Input Data in EQvis

8.4.1.1 Building Data

The building data as collected with the procedure explained in the Sect. 8.2 is been added as attributes to shapefiles of the building footprints, which were created prior to the survey. Each building point gets an attribute table where the data of the survey is stored (Genctürk 2007; Genctürk et al. 2008). The next step is to ingest the data in the EQvis platform. The following figure shows the data in the platform which then serves as the basis for all analyses performed within the test case (Fig. 8.21).

For what concerns the building structural damage, the EQvis user has to provide the inventory, the hazard characteristics, the fragility dataset (see Table 8.1), and the damage ratio dataset (see Sect. 8.5.2). These are the required information, though the user can provide some additional information to improve the result. For instance, in case required, also data concerning liquefaction could be added. This was not the case in Vienna, since the soil does not present liquefaction susceptibility.

INT_NR	STREET	STREET_NR	DWELL_USE	DWELL_PER	BUSIN_PER	NO_PEOPLE	BLDG_CORN	BLDG_ADJRE
421	Klosterneuburgerstraße	65	X	80	20	6		X
420	Klosterneuburgerstraße	63	X	80	20	5		X
419	Klosterneuburgerstraße	61	X	80	20	5		X
418	Gerhardusgasse	19	X	85	15	10	X	X

Fig. 8.21 Buildings in the test area together with a small example of the attribute table

Table 8.1 Fragility curves used in Vienna test case for RC and masonry buildings

RC buildings	
Fragility curves	*IMT*
Borzi et al. (2007) – RC – 8 storeys-seismically designed (c = 10 %)	PGA
Borzi et al. (2007) – RC – 4 storeys-seismically designed (c = 10 %)	PGA
Erberik (2008) – RC – low rise bare frame LRBR	PGV
Erberik (2008) – RC – mid-rise bare frame MRBR	PGV
Erberik (2008) – RC – mid-rise infilled frame MRIR	PGV
Erberik and Elnashai (2004) – RC flat slab – mid-rise infilled frame MRINF	Sd
Kappos et al. (2003) – RC3.1-HR-HC	PGA
RISK-UE (2003) – RC moment frame-HR-HC-UTCB hybrid approach	Sd
RISK-UE (2003) – RC moment frame – LR-HC-UTCB hybrid approach	Sd
RISK-UE (2003) – RC moment frame – MR-HC-IZIIS approach	Sd
Vargas et al. (2010) – RC – 8 storeys	Sd
Masonry buildings	
Fragility curves	*IMT*
Borzi et al. (2008) – MA brick – high percentage voids – 2 storeys	PGA
LESSLOSS (2005) – adobe and rubble stone – 8–15 storeys – Lisbon	Sd
RISK-UE (2003) – M12-HR-UNIGE approach	Sd
Borzi et al. (2008) – MA brick-low percentage voids – 4 storeys	PGA
RISK-UE (2003) – M12-LR-UNIGE approach	Sd

Table 8.2 Fragility curves used in Vienna test case for the railway network

Railway tunnels	
Fragility curves	*IMT*
ALA-2001 Poor Quality Rock Tunnel	PGA
ALA-2001 Poor Quality Alluvial Soil Tunnel	PGA
ALA-2001 Good Quality Rock Tunnel	PGA
ALA-2001 Good Quality Alluvial Soil Tunnel	PGA
SYNER-G Rectangular Tunnel Soiltype B	PGA
SYNER-G Rectangular Tunnel Soiltype C	PGA
SYNER-G Circular Tunnel Soiltype B	PGA
SYNER-G Circular Tunnel Soiltype C	PGA
Railway bridges	
Fragility curves	*IMT*
SYNER-G Bridge Abutment 6 m Soiltype C	PGA

8.4.1.2 Railway Data

As mentioned in the introduction, the Vienna railway network is a complex system consisting of many components. For a first attempt, in the deterministic analysis we considered only the most critical elements of the networks i.e. the railway tunnels and the railway bridges. The fragility functions used are listed in Table 8.2. The railway infrastructure is presented in Fig. 8.22.

8.4.1.3 Road Network Data

The road network that crosses Brigittenau districts connects the north-east part of the city (the part that is located on the east side of the Danube) with the west side of the city that is also the neuralgic centre. Therefore the network consists of main roads (with four or more ways), road bridges that allow the connection east-west side of the river and additional small roads in the inner part of the district for internal displacements.

Table 8.3 shows the fragility functions used for the assessment roads bridges. In this test case, SYNER-G fragility functions have been used for the vulnerability assessment (Pitilakis et al. 2014).

Comparing Fig. 8.22 bottom with Fig. 8.23 bottom, we can note that some of the Danube bridges are only for railway network.

8.5 EQvis Deterministic Analysis: Output Data

8.5.1 Seismic Hazard

Two deterministic cases were simulated: a 1950 historical earthquake located in Neulengbach with a moment magnitude of 6 and a "method testing" earthquake

Fig. 8.22 Railway tunnels (*top*) and bridges (*bottom*) in the Brigittenau district

Table 8.3 Fragility curves used in Vienna test case for the road network

Road bridges	
Fragility curves	*IMT*
SYNER-G Bridge Abutment 6 m Soiltype C	PGA
SYNER-G Bridge Abutment 7.5 m Soiltype C	PGA

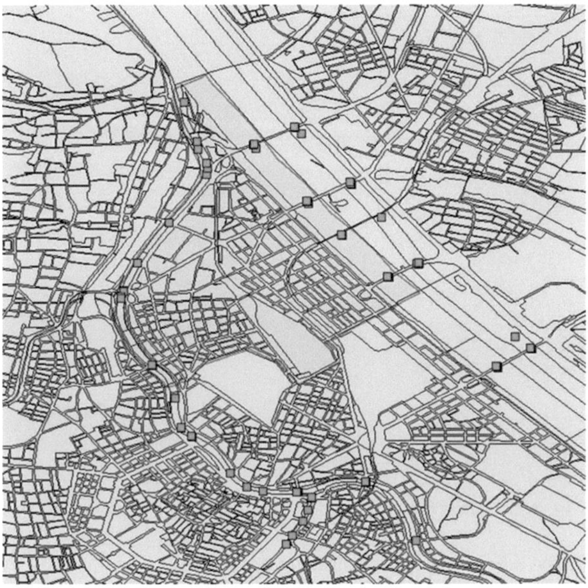

Fig. 8.23 The road bridges in the test area

Table 8.4 Earthquakes created for the simulations

	Neulengbach	Method testing
Magnitude	6	7
Longitude	15.909722	16.543582
Latitude	48.200278	48.0366424

with a moment magnitude of 7. Table 8.4 gives the characteristics of the earthquakes produced for the simulations. Campbell and Bozorgnia (2006) NGA attenuation functions (Campbell and Bozorgnia 2006) are used.

In the framework of the SEISMID FP7 project, VCE has performed detailed studies on the soil characteristics in the Vienna basin. The result of the studies and measurements performed is a very detailed microzonation. In particular, very detailed results are available for the 20th districted where an extensive campaign of measurements was performed. The results are organized in the map in Fig. 8.24: two soil types (soil classes B and C) are identified according to EC 8 (CEN 2004).

8.5.2 *Results*

The 6.0 moment magnitude Neulengbach earthquake and the 7.0 moment magnitude earthquake produce different values of Peak Ground Acceleration and hence different damage scenarios.

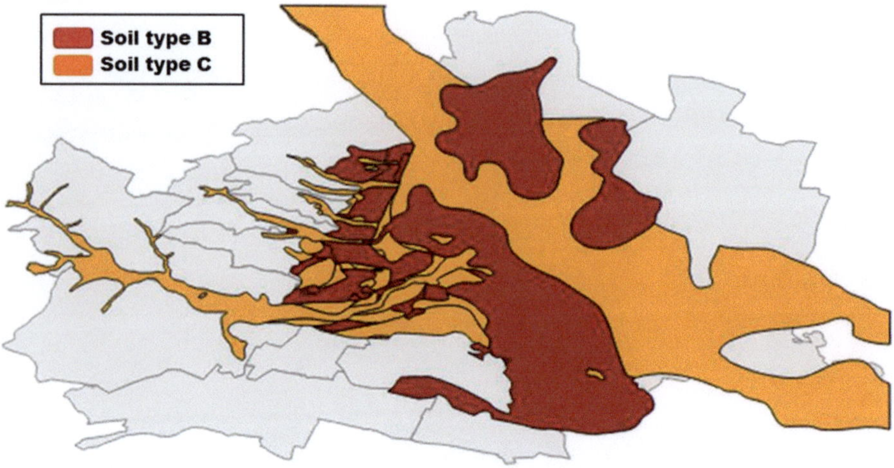

Fig. 8.24 Soil types around the test area

Table 8.5 Damage ratios used in the test case

	Insignificant	Moderate	Heavy	Complete
Buildings	0.005	0.155	0.55	0.9
Roadway	0.005	0.08	0.25	1
Railway	0.005	0.08	0.25	1
Road tunnel	0.005	0.08	0.25	1
Railway tunnel	0.005	0.08	0.25	1

We decided to display the damages scenarios in two different ways: for the Neulengbach earthquake, we considered the probability of reaching the damages state "slight", which is the first damage state in the fragility curves used.

In the case of the "method testing" earthquake the "mean damage" as derived in Eq. 8.1 is considered.

$$meandamage = g_i * i + g_m * m + g_h * h + g_c * c \tag{8.1}$$

where i is the probability of reaching the damage state "slight", m is the probability of reaching the damage state "moderate", h is the probability of reaching the damage state "heavy" and c is the probability of reaching the damage state "complete". The factors before the probabilities are called "damage ratios" and can be specified by the user when ingesting the data. The damage ratios used in this test case are written in Table 8.5.

Fig. 8.25 Building damage for the "method testing" earthquake

8.5.3 Results for the 7.0 Magnitude "Method Testing" Earthquake

EQvis computes and visualises the damage scenario at a very high resolution i.e. at building scale. The user can quickly look at the results for each building and filter them. Each building is characterized by a very detailed description of the contents as described in the previous chapters.

The distribution of damage to the buildings shows that the building stock is very homogenous (Fig. 8.25). There are very few building collapses, some heavily damaged buildings but the majority of the building stock remains in good condition.

The damage to the railway tunnels is very low compared to the building damage (Fig. 8.26). There is only one tunnel with a mean damage of 0.05, a very low level. It was expected that an earthquake of this magnitude and distance will not produce major damages to tunnels in general.

The maximum mean damage to railway bridges is 0.11 which assumes certain relevance. As expected the bridges closer to the epicentre as well as the bridges with poor soil conditions have the highest values of damage.

The damage to road bridges is similar to the damages to the railway bridges (Fig. 8.27). As before, the mean damage increases towards southeast.

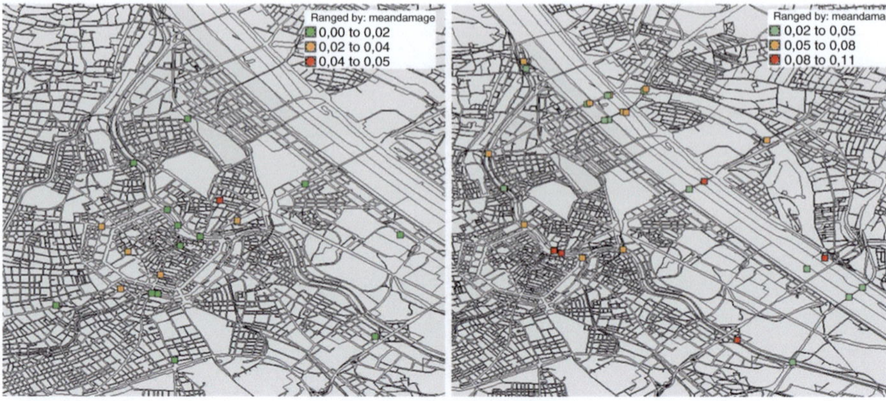

Fig. 8.26 Railway tunnel (*left*) and railway bridge (*right*) damage for the "method testing" earthquake

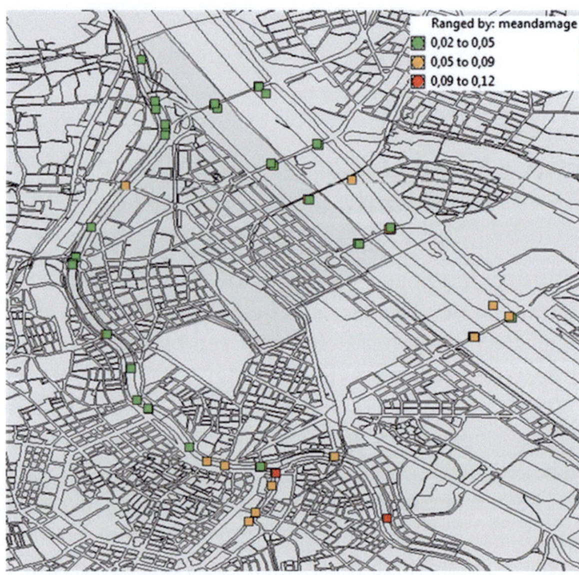

Fig. 8.27 Road bridge (*right*) damage for the "method testing" earthquake

8.5.4 Results for the "Neulengbach" Earthquake

This case reproduces an actual earthquake that occurred in 1590. There are very few articles and data about the consequences of this earthquake, but damages to some building in Vienna were reported in the historical annals.

Fig. 8.28 Building damage for the "Neulengbach" earthquake

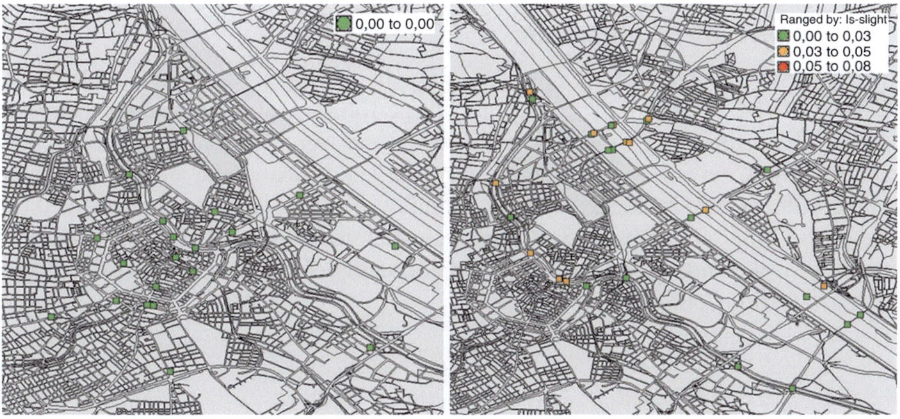

Fig. 8.29 Railway tunnel (*left*) and bridge (*right*) damage for the "Neulengbach" earthquake

The simulation confirms cases of potential failure for some buildings. The maximum mean damage is around 0.14 with moderate damage probabilities up to 0.34 (Fig. 8.28).

The damages to road bridges, railway bridges and tunnels are very low. (Figs. 8.29 and 8.30) All the figures show the probability of reaching the damage state "slight" and the maximum value is 0.18.

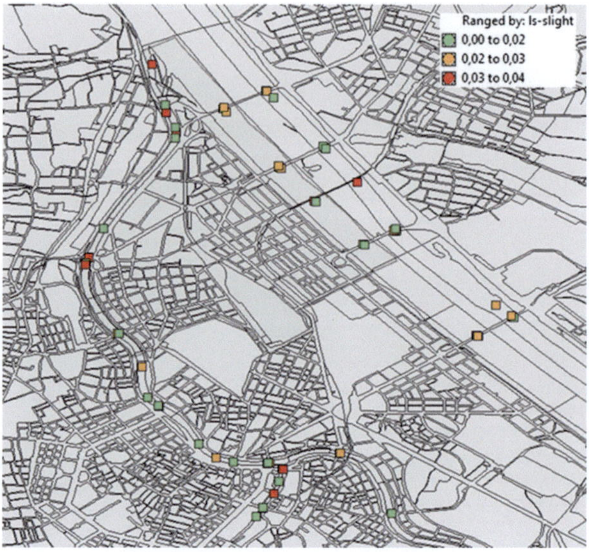

Fig. 8.30 Road bridge damage for the "Neulengbach" earthquake

8.5.5 Probabilistic Analysis with the SYNER-G Prototype Software: Input Data

The SYNER-G prototype software allows performing of probabilistic analysis: hazard characteristics, buildings inventory, water supply system and road and electric power network are the required input. The probabilistic analysis is based on the Monte Carlo method, selecting a minimum value for the covariance of 0.02 and performing 10,000 runs. Each run is characterized by a different location and intensity of the earthquake, producing consequently different scenarios of damage (Duenas-Osorio 2005).

8.5.5.1 Seismic Hazard

For the seismic hazard input, two seismic zones with $M_{min} = 5.5$ and $M_{max} = 7.5$ are selected based on the results of the SHARE European research project (Giardini et al. 2013). Figure 8.31 shows the active seismic zones that could affect the Vienna site.

We used Akkar and Bommer (2010) ground motion prediction equation, choosing the peak ground acceleration as primary Intensity Measure and area fault as source model. Expected values of magnitude can vary in the interval 4.8–6.2, according to the historical seismicity of the zone.

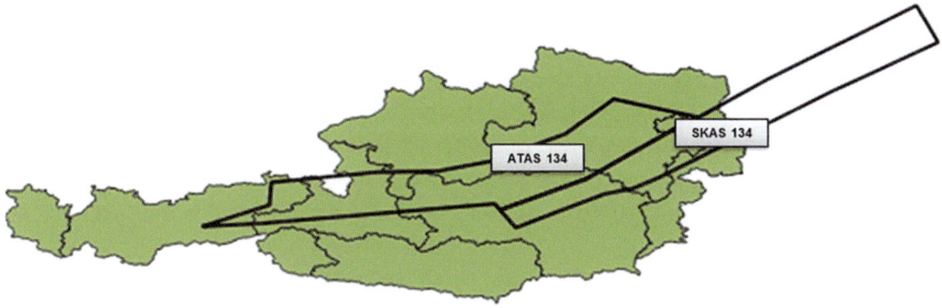

Fig. 8.31 Seismic zones that could affect Vienna site based on the SHARE results (Giardini et al. 2013)

Fig. 8.32 Masonry and reinforced concrete buildings distribution in Brigittenau districts and sub-districts (*red* are the reinforced concrete buildings, *blue* the masonry buildings)

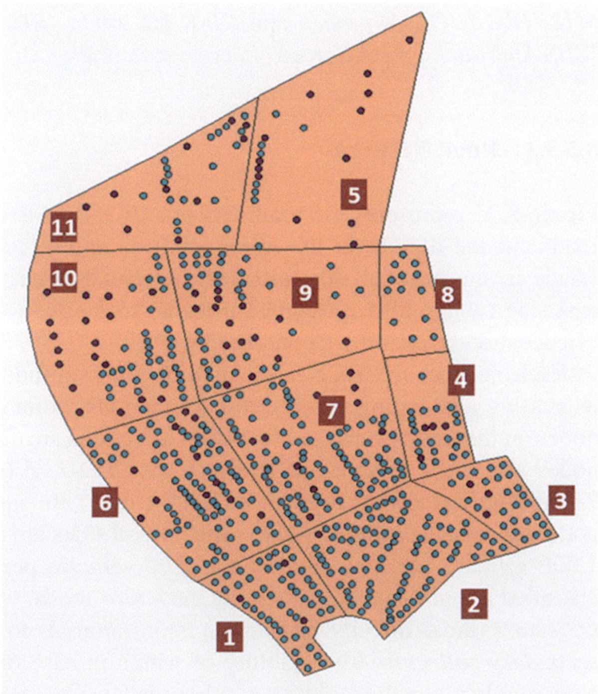

8.5.5.2 Buildings

Brigitttenau district has been divided into two land use zones, one in the north and one in the south. Three main sub-city districts are also identified: for each of them, general information concerning the buildings and their inhabitants (as respectively average building height and employment rate) are given as input.

In addition, 11 census tracts have also been identified (see Fig. 8.32). While in the deterministic analysis performed with EQvis, buildings have been input one by one, each one with its own characteristics, in probabilistic analysis, the buildings have been grouped into zones (census tracts) and for each of those zones, the structural features of the buildings have been statistically classified (Wen et al. 2003).

From the statistical analysis, it was determined that 70 % of the buildings in the district are masonry buildings, the remaining 30 % are reinforced concrete ones. All the sub-districts have a preponderance of masonry buildings; only in the 5th sub-districts the percentage of reinforced concrete buildings is greater than the one of masonry buildings.

Then, for each building typology, the more appropriate fragility curve has been selected (Table 8.3).

The final input gives for each census tract the percentage of buildings associated to the fragility curve selected.

The most used fragility function of masonry building is the *RISK-UE2003 – M12-HR-UNIGE Approach* and *RISK-UE2003 – RC Moment Frame – MR-HC-IZIIS Approach* for reinforced concrete structures.

8.5.5.3 Road Network

Figure 8.33 represents the road network (RDN) in Brigittenau district. Two main roads cut the district in the north-south direction (Jägerstraße and Brigittenauer Lände in the western side, along the Donau Kanal). Wallensteinstraße links the east side (where also a freight harbor is) to the west side of the city through the Friedensbrücke over the Donau Kanal.

Each node of the RDN is defined by its longitude and latitude; each side by its starting and ending nodes. From a functional point of view, starting and ending nodes on the north-south and east-west directions are defined as *external* nodes; the nodes where the main roads intersect are *CBD-TAZ* (*Central Business District – Traffic Analysis Zones*) type nodes; all the other are simple *intersection* nodes. All nodes are considered as *not-vulnerable*. Road sides are divided in *principal* (around 1,000 vehicles per hour) and *minor* (600 vehicles per hour). *Principal* roads are classified as *major arterials*; among the *minor* roads, we distinguished the *primary collectors* (those directly linking the major arterials to the smallest roads) and the *secondary collectors* (the viability of which in case of extensive collapses would not strongly affect the viability of other roads). The majority of road sides have two traffic lanes (*roadsegmentA*); Brigittenauer Lände has four traffic lanes (therefore considered as a *roadsegmentB*). All the sides are considered as *vulnerable*.

For each road it is also given its width, the distance with the adjacent buildings, specifying also if there are buildings on both sides or only on one side. The site characterization is expressed in terms of V_{s30} values (at nodes and sides) and site class.

Neither tunnel, nor bridge is in the part of the district analysed.

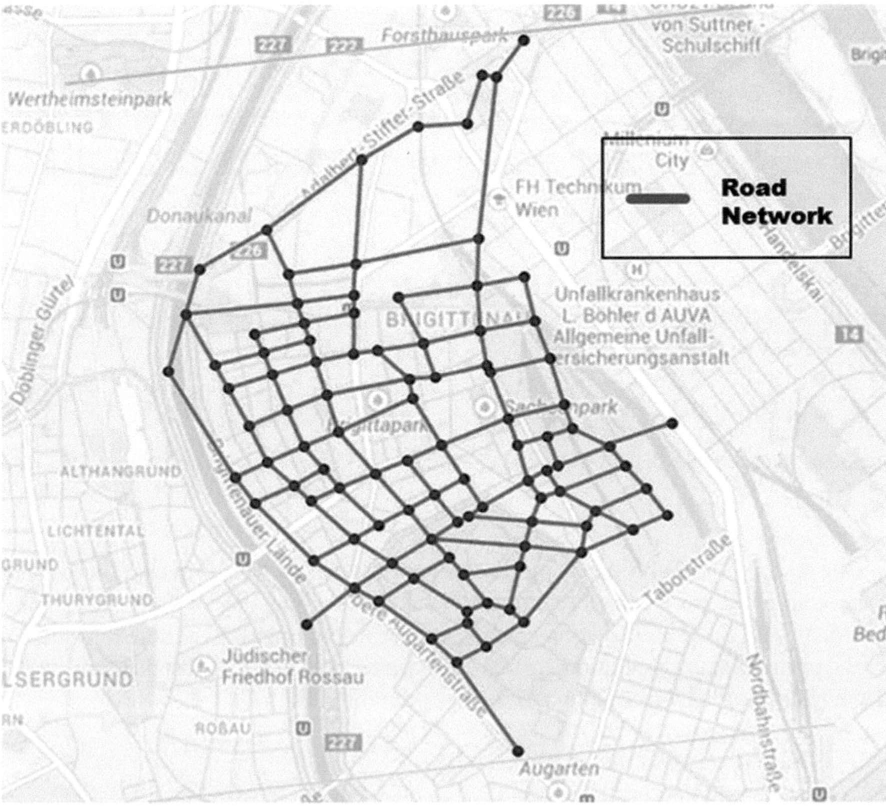

Fig. 8.33 Road network in Brigittenau

8.5.5.4 Water Supply System

Figure 8.34 (left) represents the Water Supply System (in orange) overlaid to the Road Network: the Water Supply System mostly follows the Road Network (with some exceptions). Three external points (one on the north, one on the west and one on the south-east) represent the *constanttank* nodes that supply the water to the entire district. No vulnerability is assigned to the nodes, while all sides are considered vulnerable.

Sides that deliver the water from the supply-nodes have bigger pipes diameter (1,200 mm); the other sides have smaller diameter (600 mm). Only 2 diameter sizes are present. All the pipes are in *castIron* and lay 2 m under the ground level. Also here, the site characterization is expressed in terms of V_{s30} values (at nodes and sides) and site class. The fragility functions of ALA (2001) are used for the vulnerability analysis of pipelines.

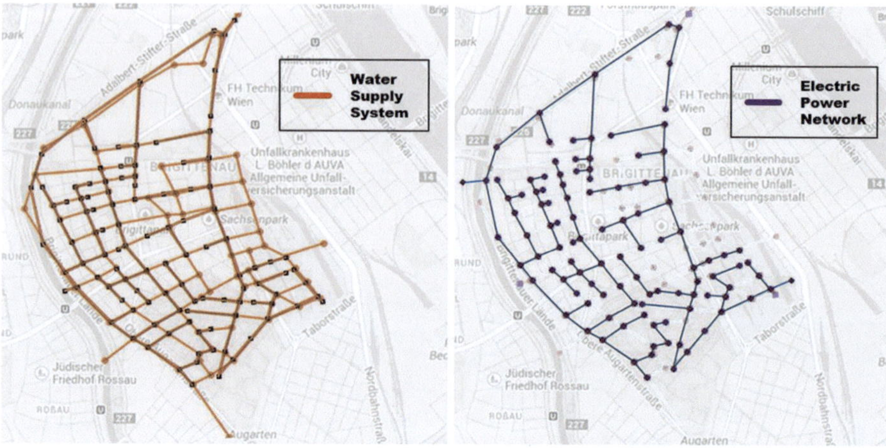

Fig. 8.34 Water supply system (*left*) and electric power network (*right*) overlying to the road network

Due to the configuration of the tested area (an island in a biggest context that is the whole city of Vienna), redundancy and interdependency of the water supply system as well as the electric power network (see next paragraph) as here reproduced for the probabilistic calculation are not fully represented. This affects the correctness of the results, at least from a quantitatively point of view.

8.5.5.5 Electric Power Network

The Electric Power Network follows the layout of the Water Supply System (Fig. 8.34 right). Two *generator* nodes are identified: one on the west side of the district where the thermal waste treatment plant of Spittelau is; the other one on the east side. The network lay underground and has a voltage of 230 kV. Also here, the site characterization is expressed in terms of V_{s30} values (at nodes and sides) and site class.

For the vulnerability analysis of the electric power transmission stations, the fragility curves proposed in the SRM-LIFE (2007) research project are used, which are provided in terms of peak ground acceleration (PGA). The fragility curves for transmission substations are classified in three classes (open, mixed and closed-type).

Two set of 10,000 runs have been performed: the first simulation considers interdependency among electric power network and water supply system, the second instead considers the two systems not dependent from the other (Bompard et al. 2011).

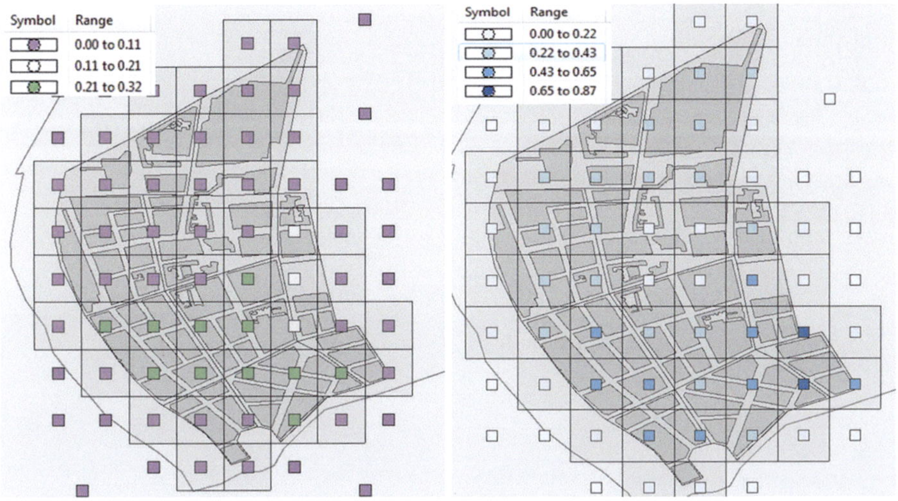

Fig. 8.35 Average building collapse (*left*) and building yielding distribution (*right*): higher values correspond to more extend level of collapse/yielding

8.5.6 *Probabilistic Analysis with the SYNER-G Prototype Software: Output Data*

8.5.6.1 Average Results

In the output of the probabilistic calculation the case study area is subdivided into cells and calculations are performed for each cell. Cell dimension is approximately 100 × 100 m. The results reported below refer to the case which interdependency is considered among the water supply system and the electric power network. In particular in what follows we report the data obtained by averaging the results of each run over the total number of runs. This implies that damage level (for buildings, roads, water supply system, and electric power network) spans in the range 0–1, while deaths and injured average (being obtained as sum of affected persons divided by 10,000) can have a different range. Please note that the range in all of the figures can have a different meaning. It is always explained in the figure captions.

8.5.6.2 Buildings

Figures 8.35 and 8.36 present respectively the damage distribution and the affected persons in the area of interest. Biggest damage level and death/injured persons are mainly concentrated in the south zone of the district where there are almost only masonry buildings.

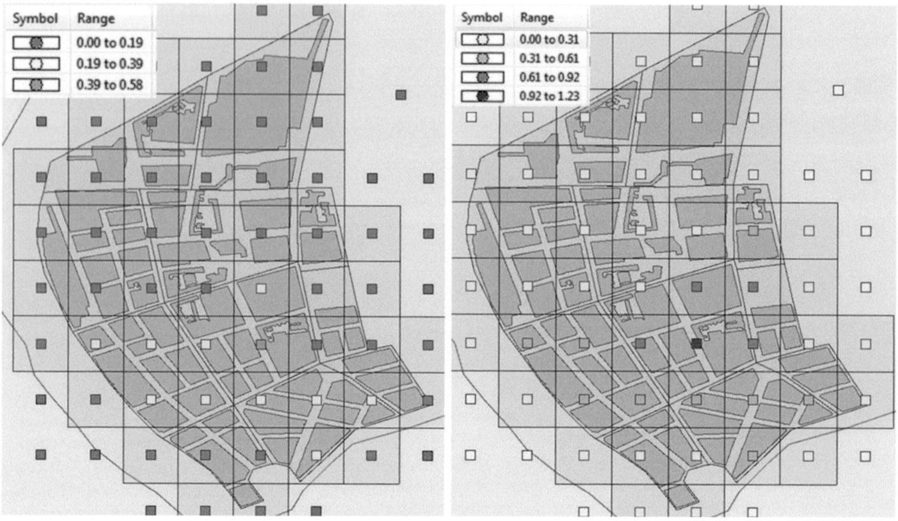

Fig. 8.36 Average death (*left*) and injured (*right*) distribution

Analyzing the mean annual frequency of exceedance and the moving average (Fig. 8.37) one can obtain:

Mean annual frequency of exceedance – deaths – 500 years return period:

$0.7 * 10^{-3} * 35{,}402$ (inhabitants) = 24 (dead persons)

Moving average – deaths – average over all runs:

$1.1 * 10^{-4} * 35{,}402$ (inhabitants) = 4 (dead persons)

The earthquake that corresponds to death toll with 500 years return period determines expected fatalities of 24 while over 10,000 runs average death persons tends to the value of 4.

Also, with reference to Fig. 8.38 one can obtain:

Mean annual frequency of exceedance – injured persons – 500 years return period:

$1.9 * 10^{-3} * 35{,}402$ (inhabitants) = 67 (injured persons)

Moving average – injured persons – average over all runs:

$3 * 10^{-4} * 35{,}402$ (inhabitants) = 11 (injured persons)

8.5.6.3 Roads

Analysing the roads damage, we obtain that blocked roads are mainly concentrated in the proximity of collapsed buildings (Fig. 8.39). In particular, this applies for small roads where the debris of collapsed buildings can partially or totally block the access of the adjacent roads.

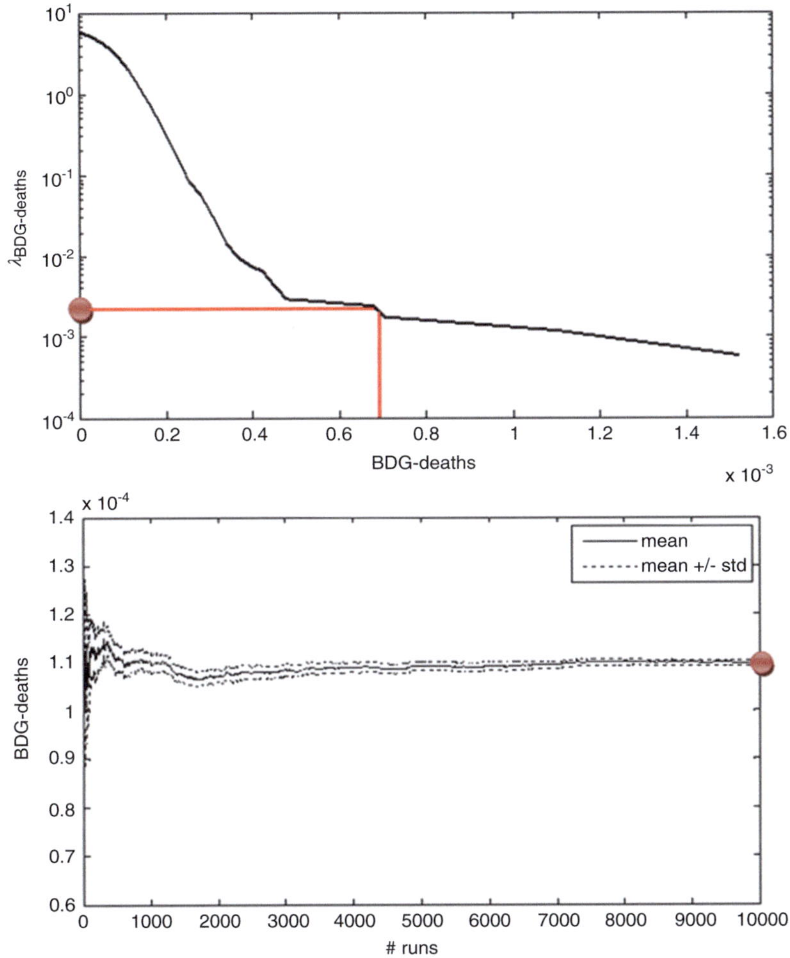

Fig. 8.37 Mean annual frequency of exceedance and moving average (death persons)

8.5.6.4 Water Supply System

Pipes and nodes of the water supply system results to be slightly affected from the earthquake and average level of damage is negligible (Fig. 8.40). The obtained result is in agreement with what observed after real earthquakes: usually only at very high magnitude the water network has registered a significant level of damage.

8.5.6.5 Electric Power Network

Also the electric power network results to be slightly damaged as shown in Fig. 8.41. There is almost no damage to the electric power network.

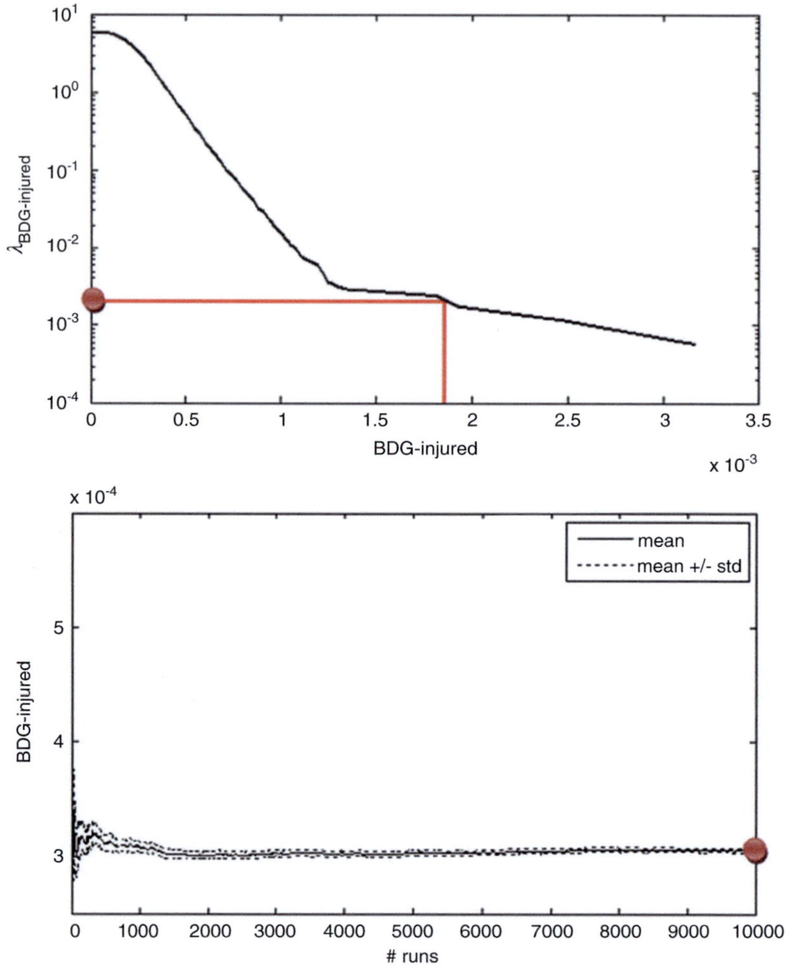

Fig. 8.38 Mean annual frequency of exceedance and moving average (injured persons)

8.5.7 *Selected Scenario*

Among the 10,000 runs, a particular scenario has been selected. It presents a 5.4 magnitude earthquake located in the south-east of Vienna, at a distance of approximately 50 km from Brigittenau district (Fig. 8.42). The selected scenario is considered meaningful since it is in the proximity of the tectonic zone of the Austrian region more prone to seismicity.

This scenario produces a PGA distribution that can reach values of 0.4 m/s^2. For the sake of simplicity, those values refer to hard rock; soil typology in each point of the calculation should be considered in order to obtain the real soil acceleration.

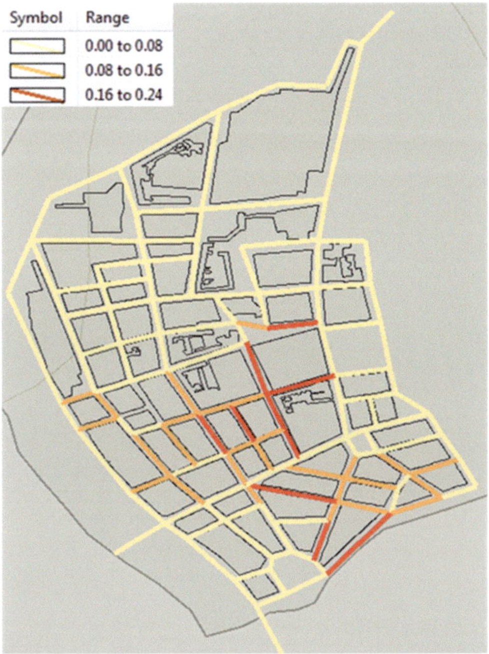

Fig. 8.39 Average blocked roads. *Darkest lines* represent lower level of usability

Fig. 8.40 Pipes broken (*left*) and non-functional nodes (*right*). The damage is completely negligible

Fig. 8.41 Average damage on the electric power network nodes

Fig. 8.42 M = 5.4 earthquake 50 km far from Brigittenau district, south-east of Vienna

8.5.7.1 Buildings

Figures 8.43 and 8.44 present respectively the distribution of collapsed and yield buildings, death and injured persons and displaced persons in case of good and bad weather conditions.

Comparing damage level and casualties, we obtain a higher number of deaths in correspondence to the collapsed buildings as it could be expected. Major damage is registered, as in the averaged results, in the south of the district where mainly masonry buildings are.

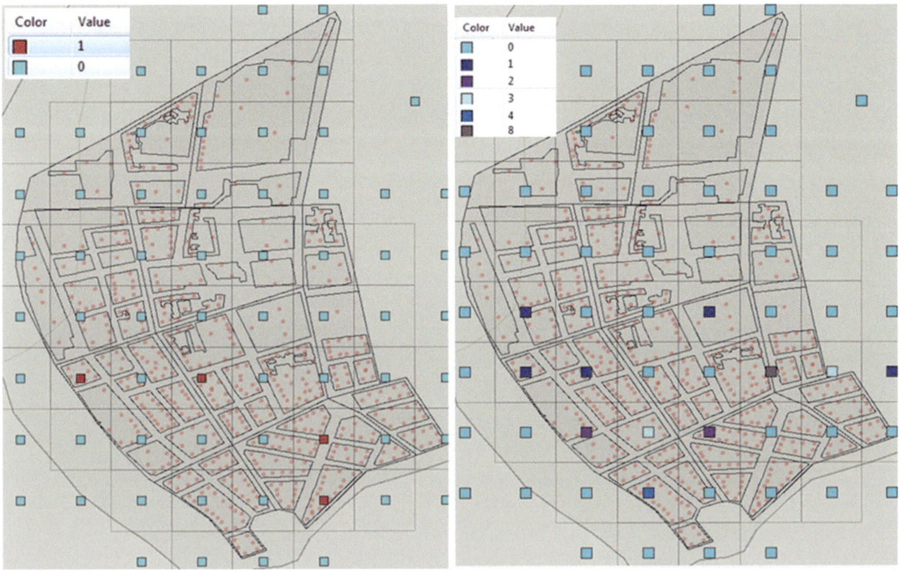

Fig. 8.43 Number of buildings collapsed (*left*) and yield (*right*) for the selected event of M = 5.4

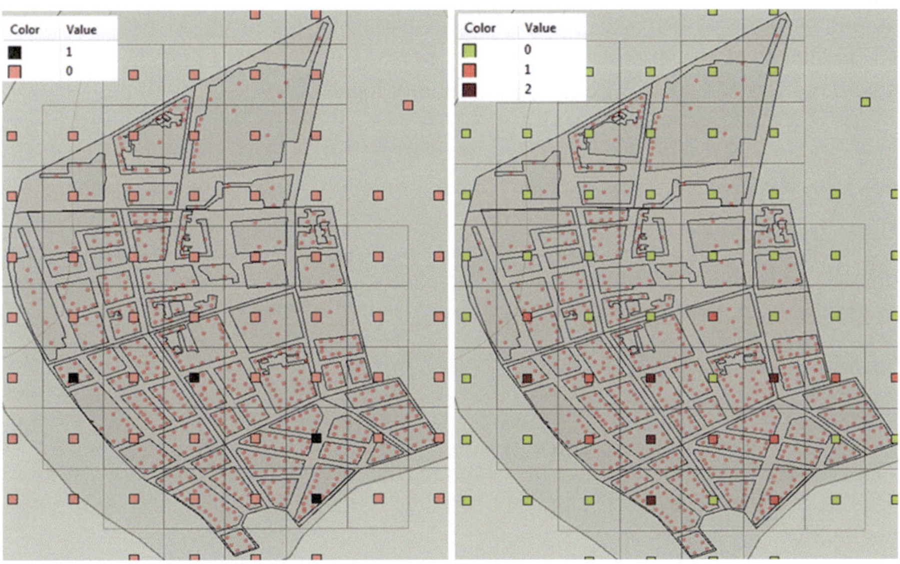

Fig. 8.44 Number of deaths (*left*) and injured (*right*) persons for the selected event of M = 5.4

Fig. 8.45 Number of displaced persons in case of bad weather (*left*) and good weather (*right*)

Figure 8.45 shows the distribution of displaced persons: the main difference among the case of bad weather and good weather is that in the first case there is an increment of the number of displaced persons in the north part of the district where reinforced concrete buildings are mainly located.

8.5.7.2 Roads

Figure 8.46 presents the damage distribution on the road network of Brigittenau district. As in Fig. 8.39, blocked roads are mainly located in the south of the district, in proximity to more vulnerable structures.

8.5.7.3 Water Supply System

The selected scenario does not produce any damage to the water supply system. This is expected considering that the average damage level obtained before was negligible.

8.5.7.4 Electric Power Network

Finally, Fig. 8.47 presents the damage level that affects the electric power network.

Table 8.6 reports the summary of damage caused by the selected event.

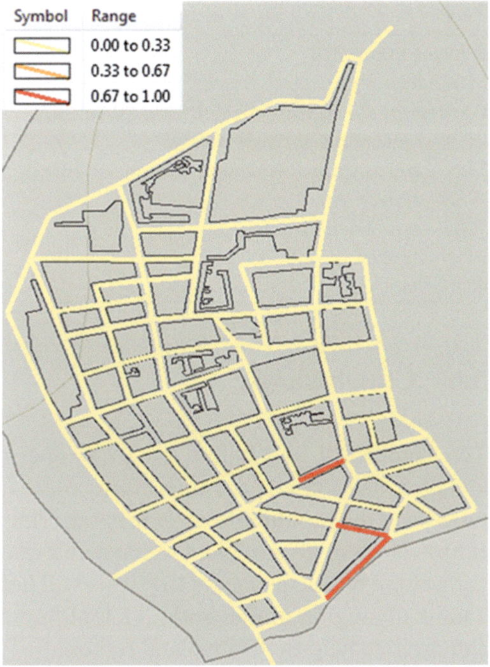

Fig. 8.46 Blocked roads left for the selected scenario

Fig. 8.47 Damage level on the electric power network for the selected scenario

Table 8.6 Data from the selected event

Selected event: 278			
M = 5.39			
Hypocenter: 17.0071, 48.0789 Depth: 10 km			
EPN – broken transmission stations	0	BDG – deaths	4
EPN – non functional demands	10	BDG – injuries	19
WSS – broken pipe	0	BDG – collapse	4
WSS – non functional demands	0	BDG – yield	27
RDN – broken	0	BDG – displaced (GW)	1,400
RDN – blocked	11	BDG – displaced (BW)	2,411

8.6 Conclusions

The Vienna test case within the SYNER-G project has brought a proof of concept and provided a number of interesting lessons. It has been proven that an assessment process at the building level is feasible with the methods developed in SYNER-G. It turned out to be sufficient to have untrained personnel performing the large data gathering exercise using the BIP. This will help implementation on a very large scale and in any region of the world. IT tools are available to support the data collection and provide help to the involved forces.

The application of the methodology to a limited area with very detailed information provides a challenge for both the software application as well as the data collection: considering the limited area means missing redundancy in particular for what concerns the network system; very detailed scale implies to collect and to handle a large amount of data and information.

Both IT tools, namely the prototype software developed in SYNER-G for the probabilistic analysis and the EQvis platform, have been applied. The probabilistic analysis accounts the systemic interdependencies whereas EQvis is able to allow a user-friendly in and output of results. Visualization plays a major role when stakeholders and officials engaged in civil protection enter the procedures. The EQvis software platform with the integrated probabilistic SYNER-G software is available as an open-source product for free download at the homepage (www.syner-g.eu).

Different earthquake scenarios have been simulated. A plausibility check on the results obtained has been performed and it has been stated that they match with the expectations of the expert community. The results in terms of buildings are rated excellent whereas the results on the utility networks are limited because of the small area involved that does not allow to fully accounting for the interdependencies and the redundancies.

The two different approaches of the software tools, namely the deterministic and the probabilistic approach, can be used together in order to help decision makers make decisions. A general overview and a full consideration of system of systems of the situation can be given by the probabilistic approach while the deterministic approach can be used as a concrete scenario and eventually as a management platform during a crisis situation.

It will be a challenge to enlarge the dataset to the entire city (180,000 buildings instead of 700 in the test area). It will bring new challenges in terms of computing power and number of interrelations to be handled. Furthermore, from the experience gathered during the SYNER-G proof of concept, it is recommended to produce an online data generation sheet to allow filling the database with the necessary information. In order to perform this exercise it will be necessary to establish a large IT project that enlarges the current boundaries of application.

Acknowledgments The authors would like to acknowledge all colleagues at VCE, especially Reinhard Scherer and Thomas Gruber, as well as all colleagues from the MAEviz team for their support.

References

Achs G et al (2010) Erdbeben im Wiener Becken. VCE, Wien

Akkar S, Bommer JJ (2010) Empirical equations for the prediction of PGA, PGV and spectral accelerations in Europe, the Mediterranean Region, and the Middle East. Seismol Res Lett 81(2):195–206

ALA (2001) Seismic fragility formulations for water systems. American Lifeline Alliance, ASCE

Bompard E, Wu D, Xue F (2011) Structural vulnerability of power systems: a topological approach. Electr Power Syst Res 81(7):1334–1340

Borzi B, Pinho R, Crowley H (2007) SP-BELA: un metodo meccanico per la definizione della vulnerabilità basato su analisi pushover semplificate, ANIDIS, Pisa (in italian)

Borzi B, Crowley H, Pinho R (2008) The influence of infill panels on vulnerability curves for RC buildings. In: Proceedings of the 14th world conference on earthquake engineering, Beijing, China

Campbell K, Bozorgnia Y (2006) Next generation attenuation (NGA) empirical ground motion models: can they be used in Europe? In: Proceedings of the first European conference on earthquake engineering and seismology, Geneva, 3–8 Sept 2006.

CEN (2004) Eurocode 8: design of structures for earthquake resistance

Clayberg E, Rubel D (2008) Eclipse plugins, 3rd edn. Addison Wesley, Upper Saddle River

Duenas-Osorio LA (2005) Interdependent response of networked systems to natural hazards and intentional disruptions. Dissertation, Georgia Institute of Technology

Erberik MA (2008) Fragility-based assessment of typical mid-rise and low-rise RC buildings in Turkey. Eng Struct 30(5):1360–1374

Erberik MA, Elnashai AS (2004) Vulnerability analysis of flat slab structures. In: 13th world conference on earthquake engineering, Vancouver

FEMA (2002) Rapid visual screening of buildings for potential seismic hazards, FEMA154, 2nd edn. Applied Technology Council, Redwood City

Flesch R (1993) Baudynamik band 1. Bauverlag, Wien

Genctürk B (2007) Improved fragility relationships for populations of buildings based on inelastic response. MS thesis, University of Illinois at Urbana-Champaign

Genctürk B, Elnashai AS, Song J (2008) Improved fragility relationships for populations of buildings based on inelastic response. In: Proceedings of the 14th world conference on earthquake engineering, Beijing, China, 12–17 Oct 2008

Giardini D, Woessner J, Danciu L, Crowley H, Cotton F, Grünthal G, Pinho R, Valensise G, Akkar S, Arvidsson R, Basili R, Cameelbeeck T, Campos-Costa A, Douglas J, Demircioglu MB, Erdik M, Fonseca J, Glavatovic B, Lindholm C, Makropoulos K, Meletti C, Musson R, Pitilakis K,

Sesetyan K, Stromeyer D, Stucchi M, Rovida A (2013) Seismic Hazard Harmonization in Europe (SHARE). Online data resource, http://portal.share-eu.org:8080/jetspeed/portal/. doi:10.12686/SED-00000001-SHARE

Kappos AJ, Nuti C, Sucuoglu H (2003) Seismic assessment and retrofit of RC buildings: case studies. fib Bulletin No. 24, May 2003, pp 251–306

LESSLOSS (2005) LESSLOSS deliverable report D84 – report on building stock data and vulnerability data for each case study

MAE (2013) MAEviz, developed by the Mid-America Earthquake Center and the National Centre for Supercomputing Applications. http://mharp.ncsa.illinois.edu/

McAffer J, Lemieux JM, Aniszczyk C (2010) Eclipse rich client platform, 2nd edn. Addison Wesley, Upper Saddle River

Mid-America Earthquake Center (2009) Impact of new Madrid seismic zone earthquakes on the Central USA, MAE Center report no. 09-03

Pitilakis K, Crowley H, Kaynia A (eds) (2014) SYNER-G: typology definition and fragility functions for physical elements at seismic risk, vol 27, Geotechnical, geological and earthquake engineering. Springer, Dordrecht. ISBN 978-94-007-7871-9

RISK-UE (2003) An Advanced approach to earthquake risk scenarios with applications to different European towns. WP4: vulnerability of current buildings risk-UE 2003, Greece

Schäfer D, Pietsch M, Wenzel H (2013) EQvis: a consequence based risk management software tool. In: Proceedings of the 5th international conference on structural engineering, mechanics and computation, University of Cape Town, 2–4 Sept 2013

SRMLIFE (2007) Development of a global methodology for the vulnerability assessment and risk management of lifelines, infrastructures and critical facilities. Application to the metropolitan area of Thessaloniki. Research project, General Secretariat for Research and Technology, Greece

Vargas YF, Pujades LB, Barbat AH (2010) Probabilistic assessment of the global damage in reinforced concrete structures, 14ECEE, Ohrid

Wen YK, Ellingwood BE, Bracci J (2003) Vulnerability function derivation for consequence-based engineering. Mid-America Earthquake Center report

Chapter 9
Application to L'Aquila Gas Network

Simona Esposito and Iunio Iervolino

Abstract This chapter, after an introduction presenting the general framework for the seismic risk assessment of a gas network according to the SYNER-G methodology (Chap. 2), describes the case study of L'Aquila (central Italy) gas distribution system, a 621 km pipeline network managed by Enel Rete Gas s.p.a. and operating at medium- and low-pressure. Subsequently, the main features regarding the implementation of the application study within the SYNER-G framework are reported, and the process for the seismic performance characterization is summarized. Then, the risk analysis of the system is described, and results in terms of connectivity-based performance indicators are presented.

9.1 Introduction

This study presents the probabilistic seismic risk assessment of a gas distribution system. A gas distribution system comprises two main categories of components: (i) a number of point-like facilities (reduction stations and groups where gas is reduced or simply measured); (ii) pipelines constituting the distribution network.

The causes of earthquake damage to components of gas distribution systems include large permanent soil deformations produced by fault displacements, landslides and liquefaction (O'Rourke and Liu 1999), as well as ground shaking associated with traveling seismic waves. The latter case is often referred to as transient ground deformation (TGD), while the former is generally referred to as permanent ground deformation (PGD); i.e., the *geotechnical hazard*.

S. Esposito (✉) • I. Iervolino
Dipartimento di Strutture per l'Ingegneria e l'Architettura, Università degli Studi di Napoli Federico II, Via Claudio 21, 80125 Naples, Italy
e-mail: simona.esposito@unina.it; iunio.iervolino@unina.it

K. Pitilakis et al. (eds.), *SYNER-G: Systemic Seismic Vulnerability and Risk Assessment of Complex Urban, Utility, Lifeline Systems and Critical Facilities*, Geotechnical, Geological and Earthquake Engineering 31, DOI 10.1007/978-94-017-8835-9_9,

Ground shaking usually affects wide geographical areas and can produce well-dispersed damage. Damage induced by permanent ground deformation typically occurs in isolated and localized areas and results in high damage and consequent repair rates, varying in relation to the amount, geometry, and spatial extent of the zone subjected to PGD.

This application study was aimed at evaluating, in a complete performance-based earthquake engineering framework, the seismic risk of a gas distribution network. The work includes the probabilistic characterization of seismic input, the evaluation of the vulnerability of network's components, the analysis of the system's performance measures, and finally the probabilistic simulation for risk assessment.

As an illustrative, yet real, application, the mid-pressure part of the L'Aquila (central Italy) gas distribution system is considered. In particular, the selected network is characterized by three reduction stations connecting the network to the high-pressure nationwide network, more than 200 km of pipelines, either made of steel or high density polyethylene (HDPE) pipes, and about 200 reduction groups.

In fact, detailed information about the system was available for this network, including performance in the 2009 M_w 6.3 earthquake (Esposito et al. 2013a), due to a dedicated partnership with the network's operator (*Enel Rete Gas s.p.a.*). For the purposes of seismic risk assessment, a single earthquake source is considered, the Paganica fault (Pace et al. 2006); it is beneath the region served by the network and it is believed to have generated the 2009 L'Aquila earthquake.

Seismic performance is expressed in terms of probability of exceedance of service disruption levels, measured by connectivity-based indicators, given the occurrence of an earthquake on the considered source.

The chapter is structured such that Sect. 9.2 presents the general framework for the seismic risk assessment of a gas network according to the SYNER-G methodology. Section 9.3 describes the case study and provides details on the implementation in the prototype SYNER-G software. Subsequently, the analysis of the system is carried out at the connectivity level, and results in terms of performance indicators are, finally, presented in Sect. 9.4.

9.2 Methodology

In this section the SYNER-G framework for the risk assessment of spatially distributed systems is recalled focusing on gas distribution system and its components (namely pipelines and stations). In particular, a concise summary of: the methodology with respect to seismic hazard analysis (Chap. 3), fragility functions of the vulnerable elements (Gehl et al. 2014), and performance analysis (Chap. 5), is provided in order to facilitate the understanding of the application study.

9.2.1 Seismic Hazard Analysis

The analysis of seismic hazard (in terms of both TGD and PGD) for spatially distributed systems, presents different challenges with respect to those addressed by consolidated tools adopted for hazard characterization of point-like structures.

First, because this kind of systems is extended in space, a key difference is that the seismic hazard has to be evaluated jointly for all the locations of the system components. Large vectors of ground motion intensity measures (*IMs*), usually expressed in terms of peak parameters (e.g., peak ground acceleration, PGA, or peak ground velocity, PGV) should be estimated through ground motion prediction equations (GMPEs), but also accounting for spatial correlation. In fact, if probabilistic assessment of ground motion intensity at two or more sites, at the same time, is of concern, then a spatial correlation model accounting for the statistical dependencies between intra-event residuals of *IMs* as a function, for example, of inter-site separation distance, is needed. This serves to model the joint probability density function of *IMs* at all locations. This means that seismic hazard has to be represented in terms of random fields (e.g., Esposito and Iervolino 2011, 2012).

Furthermore, the performance of spatially distributed systems may be conditional upon the failure of many components each of which is sensitive to different *IMs*. In particular, some elements of a gas system, such as reduction stations, have fragility curves typically expressed in terms of PGA, while fragility of pipelines may be a function of PGV. Each *IM* is spatially correlated, but the seismic input assessment has to take into account the possibility of the existence of a cross-correlation between *IMs*, in order to model the joint distribution of different random fields. To address this issue, the *conditional hazard* approach may be considered (Iervolino et al. 2010).[1] It consists of obtaining the conditional distribution of a secondary intensity measure, IM_2, at a site (e.g., PGV), given the occurrence of a primary intensity measure, IM_1, at the same site (e.g., PGA) for which a spatial correlation model is available (see Chap. 3 for more details).

Finally, geotechnical hazards associated with earthquakes (i.e., liquefaction, landslide and co-seismic rupture) may have a significant impact on lifelines, particularly those with buried elements. An assessment of seismic risk of spatially distributed elements must, therefore, incorporate PGD hazard in a manner that is also consistent with the TGD hazard. There are many models available that have the intent to relate PGD, and the probability of occurrence of each geotechnical hazard, to the intensity of ground motion (typically expressed in terms of PGA). In this context, the need of practical implementation of PGD for large areas is addressed following the HAZUS (FEMA 2004) approach that requires limited information about the geotechnical characterization of the region. In fact, the HAZUS approach has been slightly refined within SYNER-G, considering recent semi-empirical

[1]The simplest application of conditional hazard requires site-specific correlation coefficients between *IMs* and acceptability of the hypothesis of joint normality of the logarithms of the primary and secondary *IMs*.

models for the computation of the permanent ground deformation, and including the probabilistic distribution of model residuals (again, details are given in Chap. 3).

9.2.2 *Fragility Functions*

To estimate earthquake damage for a natural gas distribution system, given ground shaking (or ground failure), earthquake intensity parameters have to be related to the seismic impact on the system's components via fragility functions. In particular, for point-like components (i.e., stations) these relations typically provide the probability of reaching or exceeding some *damage state* given the intensity. This applies to the aboveground components of a gas distribution network, while for pipelines the fragility models usually consist of a seismic-intensity-dependent rate, providing the number of damages (e.g., *leaks* or *breaks*) per unit length.

9.2.2.1 Buried Pipelines

In the case of pipeline components, fragility curves available in literature are usually based on empirical data collected in past earthquakes. Empirical fragility functions are mostly based on the recorded number of repairs collected from field crews (e.g., ALA 2001). As a result, all fragility relations for pipelines are given in terms of the repair rate, R_R, per unit length of pipe. Further factors affecting the vulnerability of pipelines, usually accounted for in the formulation of the repair rate, are material, diameter, or connection type. Then, using a Poisson probability distribution and R_R as its parameter, one can assess the probability of having any number of damages in a pipe segment, given the local intensity.

As mentioned above, buried pipelines are sensitive to permanent ground deformation, in addition to TGD due to seismic wave propagation. Among the various seismic parameters used to correlate the ground shaking effects to the damage suffered by buried pipelines, PGV is often preferred (O'Rourke et al. 1998). Regarding ground failure effects, permanent ground deformation (PGD), a measure of displacement, is used as the demand descriptor.

According to HAZUS, two damage states may be considered for pipelines: leaks and breaks, with the type of damage depending on the type of hazard. In particular, when a pipe is damaged due to ground failure, it is assumed that the proportions of leaks and breaks are 0.2 and 0.8, respectively; whereas for ground shaking, leaks and breaks relative proportions are 0.8 and 0.2, respectively.

9.2.2.2 Stations

In a gas distribution system, three different types of stations may exists: (i) metering/pressure reduction stations (M/R stations) that contain metering equipments for monitoring and measuring the gas flow, and reduction lines for the reduction of the gas pressure (from high to medium) before its distribution into the pipe system; (ii) reduction groups, where the gas pressure is reduced (from medium to low) as required for the gas to arrive to the end-user; and (iii) metering stations that are only flow measurement points.

Considering that stations comprise the shelter and the equipment inside, they may be classified with respect to different features such as building typology and/or presence of anchored or unanchored subcomponents. Although in literature no fragility curves are available for, at least, (i), some authors (e.g., Chang and Song 2007; Song and Ok 2009) assume that they can be characterized by the same fragility functions of compressor stations. Damage states and fragility curves for compressor stations are usually defined as a function of PGA (FEMA 2004).

9.2.3 Vulnerability of the System and Performance Indicators

9.2.3.1 Gas Network Model and Analysis

As mentioned in Chap. 5, the seismic performance of a gas network (and of lifeline networks in a general sense) may be measured according to two categories of indices: (1) those based on connectivity analysis that allows assessment of serviceability in terms of the aggregate functionality of facilities (nodes) composing the system; e.g., the number of distribution nodes which remain accessible from at least one supply node after the earthquake; (2) those based on capacitive analysis that include consideration of flow equations used to compute flows from sources to sinks (i.e., distribution nodes), based on the damages sustained by the network components.

Connectivity analysis requires a simple description of the network, defined as a collection of nodes (i.e., stations) and lines (i.e., pipes) joining all or some of these points. Connectivity analysis tools are limited to those of graph theory (e.g., Ching and Hsu 2007). These algorithms are applied on the network after removing the parts of the system that failed in the seismic event.

In flow-based analysis, the network's performance is measured by evaluating the actual flow delivered to end users, as a function of pressure at demand nodes. For the purpose of calculating pipe flow and nodal pressure, before and after the seismic event, it is necessary to consider flow equations (the application of flow equations is required for the calculation of the pressure drop along the network) and a method to solve the network analysis problem (see Osiadacz 1987, for a discussion).

9.2.3.2 Performance Indicators

Depending on the goal of the analysis (connectivity- or flow-based) different performance indicators (PIs) may be evaluated. PIs are able to quantify the degree to which the system is able to meet established specifications and/or customer requirements following an earthquake event. For a gas distribution network two possible system-level PIs that may be used for a connectivity analysis are: the Serviceability Ratio (SR) and the Connectivity Loss (CL).

The first index, originally defined by Adachi and Ellingwood (2008) for water supply systems, is directly related to the number of distribution nodes in the utility network, which remain accessible from at least one supply facility following the earthquake. It is computed as in Eq. 9.1, where w_i is a weighting factor assigned to the distribution node *i* (i.e., customers related to the demand node or nominal flow of the distribution node), X_i represents the functionality of *i-th* demand node, which is modeled as the outcome of a Bernoulli trial ($X_i = 1$ if the facility is accessible from at least one supply facility and zero otherwise), and *n* is the number of distribution nodes.

$$SR = \sum_{i=1}^{n} (w_i \cdot X_i) \Big/ \sum_{i=1}^{n} w_i \tag{9.1}$$

The second index, originally defined by Poljanšek et al. (2012), was adapted for the purpose of this study and it is expressed in Eq. 9.2. In particular, it is related to the number of the demand nodes connected to the *i*-th source in the original (undamaged) network and then in the damaged network, $N^i{}_{demand,dam}$. In Eq. 9.2, $\langle\rangle$ denotes averaging over all source nodes.

$$CL = 1 - \left\langle \frac{N^i_{demand,dam}}{N^i_{demand,orig}} \right\rangle_i \tag{9.2}$$

For a more extended discussion on other possible PIs see Chap. 5 and Esposito (2011).

9.3 Application to L'Aquila Gas Network

9.3.1 The Case Study

In the L'Aquila region (central Italy) the gas is distributed via a 621 km pipeline network, 234 km of which with gas flowing at medium pressure (2.5–3 bar), and the remaining 387 km with gas flowing at low-pressure (LP) (0.025–0.035 bar).

The medium-pressure (MP) distribution network is connected to the high-pressure transmission network through three M/R stations [referred to as Re.Mi stations ("*stazioni di Regolazione e Misura*" in Italian)]. The three M/R stations are cased in one-story reinforced concrete structures with steel roofs (Fig. 9.1) hosting internal regulators and mechanical equipment (heat exchangers, boilers and bowls), where the gas undergoes the following processes: (1) gas pre-heating; (2)

a b

Fig. 9.1 M/R reduction stations in Onna (L'Aquila, Italy): (**a**) external view; (**b**) internal view (Esposito et al. 2013a)

Fig. 9.2 One of the 300 RGs housed in a metallic kiosk (Esposito et al. 2013a)

gas pressure reduction; (3) gas odorizing; (4) gas pressure measurement. Pipelines of medium and low-pressure distribution network are either made of steel or HDPE according to the pressure level.

The transformation of the MP into the LP is operated via 300 Reduction Groups (RGs) that are buried, sheltered in a metallic kiosk or housed within/close to a building (Fig. 9.2).

Several demand nodes (referred to as IDU, *Impianto di Derivazione Utenza* in Italian), consisting of buried and above-ground pipes and accessory elements, allow the supply of natural gas to utilities from LP network. For users such as industrial facilities, the demand node IDU is located along the MP network.

Close collaboration with the network operator has allowed the characterization of the system, necessary for the evaluation of gas system seismic performance. A geographic information system (database) was jointly developed and contains data on system physical and operational characteristics.

For the evaluation of seismic performance within this study, the medium-pressure portion of the L'Aquila gas system was selected. In particular, the selected part (shown in Fig. 9.3) is characterized by 3 M/R stations, 209 RGs, and pipelines either made of steel or HDPE.

9.3.2 Implementation of the Gas Distribution System Within the SYNER-G Framework

This section provides additional details regarding the software implementation of the case study with respect to the general description reported in Chap. 5 for gas systems. For the purposes of the application study, the SYNER-G computational framework (i.e., the SYNER-G prototype software) was equipped with the *GAS* class. The gas distribution system is modeled as an undirected graph, hence, the *GAS* class is considered a subclass of the *Undirected* abstract class. As shown by the class diagram in Fig. 9.4, the network is composed of nodes and link/edges. As a consequence, the *GAS* class is the composition of the *GASedge* and *GASnode* abstract classes, the first of which is the generalization of the *PipeGAS* class, while the second is the generalization of the *GASdemand*, *GASsource* and *Joint* classes. The *Joint* class represents all nodes used to reproduce the geometry of the system, the *GASsource* class represents M/R stations that are used to connect the distribution mid-pressure network to the high-pressure transmission lines, and the *GASdemand* is the generalization of *IDU* class and *RG* class. The *IDU* class represents the nodes directly connected with customers in the low-pressure network, while the *RG* class represents reduction groups that are considered as final nodes when only the mid-pressure network is analyzed. The definition of these edge typologies, together with associated fragility models are given in Gehl et al. (2014).

All the attributes and methods of the *GAS* class and its subclasses (listed in Esposito and Iervolino 2013) were defined in order to evaluate the state of the network and of each component of the gas system. Attributes refer to properties that describe the whole system and each component. They can be listed as:

1. *List of pointers (e.g., all the connections between objects);*
2. *Gas network global properties (e.g., the number of nodes, or edges);*

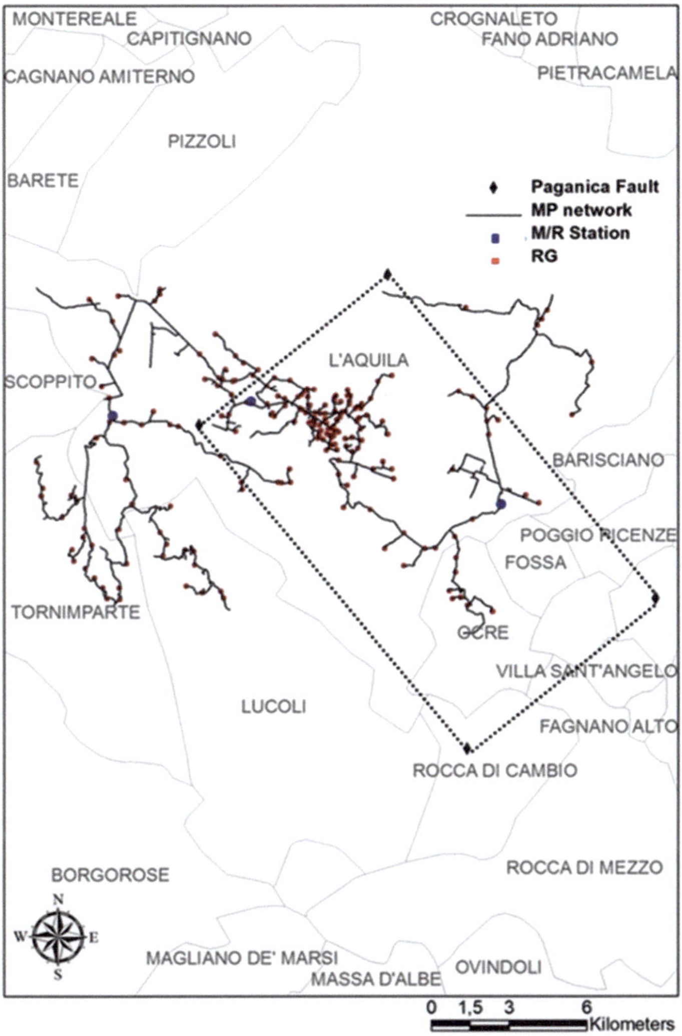

Fig. 9.3 Application network (L'Aquila, central Italy)

3. *Edge and node properties (e.g., the length of the edge or the position of a node, or the corresponding landslide susceptibility);*
4. *Properties that record the state of the GAS for each event;*
5. *Properties that store the global performance of the GAS at the end of simulation.*

Methods refer to functions used to evaluate the state of the network or of each component of the system. For example, methods include functions to evaluate the

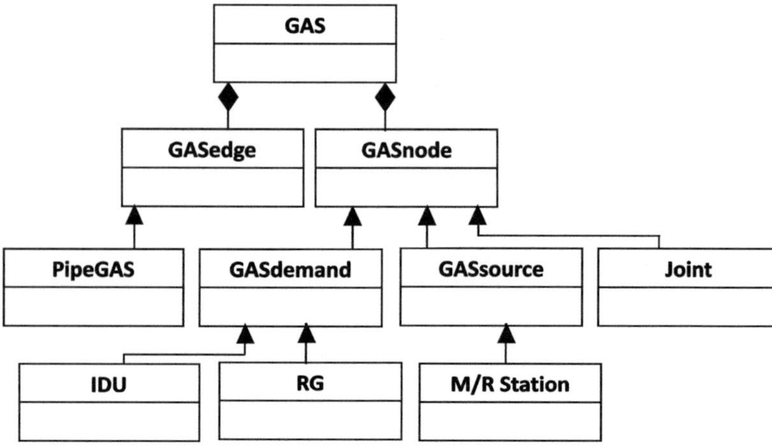

Fig. 9.4 Class diagram for the gas distribution network

accessibility of demand nodes, based on the network damage for the generic event, or the damage state of links and nodes (if they are considered vulnerable) according to a fragility function selected for these elements.

In order to cope with the limitations of the prototype MATHWORKS-MATLAB® implementation of the SYNER-G framework, it was decided to simplify the analysis reducing the amount of data, without compromising the nature of the study; i.e., the application to a real case. Therefore, a data reduction process was performed by means of: (i) removal of all *dead ends* (i.e., pipes that are not carrying gas to stations or end users); (ii) simplification of the geometry merging adjacent pipes with the same geometrical and material properties. The resulting network is composed of 602 nodes (3 sources, 209 RGs and 390 joints) and 608 links.

9.4 Analysis and Results

As mentioned, a simulation-based connectivity analysis was the subject of this chapter. Considering that the mission of a gas network at mid-pressure is to deliver gas to reduction groups, the network's performance was assessed evaluating the availability of end nodes (RGs) of the L'Aquila system.

Both TGD and PGD hazards were addressed. Pipelines and M/R stations were considered as the only vulnerable elements within the network, and the risk assessment was performed in terms of two performance indicators (SR and CL).

In the following sub-sections a detailed description of the methodology and tools adopted is provided.

9.4.1 Procedure

The process to compute the seismic performance, in terms of CL and SR is essentially divided into eight separate stages:

1. Simulation of the event on the considered seismic source.

The Paganica fault (normal fault type) was used as source for the generation of characteristic earthquakes of moment magnitude M_w 6.3 and return period equal to 750 years (Pace et al. 2006). Data on geometric source model used herein can be found in Chioccarelli and Iervolino (2010).

2. Simulation of random field of the primary *IM* at bedrock (PGA_r).

The strong ground motion for the primary *IM* was evaluated using a GMPE computed on a regular grid of points discretizing the region covered by the gas network. The regular grid that discretizes the region occupied by the network was identified based on the correlation structure of the primary *IM* intra-event residual; i.e., a grid able to represent correlation of *IMs* in one earthquake. In this case, 1 km grid spacing was selected.

3. Conditional simulation of the cross-correlated ground motion for secondary *IM* at bedrock (PGV_r).

As mentioned in Sect. 9.2 and discussed in Chap. 3, the primary *IM* is chosen as an intensity measure for which a spatial correlation model is available, and it is used to generate the realization of a random field in each event, and to obtain the secondary *IM* for each site of interest via the *conditional simulation* approach.

For this case study, PGA was identified as primary *IM*, which is needed in the fragility function for stations. On the other hand, since pipelines are sensitive to PGV (i.e., their fragility models are expressed in terms of this parameter), the latter was selected as the secondary *IM*.

The GMPE used for the evaluation of strong motion is that by Akkar and Bommer (2010) and spatial variability was modeled using correlation models provided by Esposito and Iervolino (2011).

For each event the primary *IM* was calculated for each site of the grid and then interpolated at vulnerable sites by a distance-based interpolation. The resulting ground motion intensities correspond to rock sites (PGA_r). Then, at each site, the realization of the logarithm of PGV at bed-rock (PGV_r) was obtained sampling a Gaussian variable whose parameters are conditional on PGA_r.[2]

[2]To this aim, assuming the joint normality between the two *IMs*, the correlation coefficient between PGA_r and PGV_r was specifically estimated starting from the dataset used for the Akkar and Bommer (2010) GMPE.

4. Amplification due to local site conditions to get PGA_s and PGV_s that are the *IMs* at the surface.

To account for local site conditions GMPE-based amplification factors were considered. To this aim each site of the network was characterized according to the site classification scheme adopted by the employed GMPE, starting from geological analysis of the region described in the next paragraph. This allows to obtain PGA_s and PGV_s (see Esposito et al. 2013b).

5. Simulation of displacement consequential to PGD.

Regarding the PGD hazard, the landslide potential of L'Aquila region, according to the HAZUS (FEMA 2004) procedure was evaluated (no significant liquefaction potential was found in the region, and co-seismic surface ruptures were neglected). A landslide-susceptibility map of L'Aquila region, based on the lithological groups, slope angles, and ground-water conditions, was specifically obtained for the purposes of this study, starting from the methodology formulated by Wilson and Keefer (1985), and described in Chap. 3.

More than 40 different outcropping formations were detected in the region of interest, starting from 1:50,000 scale ISPRA geological maps (http://www.isprambiente.gov.it).

A slope angle map was generated from topographic data and six slope classes were defined: 3–10, 10–15, 15–20, 20–30, 30–40, >40°. In particular, starting from a topographic map 1:25,000 (Istituto Geografico Militare, http://www.igmi.org/prodotti/cartografia/carte_topografiche) a digital elevation model of the studied area was obtained.

Bounding groundwater conditions were assumed, by considering either dry (groundwater below the depth of the sliding surface) or wet conditions (groundwater level at ground surface).

Finally, a critical acceleration (*yield coefficient*) value, k_c, ranging from 0.05 g (most susceptible) to 0.6 g (least susceptible) was associated to each landsliding-susceptible category.

Overlying the slope angle, groundwater and lithology class maps, it was possible to draw a map of the landslide susceptibility, which was finally transformed into the critical acceleration map shown in Fig. 9.5, where the white color corresponds to the non-susceptible areas; i.e., those characterized by a slope angle lower than 5°. The critical acceleration value corresponds to the threshold acceleration above which slope displacement is initiated. In fact, permanent displacements either occur or not in a susceptible deposit, with probability P^* in those cases in which PGA_s exceeds k_c. A different P^* (corresponding to the map area proportions showed in Table 3.3, Chap. 3) is associated to each landsliding-susceptible category. The resulting displacement is finally calculated via the Saygili and Rathje (2008) empirical model.

6. Computation of repair rate, R_R, for pipelines.

As mentioned, to estimate the earthquake-induced damage, *IMs* were related to system component damage via fragility models. For buried pipelines, the Poisson repair rates functions of PGV_s and PGD proposed in ALA (2001), for different pipe

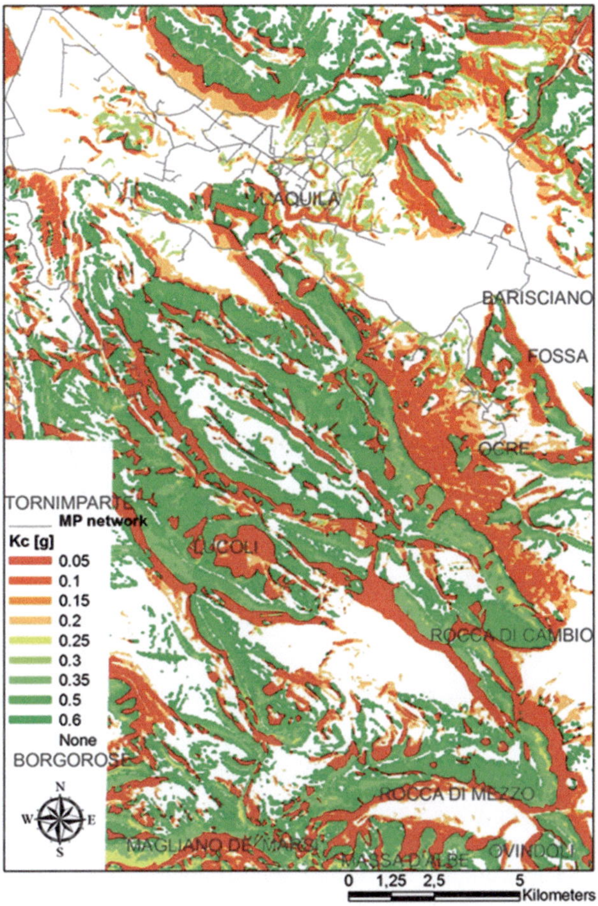

Fig. 9.5 Critical acceleration map for L'Aquila region where the network deploys

material (steel and HDPE) and diameter, were selected according to the analysis of damage occurred on the gas network following the 6th April 2009 L'Aquila earthquake (Esposito et al. 2013a). These relations are expressed in Eqs. 9.3 and 9.4 where R_R is expressed in *1*/km, PGV_S and PGD are given in cm/s and m, respectively; K_1 and K_2 represent the modification factors according to pipe material and diameter.

$$R_R = K_1 \cdot 0.002416 \cdot PGV_s \tag{9.3}$$

$$R_R = K_2 \cdot 11.223 \cdot PGD^{0.319} \tag{9.4}$$

At each location, and in each event (i.e., simulation run), the repair rate is equal to the largest value calculated as a function of PGV_s and PGD, and is then multiplied by 0.2 and 0.8, respectively; since only breaks are considered in the connectivity analysis. In fact, damage states selected for the evaluation of seismic vulnerability are chosen on the basis of the objective of the analysis; i.e., the performance indicators. In this case, the system is considered functional if demand nodes (reduction groups) continue to provide gas, and then if they remain accessible from at least one supply node (M/R station). To this aim it was assumed that a pipe segment cannot deliver gas when the segment has at least one break.

7. Computation of damage state for M/R stations.

While reduction groups were not considered seismically vulnerable, for the M/R stations, instead, lognormal fragility curves for un-anchored compressor stations (FEMA 2004) were adopted; the median and standard deviation (of the logarithms) are equal to 0.77 g and 0.65, respectively.

Regarding the damage state of interest for the supply node, it was assumed that it loses its connectivity when it is in *extensive* damage state.

8. Computation of connectivity-based performance indicators.

As mentioned, the quantitative measure of the functionality of the gas network is given by performance indicators. Herein the SR and the CL were considered. In particular, for SR, the weighting factor considered is represented by the nominal flow (m^3/h) of the demand node.

9.4.2 Results

A Monte Carlo simulation was carried out in order to evaluate the probability of exceeding predefined levels, *u,* of performance (i.e., the complementary cumulative distribution function, CCDF), given the occurrence of an earthquake on the fault.

The number of runs of the simulation was defined in order to yield stable estimates of the probability of exceeding the considered PI.

Results indicate that the expected value of connectivity loss given the occurrence of an earthquake is 0.66; i.e., it is expected that the average reduction in the ability of demand nodes to be connected to M/R stations is of 66 % when a 6.3 event occurs on the Paganica fault. While for the SR indicator, it is expected that the 68 % of demand nodes receive gas. Figure 9.6 shows the moving average, μ, curves as well as the $\mu + \sigma$ and $\mu - \sigma$ (moving average plus/minus moving standard deviation) curves for the two performance indicators, and the probability of exceedance of the two PIs.

It may be observed from the 'steps' in the curve that the connectivity loss is characterized by a multi-modal distribution. This different behavior may be due to the different definitions of the two performance indicators and the network configuration specific to the application case.

Finally, the influence of modeling of spatial correlation of intra-event residuals on risk assessment was also investigated. In particular, the risk assessment was performed also assuming in the first case a correlation coefficient equal to zero; i.e., considering intra-event residuals uncorrelated. Minor differences with respect

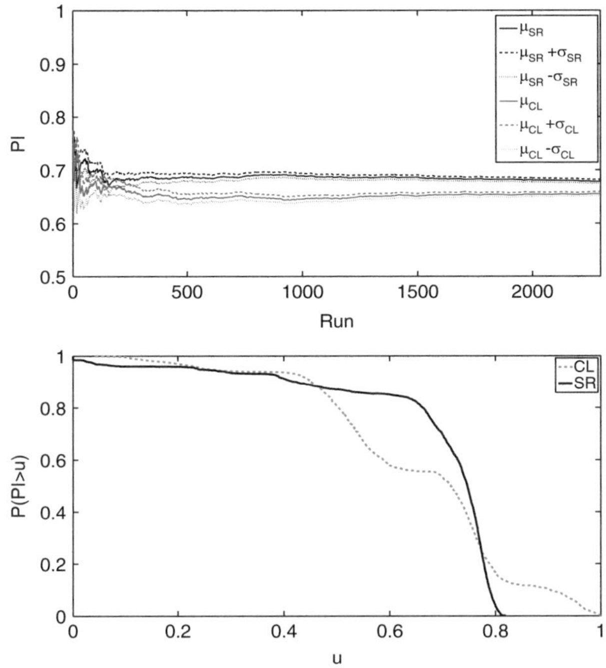

Fig. 9.6 Moving average μ, $\mu + \sigma$ and $\mu - \sigma$ curves (*top*) and CCDF for CL and SR (*bottom*)

to results in Fig. 9.6 indicate that spatial correlation has a relatively small impact on risk evaluation of the selected system and with reference to the considered performance indicators (see Esposito et al. 2013b, for more details).

9.5 Conclusions

This chapter summarized the application of the SYNER-G framework to a real case study: L'Aquila (central Italy) gas distribution network. The study employed probabilistic seismic and geotechnical hazard analysis, empirical relations to estimate pipeline response, fragility curves for the evaluation of vulnerability, and connectivity performance indicators to characterize the functionality of the network.

In fact, special emphasis was put on the medium-pressure part of the L'Aquila gas system for which detailed information was retrieved. In particular, the selected network was characterized by 3 M/R stations, 209 reduction groups, and pipelines either made of steel or HDPE.

The principal result of the risk analysis was the probability of exceeding a set of performance levels, which can be transformed in the annual rate of their exceedance, given the occurrence of a characteristic earthquake on the fault beneath the region where the system deploys.

In particular, earthquakes were generated using as source the Paganica fault and considering characteristic earthquakes of moment magnitude *Mw* 6.3. Strong ground motion intensities for the primary TGD intensity measure were evaluated though an European GMPE and an European spatial correlation model on a regular grid defined based on the correlation structure of PGA; since fragilities of gas pipelines are often expressed in terms of PGV, the latter was selected as the secondary *IM*. For each site, the secondary *IM* was determined, in a probabilistically consistent manner, via the conditional hazard approach. The resulting intensities correspond to rock sites. To account for local site conditions GMPE-based amplification factors were applied.

Regarding geotechnical hazards, often resulting the most effort-demanding issue in this kind of problems, the landslide potential of L'Aquila region according to the HAZUS procedure was performed.

To estimate earthquake-induced damage for buried pipelines, repair rate functions of PGV_s and PGD were selected for each pipe typology and diameter. Reduction groups were not considered seismically vulnerable. For the M/R stations, instead, a lognormal fragility curve for un-anchored compressor stations was adopted. Damage states considered for the evaluation of seismic vulnerability are strictly related to the objective of the analysis. In this case a connectivity analysis was performed; i.e., the system is considered functional if demand nodes (reduction groups) continue to provide gas after the earthquake, and then if they remain accessible from at least one supply node (M/R station). To this aim it was assumed that a pipe segment cannot deliver gas when the segment has at least one break, while for the supply node it was assumed that it loses its connectivity when it is in extensive damage state.

The adaptation of two connectivity performance indicators (serviceability ratio and connectivity loss) was considered to include damage of stations and distributing elements into the risk assessment for the system.

The study finally proves feasibility of simulation-based seismic risk assessment of gas distribution networks in a performance-based earthquake engineering framework.

Acknowledgments Authors want to acknowledge the network operator (Enel Rete Gas s.p.a.) for kindly providing data for the characterization of the case study.

References

Adachi T, Ellingwood BR (2008) Serviceability of earthquake-damaged water systems: effects of electrical power availability and power backup systems on system vulnerability. Reliab Eng Syst Saf 93(1):78–88

Akkar S, Bommer JJ (2010) Empirical equations for the prediction of PGA, PGV and spectral accelerations in Europe, the Mediterranean region and the Middle East. Seismol Res Lett 81(2):195–206

ALA (2001) Seismic fragility formulations for water systems: Part 1 – Guidelines. American Lifeline Alliance, ASCE, Washington, DC

Chang L, Song J (2007) Matrix-based system reliability analysis of urban infrastructure networks: a case study of MLGW natural gas network. Paper presented at the 5th China-Japan-US trilateral symposium on lifeline earthquake engineering, Haikou, China, 26–28 Nov 2007

Ching J, Hsu W-H (2007) An efficient method for evaluating origin-destination connectivity reliability of real-world lifeline networks. Comput-Aided Civil Infrastruct Eng 22(8):584–596

Chioccarelli E, Iervolino I (2010) Near-source seismic demand and pulse-like records: a discussion for L'Aquila earthquake. Earthq Eng Struct Dyn 39(9):1039–1062

Esposito S (2011) Systemic seismic risk analysis of gas distribution networks. PhD thesis, University of Naples Federico II, Naples, Italy. Advisor: I. Iervolino. Available at: wpage.unina.it/iuniervo

Esposito S, Iervolino I (2011) PGA and PGV spatial correlation models based on European multievent datasets. Bull Seismol Soc Am 101(5):2532–2541

Esposito S, Iervolino I (2012) Spatial correlation of spectral acceleration in European data. Bull Seismol Soc Am 102(6):2781–2788

Esposito S, Iervolino I (2013) Application and validation study to a gas pipeline network. Tech Report D6.5 of the SYNER-G project (http://www.vce.at/SYNER-G/)

Esposito S, Giovinazzi S, Elefante L, Iervolino I (2013a) Performance of the L'Aquila (central Italy) gas distribution network in the 2009 (Mw 6.3) earthquake. Bull Earthq Eng 11(6): 2447–2466

Esposito S, Iervolino I, d'Onofrio A, Santo A, Franchin F, Cavalieri F (2013b) Simulation-based seismic risk assessment of a gas distribution network. Comput-Aided Civil Infrastruct Eng (under review)

Federal Emergency Management Agency (FEMA) (2004) Multi-hazard loss estimation methodology-earthquake model: HAZUS MR4 technical manual, Washington, DC

Gehl P, Desramaut N, Réveillère A, Modaressi H (2014) Fragility functions of gas and oil networks. In: Pitilakis K, Crowley H, Kaynia AM (eds) SYNER-G: typology definition and fragility functions for physical elements at seismic risk, vol 27, Geotechnical, geological and earthquake engineering. Springer, pp 187–220

Iervolino I, Giorgio M, Galasso C, Manfredi G (2010) Conditional hazard maps for secondary intensity measures. Bull Seismol Soc Am 100(6):3312–3319

O'Rourke MJ, Liu X (1999) Response of buried pipelines subjected to earthquake effects. MCEER monograph no. 3, Buffalo, New York

O'Rourke TD, Toprak S, Sano Y (1998) Factors affecting water supply damage caused by the Northridge earthquake. Paper presented at the 6th U.S. national conference on earthquake engineering. EERI, Seattle, May 31–June 4 1998

Osiadacz AJ (1987) Simulation and analysis of gas network, 1st edn. D and F.N. Spon Ltd., London

Pace B, Perruzza L, La Vecchia G, Boncio P (2006) Layered seismogenic source model and probabilistic seismic-hazard analyses in Central Italy. Bull Seismol Soc Am 96(1):107–132

Poljanšek K, Bono F, Gutiérrez E (2012) Seismic risk assessment of interdependent critical infrastructure systems: the case of European gas and electricity networks. Earthq Eng Struct Dyn 41(1):61–79

Saygili G, Rathje EM (2008) Empirical predictive models for earthquake-induced sliding displacements of slopes. J Geotech Geoenviron 134(6):790–803

Song J, Ok S-Y (2009) Multi scale system reliability analysis of lifeline networks under earthquake hazards. Earthq Eng Struct Dyn 39(3):259–279

Wilson RC, Keefer DK (1985) Predicting areal limits of earthquake induced landsliding, evaluating earthquake hazards in the Los Angeles region. In: Ziony JI (ed) U.S. Geological survey professional paper, 1360, pp 317–345

Chapter 10
Application to Selected Transportation and Electric Networks in Italy

Francesco Cavalieri, Paolo Franchin, and Paolo Emilio Pinto

Abstract This chapter presents the application of the SYNER-G general methodology to two regional networks in Southern Italy: the road network of Calabria and the medium-high voltage electric power network of Sicily. Modelling and analysis of a road network, already discussed in Chap. 5, is recalled and further expanded here. The corresponding portion of the object-oriented (OO) model as implemented within SYNER-G is discussed. Then, the case study is presented, in terms of the main methodological choices, system properties (topology, vulnerability) and seismic hazard. Results of the (connectivity-only) analysis are then reported. Similarly, further aspects of the power network model implemented within SYNER-G are given before introducing and discussing the Sicily case study. The analysis of the EPN is carried out in terms of power flows.

10.1 Application to the Road Network of the Calabria Region

In the following sub-sections, the application of the SYNER-G general methodology to the Calabria road network is discussed. Methodological aspects are presented first, followed by a description of the case study and the simulation results. The study is carried out at the connectivity level, i.e. not computing the vehicle flows and congested travel times on the edges, both in the undamaged or reference state and in the damaged one.

F. Cavalieri (✉) • P. Franchin • P.E. Pinto
Department of Structural and Geotechnical Engineering, Sapienza University of Rome, Via Gramsci 53, 00197 Rome, Italy
e-mail: francesco.cavalieri@uniroma1.it; paolo.franchin@uniroma1.it; pinto@uniroma1.it

K. Pitilakis et al. (eds.), *SYNER-G: Systemic Seismic Vulnerability and Risk Assessment of Complex Urban, Utility, Lifeline Systems and Critical Facilities*, Geotechnical, Geological and Earthquake Engineering 31, DOI 10.1007/978-94-017-8835-9_10,

10.1.1 *Methodological Aspects and Implementation Within the SYNER-G Model*

10.1.1.1 Analysis of Road Networks

Specification of the SYNER-G model to transportation networks, including modelling and analysis for the purpose of seismic performance assessment, is described in Chap. 5 (Sect. 5.3.2.4). This section provides additional background to the implementation.

As stated in Chap. 5, all network systems can be analysed at different increasing levels of complexity, going from vulnerability (Level 0), through connectivity/form analysis (Level I), to flow analysis (Level II). In this chapter Level I analysis is carried out for road networks, even though Level II capabilities are already included in the model. The reason for this choice is easily explained. Traffic demand prediction in the disturbed post-seismic conditions is still an area of research in its infancy and flow analysis under pre-earthquake demand has only a conventional character and almost no utility when the focus is on the immediate event aftermath.

The most common methodology applied for Urban Travel Forecasting, called four stages transportation/land-use model or, as it is also known, the *sequential procedure*, was originally developed in the 1950s for the Detroit and subsequently the Chicago Area Transportation Studies (DATS and CATS). It goes without saying that the sequential model was not conceived as a "real-time" predictor of the state of traffic on a damaged network at discrete time instants after a seismic event. In fact, its application has only proven to be effective for relatively long term forecasting (Levinson and Kumar 1994). It is here briefly illustrated since proposals for post-disaster travel forecasting build upon it.

Typically, the region of interest is subdivided into Traffic Analysis Zones (TAZs) and *population* and *employment levels* are determined for each zone (in the original procedure by either trend or regression analysis, while in the SYNER-G model they are directly evaluated from the cell data). The four stages of the procedure are:

1. **Trip Generation**. In each zone, the frequency of *origins* and *destinations*, i.e. the number of *trips generated* from or *attracted* to each zone, are evaluated by trip purpose (passenger trips by home-to-work, home-to-shop, home-to-other, etc.; freight trips by transported goods; all is then converted to Passenger Car Units – PCU) as a function of land use, household demographics and other socio-economic factors. This stage employs so-called *Land Use Models*.
2. **Trip Distribution**. Movements between zones are evaluated, i.e. the total trips originated from zone *i* are distributed among all the other zones as a function of the *zonal demands* and *inter-zonal distances*.
3. **Mode Split**. Movements between origins and destinations are disaggregated by type of transportation or mode (bus, railway, car, airplane, etc.) depending on the availability of each mode, their respective costs and user preferences.

This stage and the previous one employ *Spatial Interaction Models*, that produce flow estimates between TAZs, in terms of origin-destination pairs, which can be disaggregated by nature, mode and time of the day. The output of the trip distribution stage is the $n \times n$ "trip table" matrix T_{ij}, also called *origin-destination* (OD) *matrix*, which displays the number of trips from each origin to each destination.

4. **Traffic Assignment**. All the estimated trips between zones, disaggregated by purpose and mode, are *loaded* on the transportation network to determine the total flows on each arc. Typical traffic assignment algorithms account for the fact that users want to minimize their travel time. The models currently in use for this stage are the *network equilibrium models*. They are based on the so-called two principles of equilibrium by Wardrop (1952):

    (a) First principle: At equilibrium all used routes from node i to node j have equal travel times, and no unused route has a lower travel time; this implies that each user non-cooperatively seeks to minimize his cost of transportation, and cannot lower his transportation cost through unilateral action. The traffic flows that satisfy this principle are usually referred to as *user equilibrium* (UE) *flows*, since each user chooses the route that is the best.

    (b) Second principle: At equilibrium the average travel time is minimal. This implies that each user behaves cooperatively in choosing his route to ensure the most efficient use of the whole system. Traffic flows satisfying Wardrop's second principle are generally deemed *system optimal* (SO) flows.

It should be noted that the four stages are to a large extent interrelated. For example when traffic flow exceeds the capacity of a specific arc, resulting in congestion and increased travel time, this, through a feedback process, may influence trip generation and distribution. The procedure is hence inherently iterative. Convergence is often measured in terms of *minimal transportation generalized cost* (e.g. distance, time, money, etc.).

As already mentioned in Chap. 5, Level II studies usually employ a static pre-earthquake origin-destination matrix, output of the first three steps in the sequential procedure, and modify only Stage 4, by evaluating flows on the damaged network. A single attempt to revise the entire procedure by accounting for changes in trip generation and attraction is the work by Chang et al. (2010) reported in Chap. 5.

Currently the SYNER-G model implementation has the full sequential procedure as briefly outlined above, but no modification factors for trip generation and attraction to account for post-earthquake conditions. This is certainly one of the main open areas of multi-disciplinary research in disaster impact assessment and management.

10.1.1.2 RDN Within the SYNER-G Model

As described in Chap. 2 and further specified in Chap. 5, the SYNER-G object oriented (OO) model of the Infrastructure and the Hazards acting upon it consists

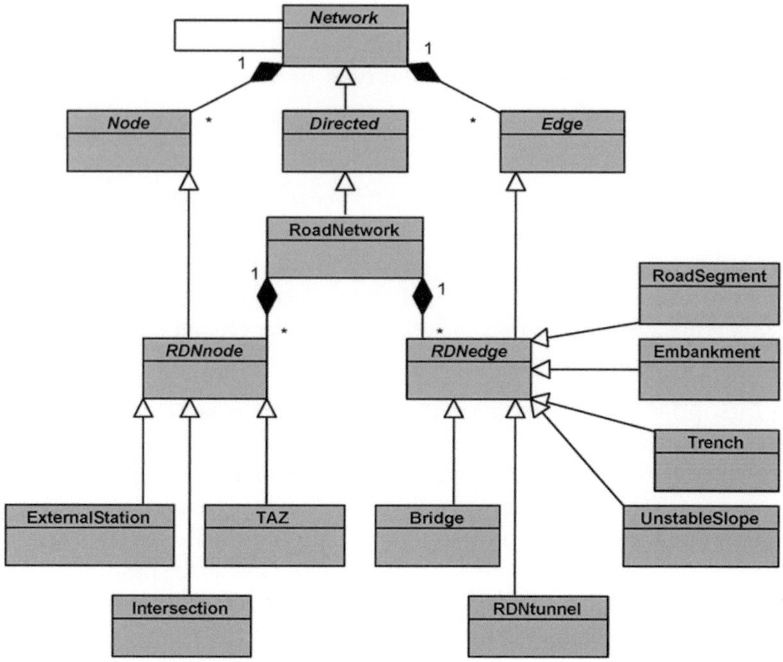

Fig. 10.1 Class diagram for the RDN class

of a set of inter-related classes. Figure 10.1 illustrates the RDN portion of the class diagram describing the model.

The RDN is modelled as a *directed* graph, i.e. a graph in which all edges have a travelling direction, from node *i* to node *j*. For this reason, the RDN class is a subclass of the *Directed* abstract class, which in turn is a subclass of the *Network* abstract class.

Similarly to other Network classes, a road network is made up of *nodes* and *edges* connecting them. As a consequence, the RDN class is the composition of the *RDNnode* and *RDNedge* abstract classes. The first is the generalization of the TAZ, ExternalStation and Intersection classes, and itself a sub-class of the Node class. A brief explanation of these node typologies is given in Pinto et al. (2012). The *RDNedge* class is the generalization of the RoadSegment, Embankment, Trench, UnstableSlope, RDNtunnel and Bridge classes. The definition of these edge typologies, together with associated fragility models, are given in SYNER-G reference report 4 (Kaynia 2013), and in Chapters 9 and 10 of Book 1.

The RDN OO model is exemplified with the basic network in Fig. 10.2, where some of the properties related to the network, edges and nodes, in the damaged and undamaged conditions, are reported.

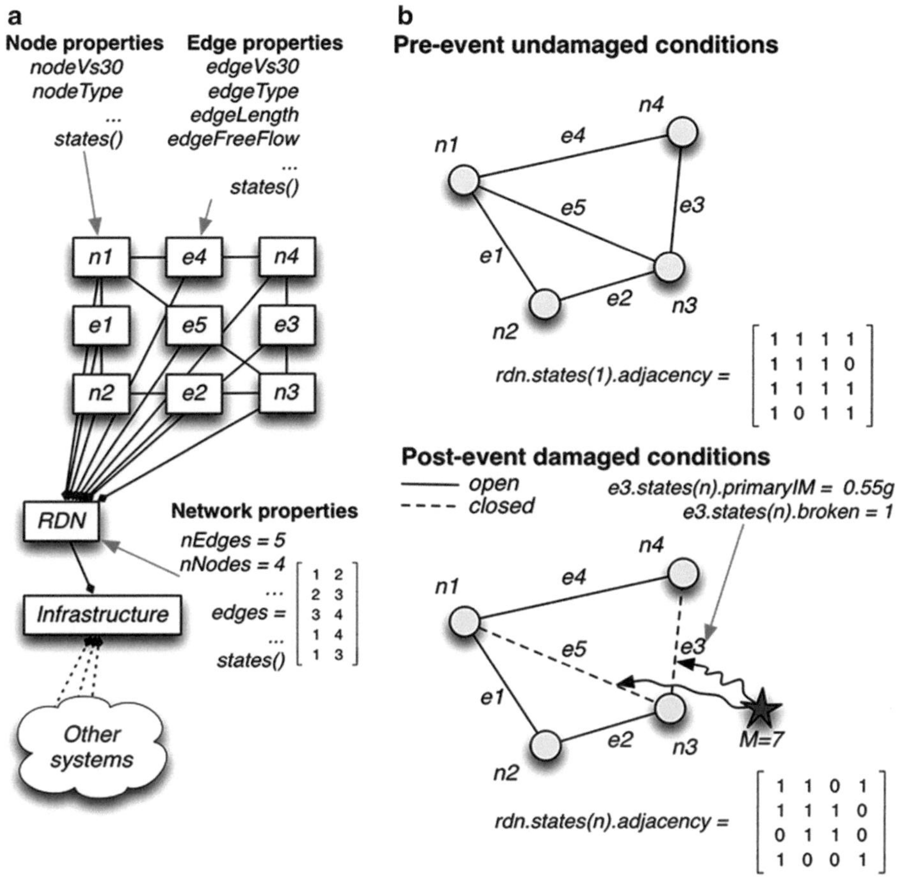

Fig. 10.2 Some properties of the RDN OO model, with reference to a basic network. (**a**) Object model. (**b**) Physical system

All the properties and methods of the RDN class and its subclasses are listed in SYNER-G reference report 6 (Pitilakis and Argyroudis 2013). They can be grouped into five lists:

1. *List of pointers (e.g. all the connections between objects in* Fig. 10.2*)*
2. *Road network global properties (variables such as the number of nodes, or edges, or the connectivity in* Fig. 10.2*)*
3. *Edge and node properties (variables such as the length of the edge or the position of a node, or the corresponding local geology, in terms of* V_{S30}*, in* Fig. 10.2*)*
4. *Properties that record the state of the RDN for each event (variable "states" in* Fig. 10.2*)*
5. *Properties that store the global performance of the RDN at the end of simulation*

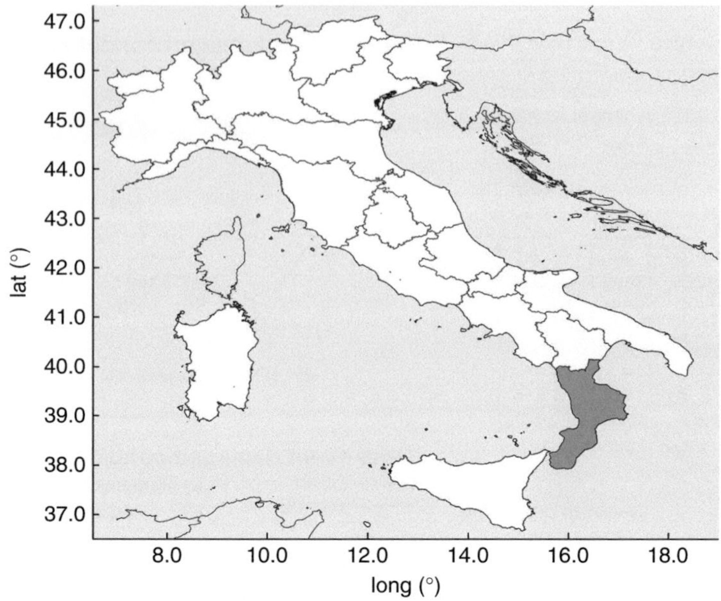

Fig. 10.3 Italy and the Calabria region, in *grey*

10.1.2 The Case Study: Road Network of Calabria (Italy)

The road network of the Calabria region, in Southern Italy (see Fig. 10.3), has been chosen as the case study. Though the region has not been affected by significant earthquakes in recent times, it is one of the most seismic areas in the country, together with the adjacent region of Sicily, which is the setting for the electric power network example to follow.

10.1.2.1 Seismic Hazard

The seismic hazard is modelled through 20 faults taken from the Italian DISS (Database of Individual Seismogenic Sources) database, employing the truncated Gutenberg-Richter recurrence model for the source activity. The faults are visualized in Fig. 10.4, while Table 10.1 reports their activity parameters.

Within the DISS source model, also used in the European SHARE (Seismic Hazard Harmonization in Europe http://www.share-eu.org/) fault database, the fault surface does not correspond to a simple rectangle (seismogenic area), but is described as a more complex polygon that can change strike according to the trace. The current fault typology used within the SYNER-G model and the prototype software implementation for shallow faults in Europe (except the Aegean, Calabrian and Cypriot subduction interfaces) is **simpleFaultGeometry**. The fault is described by a trace, a dip and an upper and lower depth (the dip direction is always 90°

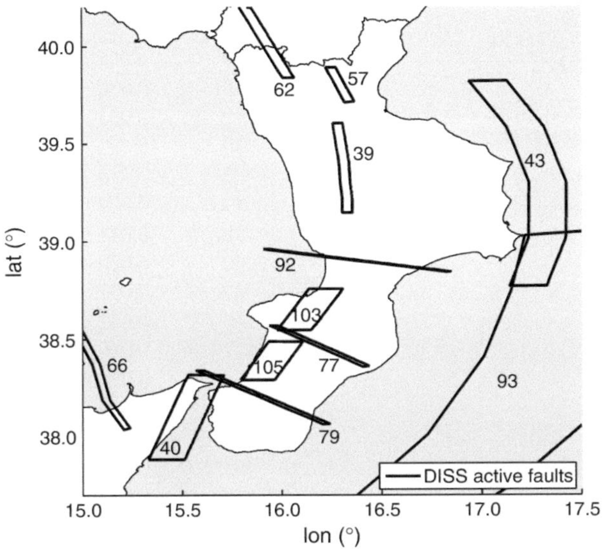

Fig. 10.4 DISS database faults affecting Calabria

clockwise from the azimuth to the last point on the trace from the first point of the trace). The rendering of an evenly-spaced discrete mesh over the fault surface is done using an adaptation of the algorithm of Mark Stirling, which is what is currently implemented in OpenSHA (Open Source Seismic Hazard Analysis http://www.opensha.org/). The fault source, as it is represented in this fashion, is equivalent to a composite source (Weatherill et al. 2011).

Figure 10.5 shows a shake field of the primary intensity measure (IM) (see Chap. 3), taken here to be Peak Ground Acceleration (PGA), computed at points of a regular grid, for a scenario M = 5.5 event on source # 39. The ground motion prediction equation (GMPE) employed in this application is that by Akkar and Bommer (2010).

During each event, as described in Chap. 2, the primary IM is then interpolated at vulnerable sites by distance-based interpolation and any IM different from the primary needed as an input to fragility model is sampled conditionally on the primary IM value. Since Permanent Ground Deformation (PGD) is needed as an input to the fragility model of road segments, the landslides and co-seismic rupture models in the geotechnical hazard module are used. No amplification has been carried out since the corresponding soil classification data were not available.

10.1.2.2 System Topology and Characteristics

The original data available for the road network of Calabria (database DBPrior10k, provided by the Cartographic Center of Calabria region, http://www.centrocartografico.it/) included the network topology, population of cities and

Table 10.1 Activity parameters of DISS database faults affecting Calabria

Fault #	α	β	M_{min}	M_{max}
29	0.058	2.303	5.0	5.8
39	0.030	2.303	5.0	6.2
40	0.063	2.303	5.0	7.0
43	0.092	2.303	5.0	6.5
53	0.054	2.303	5.0	6.0
57	0.009	2.303	5.0	6.2
58	0.041	2.303	5.0	6.8
60	0.020	2.303	5.0	6.6
62	0.027	2.303	5.0	6.4
66	0.062	2.303	5.0	6.1
77	0.016	2.303	5.0	6.6
79	0.025	2.303	5.0	6.0
87	0.019	2.303	5.0	6.9
92	0.014	2.303	5.0	6.9
93	0.414	2.303	5.0	7.1
103	0.019	2.303	5.0	6.0
105	0.015	2.303	5.0	6.6
108	0.006	2.303	5.0	5.4
110	0.014	2.303	5.0	6.2
112	0.017	2.303	5.0	6.3

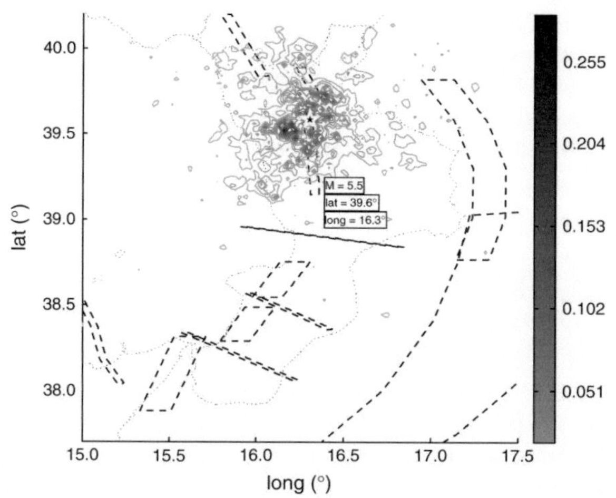

Fig. 10.5 Shake map in terms of PGA on rock at grid points, for a scenario M = 5.5 event on source # 39

suburbs served by the network, as well as the position of the hospitals of the regional health-care system. In particular, the topological data consisted of:

- 2,451 TAZs (corresponding to either municipalities or suburbs) and 3,607 intersections, for a total of 6,058 nodes;

Table 10.2 Available fragility curves for bridges

Limit state	LD		CO	
Bridge #	λ	β	λ	β
1	0.030	0.224	1.901	1.501
2	−0.481	0.212	0.662	1.604
3	0.224	0.288	7.193	1.848
4	−1.016	0.345	1.546	0.179
5	−0.009	0.156	1.722	0.167
6	−0.759	0.323	5.978	2.425
7	−0.465	0.201	1.044	0.106
8	−0.743	0.247	1.061	0.090
9	−0.564	0.270	10.941	4.149
10	−0.087	0.255	58.701	19.01
11	0.506	0.246	1.447	0.211
All others	−0.294	0.646	2.323	1.394

- 13,956 edges, divided into *main* and *secondary* roads (based on free-flow speed) and *connectors* of TAZs to the road network (these latter are used since in the available data set the TAZs' centroids have not been moved to the nearest network node);

Further available data, provided by Borzi and Fiorini (2012), included the positions and susceptibility of 9,002 landslide susceptible areas and the positions of 1,325 bridges, 11 of which are characterized with fragility data. Table 10.2 reports the parameters of such fragility curves, together with (in the last row) the mean parameters derived from an ensemble of about 400 bridges in Italy, and which have been used in the application for all the remaining bridges. The curves are in the form of lognormal cumulative distribution functions, defined in terms of the logarithmic mean λ and the logarithmic standard deviation β of the intensity measure (PGA in m/s^2). A set of two curves is provided for each bridge, corresponding to the Light Damage (LD) and Collapse (CO) limit states.

In order to cope with the limitations of the prototype MATLAB implementation of the SYNER-G model, reduction of the data set to a manageable size was deemed necessary. The data reduction process consisted of several steps:

- Removal of many TAZs, corresponding to suburbs having a very small population, aggregated into that of the corresponding municipalities. Only municipalities have been kept, reducing the number of TAZs from 2,451 to 422 (Fig. 10.6);
- Removal of bridges with total length less than 35 m (which is the upper bound value for the length of prefabricated pre-stressed concrete girders usually employed in Italy in the decades from the 1960s for ordinary viaducts; bridges with shorter length are likely to be simple single-span bridges without piers which, due to usually large seating lengths in the Italian practice, have been considered not vulnerable), resulting in a total of 521 bridges, 11 of which characterized by detailed fragility data;
- Removal of *inactive* landslides areas, resulting in a total of 2,089 areas;

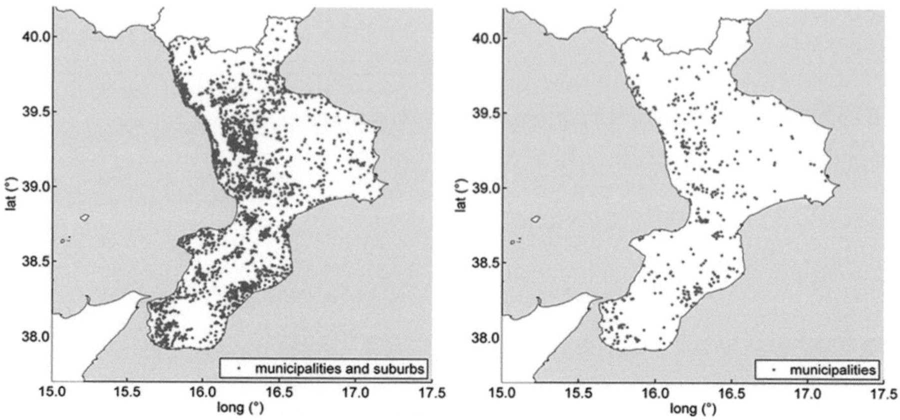

Fig. 10.6 Reduction of TAZs, aggregating suburbs into municipalities

- Displacement of TAZs to the nearest road network node and removal of corresponding connectors, not affecting network connectivity;
- Integration of bridges into the graph, since graph and bridges belonged to two different databases; bridges are considered in the software as edges; each bridge has been projected onto the road segment with the closest centroid;
- Removal of secondary roads not included in minimum paths connecting TAZs to main roads; a minimum path in this case is the path with the minimum travel time. This is of course an operation that reduces the redundancy of the network and thus influences the results;
- Removal of all dead ends resulting from the previous steps.

The results of the data reduction process can be seen in Fig. 10.7, showing a close-up on the network topology, before (left) and after (right) the reduction.

The resulting road network is composed of 2,861 nodes (422 of which are TAZs) and 5,970 edges (Fig. 10.8). Edges, that are the only vulnerable components in the network, are subdivided into road segments and bridges, with fragility models expressed in terms of PGD and PGA, respectively (see Chapters 9 and 10 of Book 1). Edges are also classified as either main roads (principal roads or highways) or secondary roads, based on their free-flow speed. Of the former 11 bridges characterized by a specific fragility function, only 9 (the "bridges with data" entry in the legend) result still included in the network, after the data reduction process. It is recalled that while within the developed model the RDN is modelled as a directed graph and all edges have a travelling direction, from node *i* to node *j*, for this particular network, all edges are two-ways roads, effectively making the graph undirected. Finally, the graph is a weighted one, with weights being the free-flow travel times of edges.

Further available data include the location of the ten public hospitals, belonging to the regional health-care system, as well as of the landslide susceptible areas (Fig. 10.9).

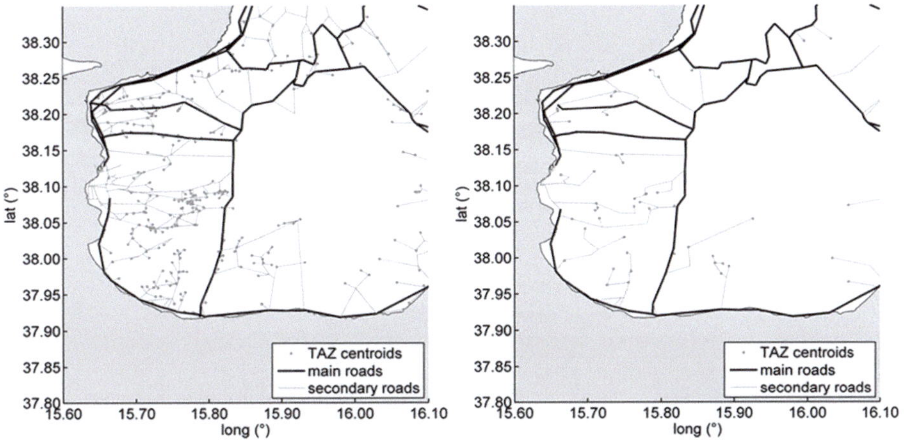

Fig. 10.7 Close-up comparing the network topology, before (*left*) and after (*right*) the data reduction process

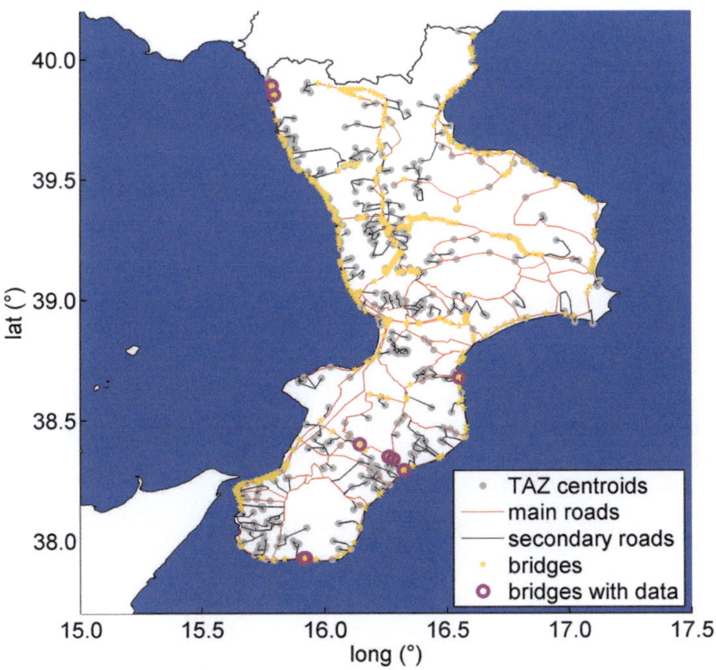

Fig. 10.8 Road network topology

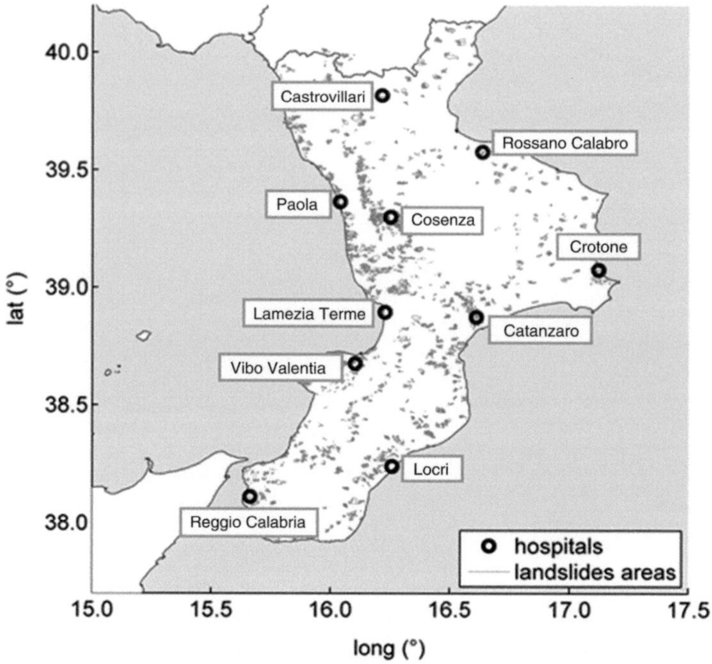

Fig. 10.9 Location of hospitals and landslides areas

10.1.3 Analysis and Results

10.1.3.1 Performance Indicators

A number of performance indicators (PIs) introduced in Chap. 5 (Sect. 5.3.2.5) have been used to measure the RDN seismic performance. Since the analysis is limited to connectivity, two global measures employed are the Simple Connectivity Loss (SCL) and the Weighted Connectivity Loss (WCL):

$$SCL = 1 - \langle N_s^i / N_0^i \rangle_i \tag{10.1}$$

$$WCL = 1 - \langle (N_s^i \cdot W_s^i) / (N_0^i . W_0^i) \rangle_i \tag{10.2}$$

where $\langle \rangle$ denotes averaging over all sink vertices, while N_s^i and N_0^i are the number of sources connected to the i-th sink in the seismically damaged network and in non-seismic conditions, respectively. With reference to an RDN, all the single TAZs, taken one at a time, are considered sinks, whereas all the remaining TAZs are sources. The weights W_s^i and W_0^i can be defined in different ways: here they are defined as the sum of the inverse of (free-flow) travel times of the single paths between the i-th sink and the sources, in the seismically damaged network and in non-seismic conditions (see Chap. 5). The statistics of these quantities have been computed from the simulation results, in particular the moving average μ and standard deviation σ, as well as the Mean Annual Frequency (MAF) of exceedance of any threshold value y. The latter has been computed as (Y denoting the generic PI):

$$\lambda_Y(y) = \sum_i \lambda_{0,i} G_{Y|i}\ (y\,|i\,) = \lambda_0 \sum_i p_i G_{Y|i}\ (y\,|i\,) = \lambda_0 G_Y(y) \tag{10.3}$$

by post-processing of the vector of sampled values of Y to first obtain the complementary (experimental) distribution function $G_Y(y)$. Then this distribution is multiplied by the MAF of all earthquakes in the region, $\lambda_0 = \sum_i \lambda_{0,i}$. The probability $p_i = \lambda_{0,i}/\lambda_0$ that, given an earthquake, it occurs on source i is respected in the complementary distribution $G_Y(y)$ where results come from events sampled in the correct proportion among the different sources.

Further PIs employed are the Minimum Travel Time to reach one of the hospitals, computed for each TAZ centroid (computed for each scenario and also averaged over the entire simulation), and the Terminal Reliability (TR), i.e. the probability that a path exists between a specific OD pair (see Chap. 5).

10.1.3.2 Results

Three types of simulation have been carried out: a plain Monte Carlo (MCS) and two improved simulations employing variance reduction techniques, i.e. Importance Sampling (ISS) and Importance Sampling with k-means clustering (ISS-KM). The latter, as explained in Chap. 2, can considerably reduce the computational effort associated with the simulation but their accuracy over different systems and a range of PIs has not been tested extensively. The ISS is carried out on magnitude and the inter- and intra-event model error terms of the adopted GMPE. Clustering to further reduce the set of events generated by ISS is done on the shake-fields with a target number of clusters equal to a few hundreds.

For the plain Monte Carlo simulation a sample size of 20,000 runs showed to yield stable estimates for all considered PIs. The results are presented in the following figures. Figure 10.10 shows the moving average μ curves for SCL (left) and WCL (right), as well as the $\mu + \sigma$ and $\mu - \sigma$ curves for the two PIs. The figure indicates that the expected value of connectivity loss given the occurrence of an earthquake is higher for WCL than for SCL (about three times larger), as expected. In fact, WCL takes into account not only the existence of a path between two TAZs, but also the increase in travel time due to the seismically induced damage suffered by the RDN. The jumps present in the plots are located in correspondence of simulation runs/samples in which at least one TAZ results disconnected from at least one TAZ, leading to SCL and WCL values larger than zero.

Figure 10.11 displays in graphical matrix form the values of TR for each pair of TAZs. The matrix, which is symmetric due to the graph being undirected (as already

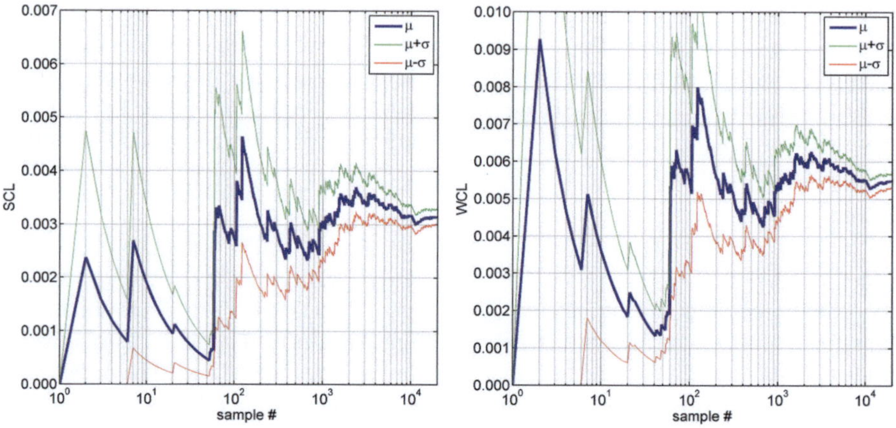

Fig. 10.10 Moving average μ, $\mu+\sigma$ and $\mu-\sigma$ curves for SCL (*left*) and WCL (*right*)

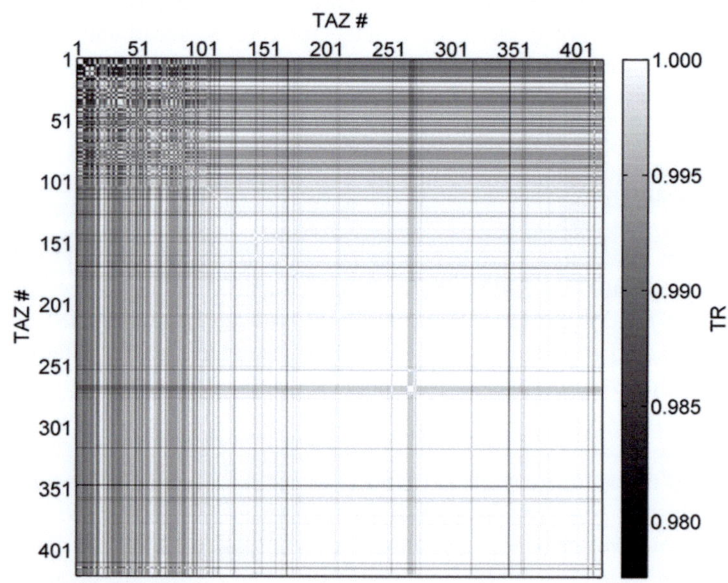

Fig. 10.11 Terminal reliability matrix

mentioned, even if the implemented RDN model is directed, this particular network is in practice undirected because there is always a pair of opposite edges between connected nodes, sharing the same vulnerability and damage state), indicates that the probability of connection is very high over the entire region, with lower reliability concentrated in the northern part of Calabria (approximately the first 100 TAZs).

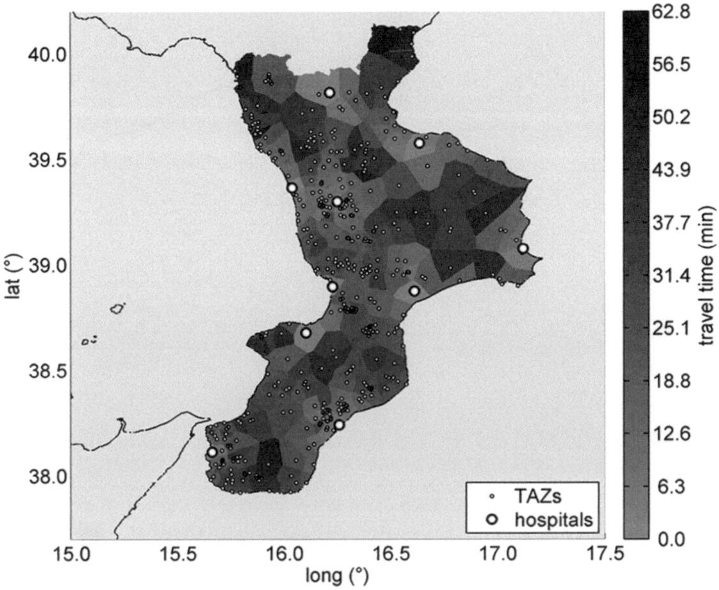

Fig. 10.12 Voronoi diagram of minimum travel time to hospitals, in non-seismic conditions

Figure 10.12 shows a Voronoi diagram of travel time to the closest hospital, in non-seismic, undamaged conditions. Such travel time, computed for each TAZ centroid, is attributed to all end-users accessing the RDN through that node; hence, a uniform colour code is used over the node tributary area (taken here equal to the Voronoi cell, collecting all points for which the node is the closest node).

The Voronoi diagram of travel time to the closest hospital is an informative representation of the impact of earthquake induced road damage. Out of the 20,000 runs of the Monte Carlo simulation, representative scenario events can be selected based on the frequency of exceedance of the corresponding SCL or WCL. For instance the Voronoi diagram in Fig. 10.14 illustrates the scenario corresponding to the 500-years value of WCL, equal to about 20 % as indicated in Fig. 10.13, which reports the MAF curves for both SCL and WCL. Figure 10.14 provides the ratio between travel time in damaged conditions (500-years event) and undamaged conditions. Overall, in most of the region the ratio equals unity, indicating no increment of travel time due to earthquake induced damage to the RDN. A ratio of up to almost five characterizes however some central mountainous parts of the region, while some zones, white coloured, experience a total disconnection from all regional hospitals.

As already mentioned, two additional more efficient simulations have been carried out. The accuracy of these variance-reduction techniques, however, has not been tested extensively over a wide range of systems and performance measures.

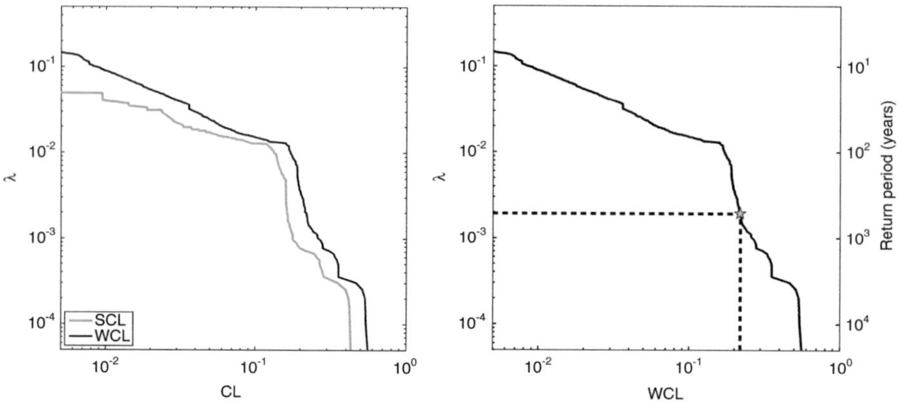

Fig. 10.13 MAF curve of SCL and WCL (*left*), with selection of a value with 500-years return period

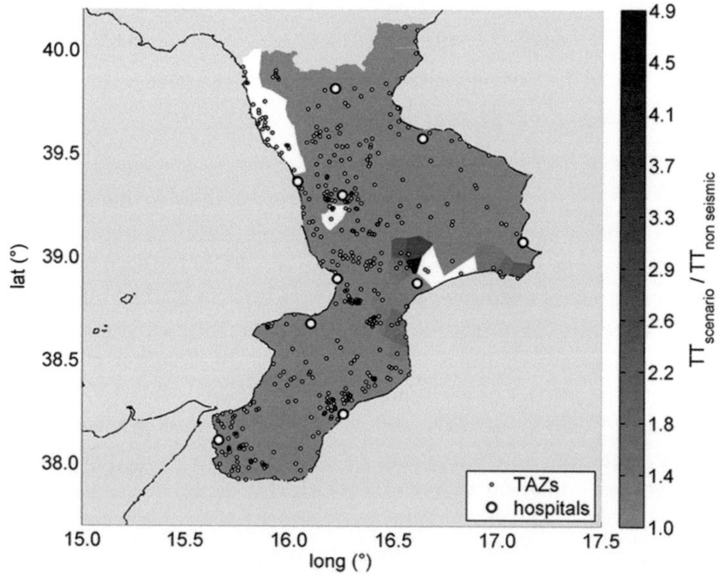

Fig. 10.14 Voronoi diagram of increment of minimum travel time to hospitals due to a return period $T_R = 500$ event

The results obtained from the Importance Sampling Simulation (2,000 runs) and the ISS enhanced with the K-means clustering (200 runs) were thus contrasted with the stable Monte Carlo results. In particular Fig. 10.15 shows the comparison of the MAFs of SCL and WCL. The match of the curves is not as good as it would be desirable, but still satisfactory. The system, on the other hand, is a road network as in the original study where these algorithms are proposed (Jayaram and

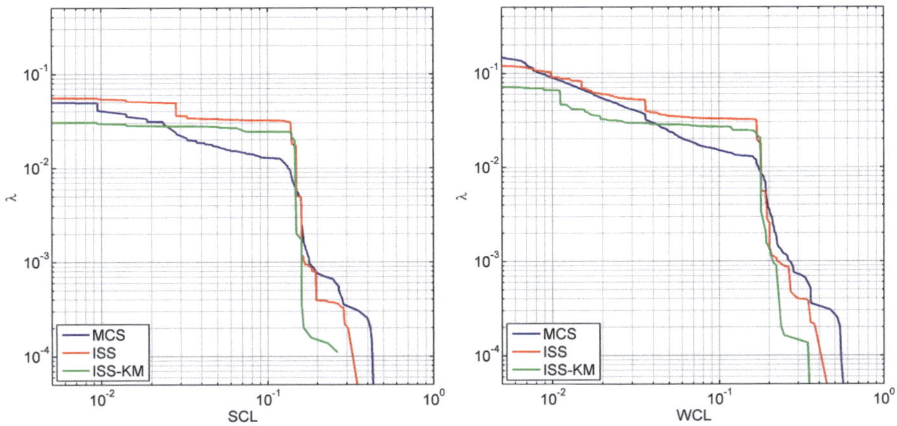

Fig. 10.15 Comparison of MAF curves obtained from MCS, ISS and ISS-KM, for SCL (*left*) and WCL (*right*)

Baker 2010). The differences, however, justify the different quality in the match. The original study uses a conventionally evaluated increase in travel time as the performance metric (static pre-earthquake OD matrix to compute flows on the damaged network), which is a more regularly behaving quantity with respect to connectivity loss measures, and validates the results coming from ISS and ISS-KM against a simulation in which only the inter- and intra-event model error terms are computed through plain Monte Carlo sampling, while the magnitudes are still obtained via IS. Further, the example used for validation was much smaller (a drastic reduction of the Bay Area network), allowing for a larger number of simulations runs to be carried out. Hence, all methods in the original reference have been run with one order of magnitude larger number of simulations runs (10^5 for MCS, 10^4 for ISS but still 10^2 for ISS-KM). All this considered, the results seem comfortably close and acceptable.

10.2 Application to the Electric Power Network of Sicily

In the following sub-sections, the application of the SYNER-G general methodology to the Sicily electric power network is discussed. Similarly to the previous section, methodological aspects are presented first, followed by a description of the case study and the simulation results. The study is carried out at the capacitive level, i.e. computing the actual power flows, voltages and currents in the network, both in the undamaged or reference state and in the damaged one.

10.2.1 Methodological Aspects and Implementation Within the SYNER-G Model

The analysis of electric power networks and its implementation within the SYNER-G model are described in Chaps. 2 and 5. Chapter 6 of Book 1 provides a selected review of the seismic fragility models used in the vulnerability assessment of these networks. According to the classification already introduced, this chapter presents a Level II analysis of the selected EPN. In analogy with the RDN, the use of a static, i.e. pre-shock demand pattern for the analysis of post-event flows is regarded as unrealistic also for EPNs or any other utility. Determination of power demands, on the other hand, does not require a complex iterative procedure such as the sequential procedure (see Sect. 10.1.1.1) but, rather, only the aggregation over tributary cells (see Chap. 2). This sum can easily account for damage to cells (displaced population, business interruption, etc.) thus leading to perturbed post-quake demand values. This capability is already implemented in the model. It is not used in the application, however, since the data set for the presented example did not contain detailed building data for the entire region of Sicily.

Figure 10.16 illustrates the EPN class diagram. In the software, the EPN is modelled as an undirected graph, i.e. a graph in which flow can occur in both directions based on voltage differences. For this reason, the EPN class is a subclass of the *Undirected* class, which in turn is a subclass of the *Network* class.

Similarly to other Network classes, the electric power network is made up of nodes and edges/lines connecting them. As a consequence, the EPN class is the composition of *EPNedge* and *EPNnode* classes, that are both abstract. The first one is the generalization of the OverheadLine and UndergroundLine classes, while the second one is the generalization of the SlackBus, PVGenerator and *LoadBus* classes. The latter is the generalization of the TransformationDistribution and Distribution classes, both of which are composed of the *Component* abstract class. This latter class is the generalization of 11 classes, one for each micro-component composing the substations.

The EPN model is exemplified with the basic network in Fig. 10.17, where some of the properties related to the network, edges and nodes, in the damaged and undamaged conditions, are reported.

All the properties and methods of the EPN class and its subclasses are listed in SYNER-G reference report 6 (Pitilakis and Argyroudis 2013). They can be grouped into five lists:

1. *List of pointers (e.g. all the connections between objects in* Fig. 10.17*)*
2. *Power network global properties (variables such as the number of nodes, or edges, or the connectivity in* Fig. 10.17*)*
3. *Edge and node properties (variables such as the length of the edge or the position of a node, or the corresponding local geology, in terms of V_{S30}, in* Fig. 10.17*)*
4. *Properties that record the state of the EPN for each event (variable "states" in* Fig. 10.17*)*
5. *Properties that store the global performance of the EPN at the end of the simulation*

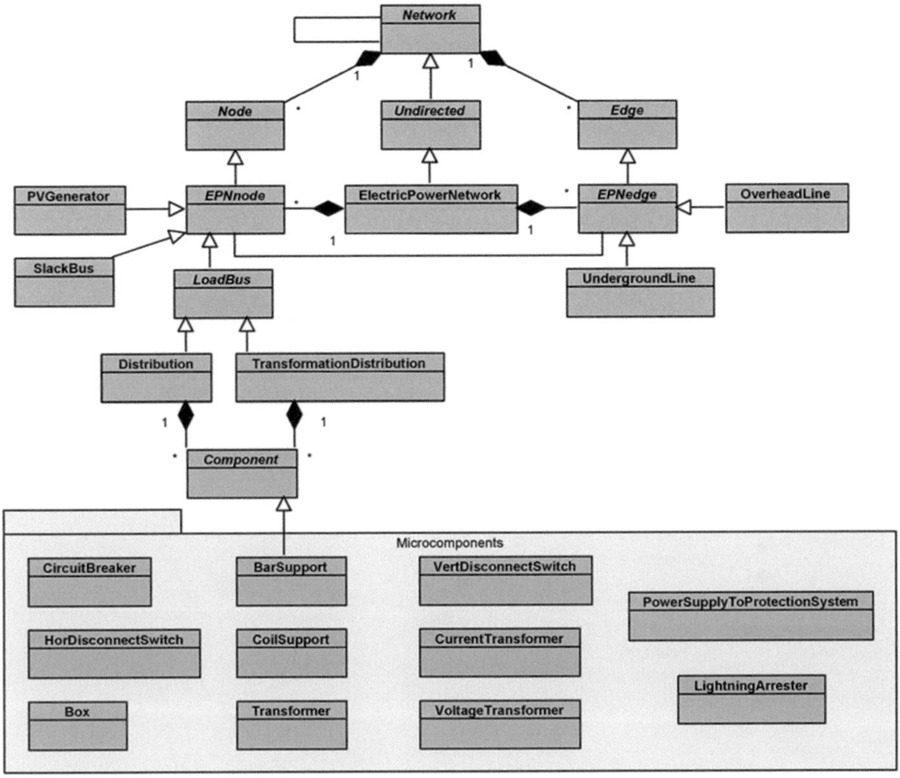

Fig. 10.16 Class diagram for the EPN class

The figure depicts a sample network with same topology of the RDN in Fig. 10.2, nodes however are differentiated in terms of type (the slack bus, as explained in Chaps. 2 and 5, is one of the generators/sources and usually coincides with that of the largest real power capacity). One important difference is that direct damage occurs in nodes rather than edges, as in the RDN. The main vulnerable elements are the components within substations, called micro-components (see Kaynia 2013, and Chapter 6 of Book 1), whereas the vulnerability of lines, especially the overhead ones, is neglected being relatively minor. It is recalled the modelling approach adopted within SYNER-G, i.e. that of describing substations in terms of their internal components. One important consequence is that within a transformation-distribution station one load bus can fail while the other remains active. This means that not all lines connected to a node are down when the substation is damaged (see example of lines converging in node n3 in the figure.). A further refinement of the model is the iterative algorithm for short-circuit propagation, that induces indirect damage in the network.

Finally, it is recalled that currently flows are determined on the portion of the damaged network that is still connected to the slack bus. This is a partially

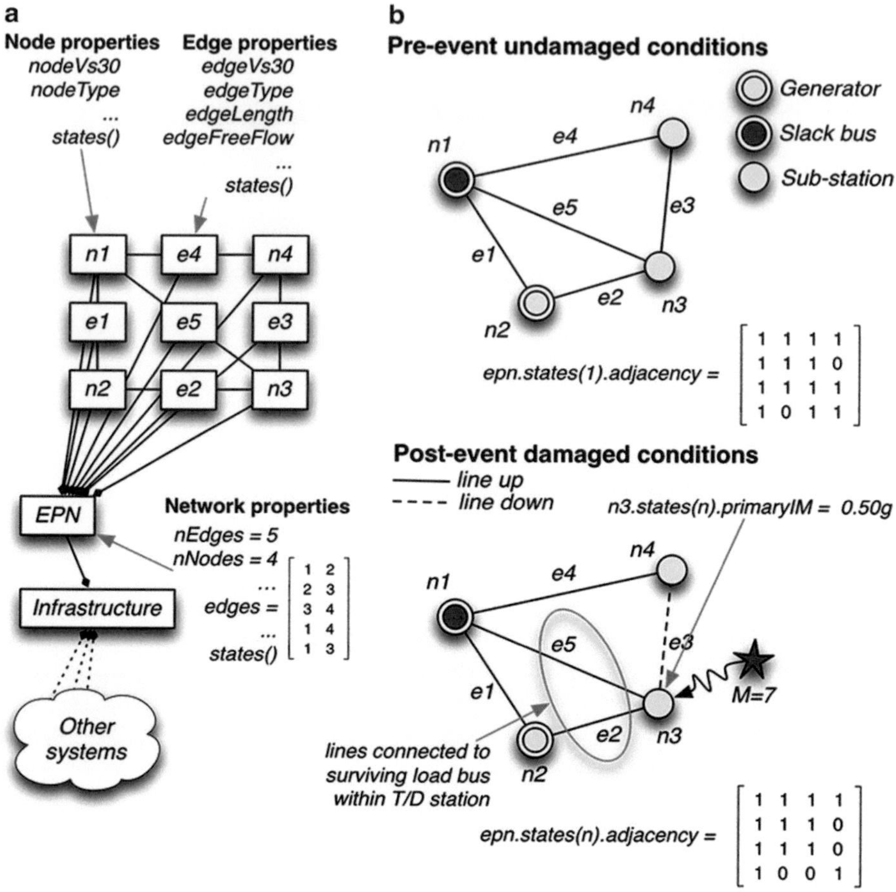

Fig. 10.17 Some properties of the EPN OO model, with reference to a basic network. (**a**) Object model. (**b**) Physical system

realistic assumption, since in reality network managers tend to restore functionality by isolating multiple sub-portions (islands) and reconnecting them step-wise. This process is not currently modelled.

10.2.2 The Case Study: Electric Power Network of Sicily

The medium-high voltage electric power network of the Sicily region, in Southern Italy (see Fig. 10.18), has been chosen as the case study. Sicily is one of the most seismic areas in the country: the last important event hitting the region is the Messina 1908 earthquake.

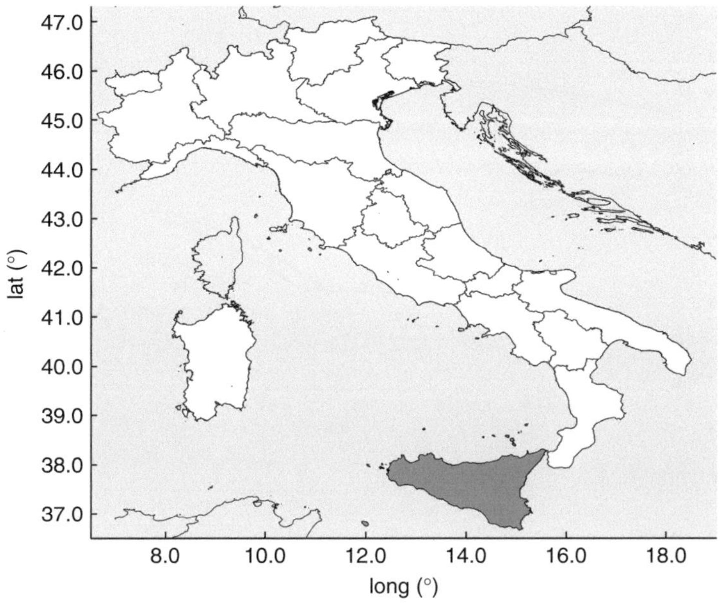

Fig. 10.18 Position of Sicily island, coloured in *grey*

10.2.2.1 Seismic Hazard

The seismic hazard is modelled through 18 faults taken from the Italian DISS (Database of Individual Seismogenic Sources) database described in Sect. 10.1.2.1. The faults are visualized in Fig. 10.19, while Table 10.3 reports their activity parameters (truncated Gutenberg-Richter recurrence model). Figure 10.20 shows a shake field of PGA, the chosen primary IM, for a scenario $M = 5.4$ event on source # 53.

10.2.2.2 System Topology and Characteristics

The available data about the system topology and characteristics are those provided by the former Italian State Electricity Board, ENEL (the network is now run under concession by another entity called TERNA, which manages only the transmission/distribution infrastructure while generation is distributed between a number of operators that compete in a recently freed market for energy). Hence, data may not be up to the date of this report. The island network is composed of 181 nodes and 220 transmission lines. The nodes, i.e. the buses, are subdivided into 175 demand or load nodes and 6 supply nodes, 5 of which are power plants and 1 is the balance node (or slack bus). Load buses, or nodes (two geographically coincident for transmission/distribution substations and one for distribution substations) deliver

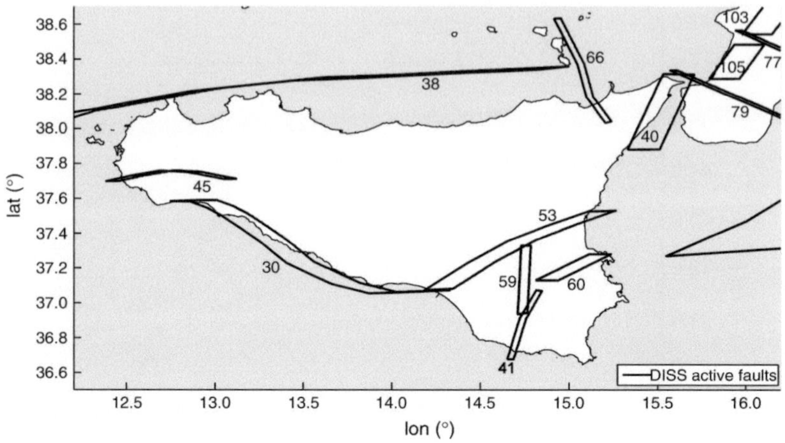

Fig. 10.19 DISS database faults affecting Sicily

Table 10.3 Activity parameters of DISS database faults affecting Sicily

Fault #	α	β	M_{min}	M_{max}
30	0.145	2.303	5.0	5.5
38	0.528	2.303	5.0	6.2
39	0.030	2.303	5.0	6.2
40	0.063	2.303	5.0	7.0
41	0.289	2.303	5.0	5.5
43	0.092	2.303	5.0	6.5
45	0.034	2.303	5.0	5.6
53	0.054	2.303	5.0	6.0
57	0.009	2.303	5.0	6.2
59	0.118	2.303	5.0	5.6
60	0.020	2.303	5.0	6.6
66	0.062	2.303	5.0	6.1
77	0.016	2.303	5.0	6.6
79	0.025	2.303	5.0	6.0
92	0.014	2.303	5.0	6.9
93	0.414	2.303	5.0	7.1
103	0.019	2.303	5.0	6.0
105	0.015	2.303	5.0	6.6

power to users. The total number of municipalities served by the network is 390 (Fig. 10.21). The balance node, generally coinciding with the generation node providing the highest power, is introduced in the adopted power-flow formulation (see Pinto et al. 2011) since power losses in the network are unknown before solving the power-flow equations. Its function is to provide the power balance between the power ingoing at generation nodes and the power outgoing from load buses, plus the power losses.

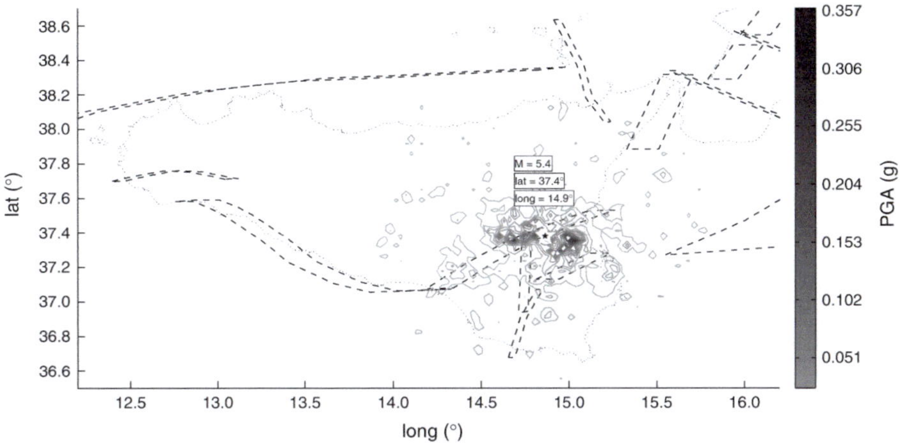

Fig. 10.20 Shake map in terms of PGA at grid points, for a scenario M = 5.4 event on source # 53

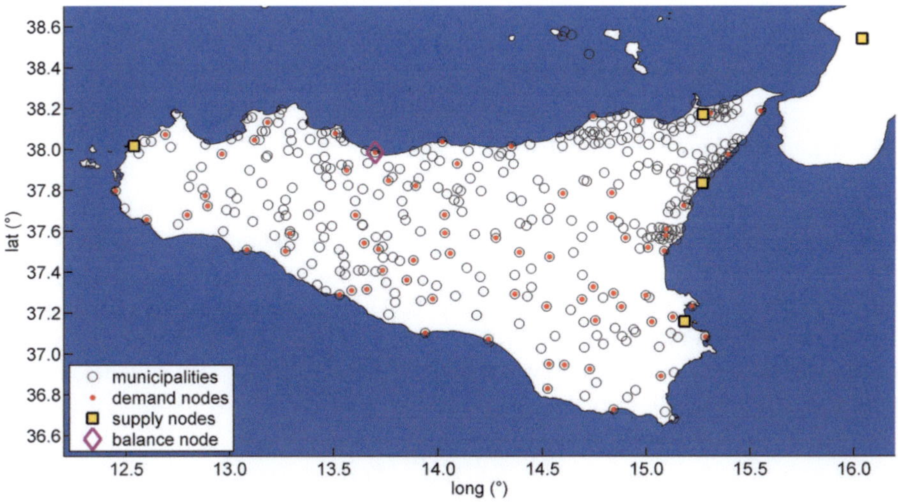

Fig. 10.21 Position of municipalities and EPN nodes

All transmission lines are overhead lines and as such they are not considered vulnerable. They are classified by voltage level into three categories, in particular 220 kV (high voltage, HV), 150 kV (medium voltage, MV) and 60 kV (low voltage, LV) lines (Fig. 10.22).

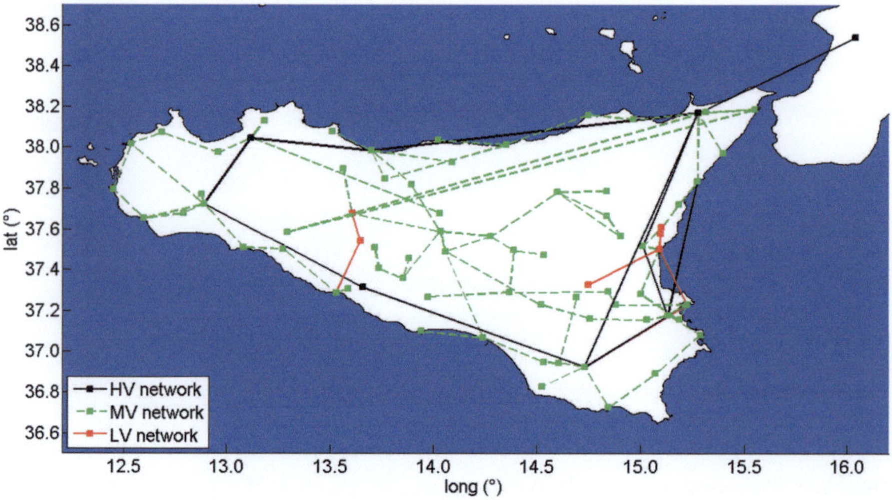

Fig. 10.22 Transmission lines, classified by voltage

10.2.3 *Analysis and Results*

10.2.3.1 Performance Indicators

A number of indicators introduced in Chap. 5 (Sect. 5.3.2.4) have been used to measure the seismic performance of the EPN. Even though the analysis addresses flows, two global connectivity-related measures are also employed: the already introduced Simple Connectivity Loss (Eq. 10.1, from now on called CL) and the Power Loss (PL) (Poljanšek et al. 2011), a weighted variant of CL which makes use of the size of power plants (in MW) to which sink vertices are still connected after the earthquake. PL is defined as:

$$PL = 1 - \langle P_s^i / P_0^i \rangle_i \tag{10.4}$$

where P_s^i and P_0^i are the sum of the real power of all the power plants connected to the i-th load bus in the seismically damaged network and in non-seismic conditions, respectively. For the case study in exam, PL yields the same values as CL, since all power plants generate the same power.

Other indicators, based on flows, are the Voltage Ratio (VR) and the System Serviceability Index (SSI). For each bus inside the substations, VR is defined as the ratio of the voltage magnitude in the seismically damaged network to the reference value for non-seismic, normal conditions:

$$VR_i = V_{i,s} / V_{i,0} \tag{10.5}$$

The voltage computation requires a power-flow analysis on the network. Hence this index expresses a functional consequence in the i-th component of the physical damage to all system components. When interactions with other systems are modelled (like more in general within the SYNER-G model), VR_i expresses the functional consequence in the i-th component of the physical damage to components of all the interacting systems, i.e. it is the value of the index that changes due to the inter- and intra-dependencies, not its definition.

SSI provides a global scalar measure of the system performance. It is defined (Vanzi 1995) as the ratio of the sum of the real power delivered from load buses after an earthquake, to that before the earthquake:

$$SSI = \frac{\sum_{i=1,\ldots,N_D} P_{i,0} \cdot (1 - R_i) \cdot w_i}{\sum_{i=1,\ldots,N_D} P_{i,0}} \cdot 100 \tag{10.6}$$

where $P_{i,0}$ is the real power delivered from the i-th load bus in non-seismic conditions, i.e. the demand. In order to compute the eventually reduced power delivered in seismic conditions, two factors are considered. The first one, $R_i = \frac{|V_{i,s} - V_{i,0}|}{V_{i,0}}$, with $V_{i,s}$ and $V_{i,0}$ the voltage magnitudes in seismic and non-seismic conditions, is the percent reduction of voltage in the i-th load bus and if $V_{i,s} < V_{i,0}$ one has $1 - R_i = VR_i$. The second factor, w_i, is a weight function accounting for the small tolerance on voltage reduction: in particular, its value is 1 for $R_i \leq 10\ \%$ and 0 otherwise. The SSI index varies between 0 and 100, assuming the value 0 when there is no solution for the power-flow analysis and 1 when the EPN remains undamaged after the earthquake. The above definition assumes that the demand remains fixed before and after the earthquake, since the index looks only at a single system, without considering the interactions of the EPN with the other infrastructure systems.

10.2.3.2 Results

As it was done also for the RDN, three types of simulation have been carried out: plain Monte Carlo (MCS), Importance Sampling (ISS) and Importance Sampling with k-means clustering (ISS-KM). MCS with 20,000 runs yield stable estimates for all considered PIs and is thus used as the benchmark for the other two simulations.

Figure 10.23 shows the moving average μ curves for CL (left) and SSI (right), as well as the $\mu + \sigma$ and $\mu - \sigma$ curves for the two PIs. The minimum sample size is strongly dependent on the chosen PI; in fact, SSI stabilises with less than 1,000 runs, whereas CL requires a much larger number of runs. The reason for this difference is that CL depends on the number of connected sources, rather than on the actual ratio of demand satisfaction at load buses. The number of connected sources is a more variable quantity, being affected by the uncertainty on short-circuit propagation, that causes a line to be turned off every time a short-circuit spreads outside a substation

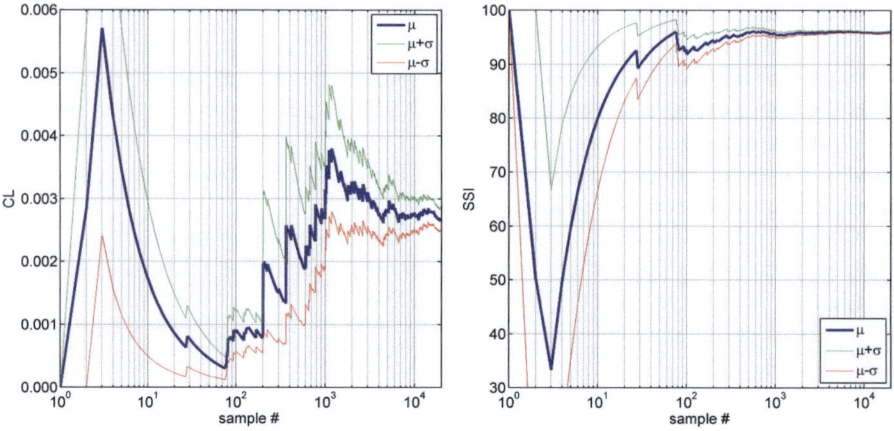

Fig. 10.23 Moving average μ, $\mu+\sigma$ and $\mu-\sigma$ curves for CL (*left*) and SSI (*right*)

Fig. 10.24 MAF curve for CL

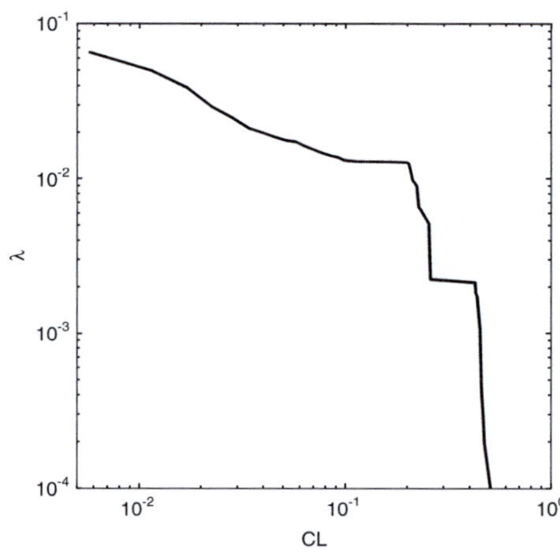

or even if it does not (as a result of the circuit breaker isolating the line). This high variability is confirmed in Fig. 10.24, which shows the MAF of exceedance for CL (the MAF curve of SSI, with values ranging in a very small interval, is not shown).

Figure 10.25 shows a Voronoi diagram of the expected values of VR, E[VR], over the whole simulation. The mean VR in each node characterizes the quality of power fed to all end-users in the area served by the node, hence a uniform colour code is used over the node tributary area (taken here equal to the Voronoi cell, collecting all points for which the node is the closest node). It can be seen that, on average, the reduction in voltage due to seismically induced damage is less than the tolerated threshold of 10 %, allowing the power demand delivery everywhere in the island, consistently with the very large values of SSI and very low values of CL.

VR maps corresponding to scenario events where damage to the EPN is severe are obtained looking at simulation samples with associated low values of SSI.

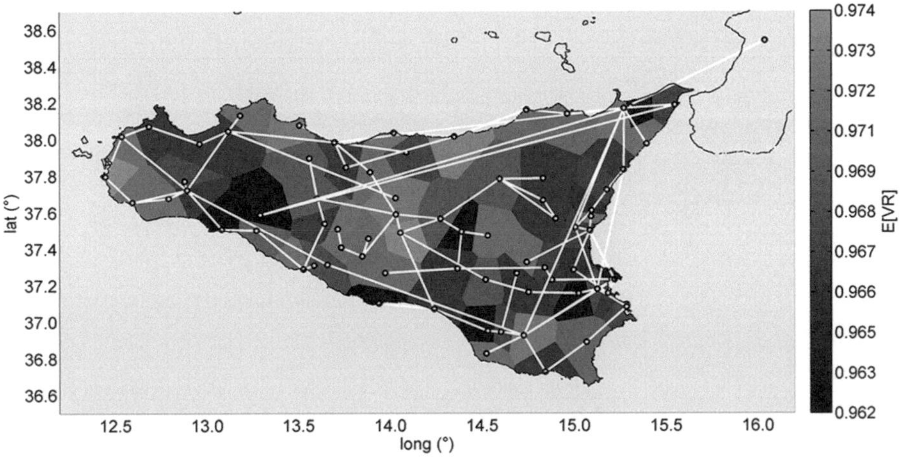

Fig. 10.25 Voronoi diagram of expected values of VR

Fig. 10.26 CDF of SSI, with selection of an SSI value with CDF = 4 %

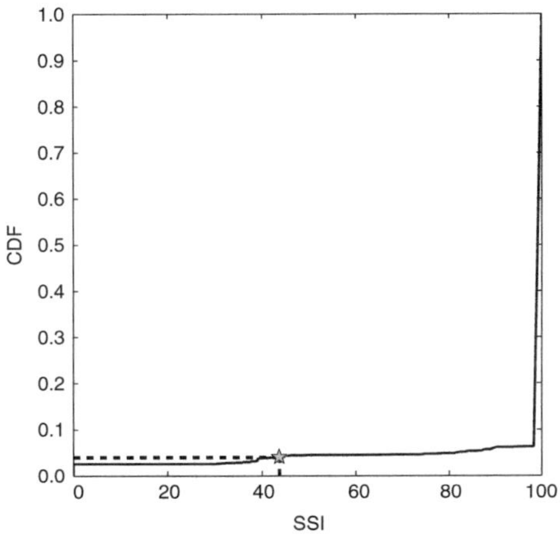

Figure 10.26 shows the cumulative distribution function (CDF) of SSI: a value of 4 % is selected from all the sampled events, and the corresponding VR map is provided in Fig. 10.27. It can be noted that large areas of the island are characterized by values ranging from 0 to about 0.9, meaning that the end-users will experience a black-out. The very high average values of VR indicate that these events are very rare.

Finally, MCS results have been taken as the reference solution and compared with those obtained by ISS (2,000 runs) and ISS-KM (200 runs). In particular, Fig. 10.28 compares moving average curves of CL (left) and SSI (right), while Fig. 10.29 compares MAF curves of the same indicators. The match is not satisfactory as it was for the RDN (Sect. 10.1), or as it has been shown to be in other applications (Jayaram and Baker 2010), although the order of magnitude is quite obviously comparable

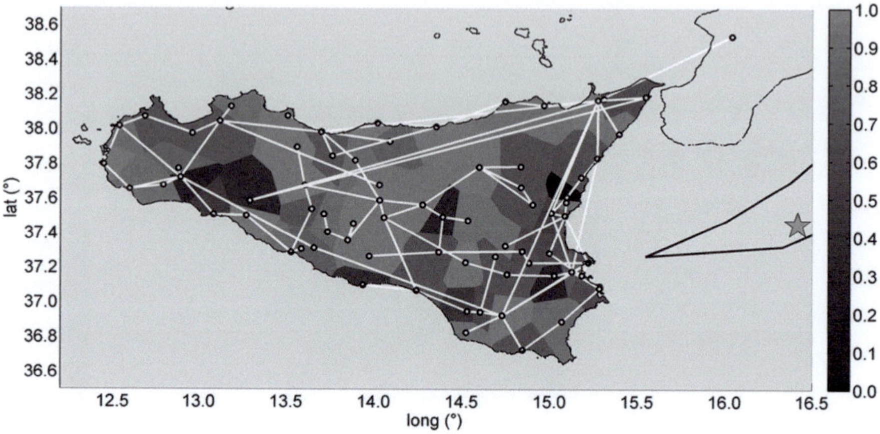

Fig. 10.27 Voronoi diagram of VR due to the selected seismic event (epicentre indicated with a *star* inside the activated fault)

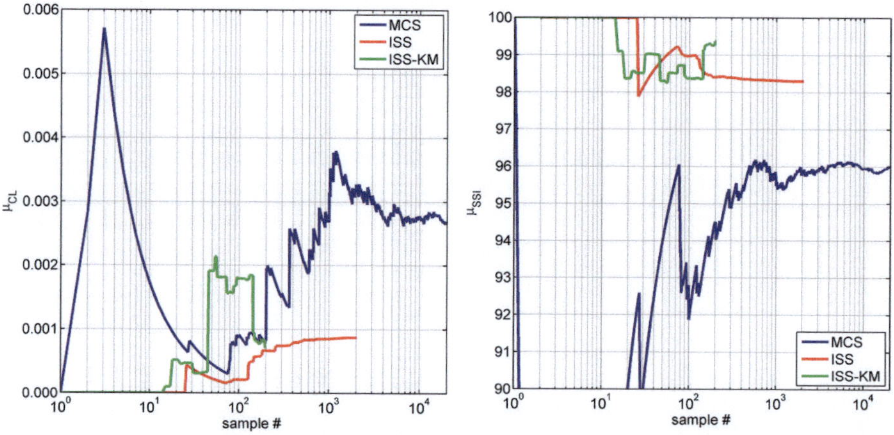

Fig. 10.28 Comparison of moving average μ obtained from MCS, ISS and ISS-KM, for CL (*left*) and SSI (*right*)

for SSI, since this quantity stabilizes with a smaller simulation number. This results hints at the need of further studies on different systems and performance measures to draw guidelines for the use of such variance reduction techniques as a practical alternative to the time-consuming plain MCS.

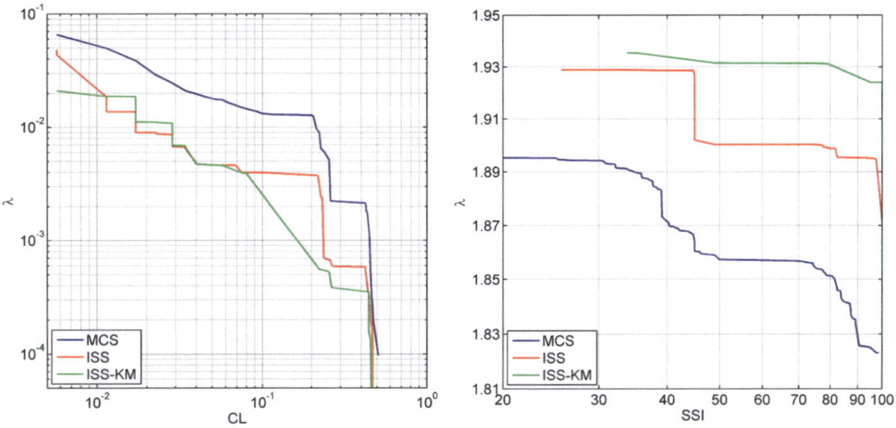

Fig. 10.29 Comparison of MAF curves obtained from MCS, ISS and ISS-KM, for CL (*left*) and SSI (*right*)

References

Akkar S, Bommer JJ (2010) Empirical equations for the prediction of PGA, PGV and spectral accelerations in Europe, the Mediterranean region and the Middle East. Seismol Res Lett 81(2):195–206

Borzi B, Fiorini E (2012) Personal communication, project DPC-Eucentre 2009–2012, subproject D4: seismic vulnerability of the national road network

Cartographic Center of Calabria region, http://www.centrocartografico.it/

Chang L, Elnashai AS, Spencer BF Jr, Song J, Ouyang Y (2010) Transportation system modeling and applications in earthquake engineering. Report 10-03, Mid-America Earthquake (MAE) Center: 1245 Newmark Civil Engineering Lab, University of Illinois at Urbana-Champaign, 205 N. Mathews Avenue, Urbana, IL, 61801

Jayaram N, Baker JW (2010) Efficient sampling and data reduction techniques for probabilistic seismic lifelines assessment. Earthq Eng Struct Dyn 39(10):1109–1131

Kaynia AM (ed) (2013) Guidelines for deriving seismic fragility functions of elements at risk: buildings, lifelines, transportation networks and critical facilities. SYNER-G reference report 4. Publications Office of the European Union, Luxembourg. doi:10.2788/19605

Levinson D, Kumar A (1994) The rational locator: why travel times have remained stable. J Am Plan Assoc 60(3):319–332

OpenSHA (Open Source Seismic Hazard Analysis) http://www.opensha.org/

Pinto PE, Cavalieri F, Franchin P, Vanzi I (2011) Systemic vulnerability and loss for electric power systems. SYNER-G deliverable report D5.2. Available online: www.syner-g.eu

Pinto PE, Cavalieri F, Franchin P, Lupoi A (2012) Systemic vulnerability and loss for transportation systems. SYNER-G deliverable report D5.5. Available online: www.syner-g.eu

Pitilakis K, Argyroudis S (eds) (2013) Systemic seismic vulnerability and loss assessment: validation studies, SYNER-G reference report 6. Publications Office of the European Union, Luxembourg. doi:10.2788/16706

Poljanšek K, Bono F, Gutiérrez E (2011) Seismic risk assessment of interdependent critical infrastructure systems: the case of European gas and electricity networks. Earthq Eng Struct Dyn 41(1):61–79. doi:10.1002/eqe.1118

SHARE (Seismic Hazard Harmonization in Europe) http://www.share-eu.org/

Vanzi I (1995). Seismic reliability of electric power networks. PhD thesis, Department of Structural Engineering and Geotechnics, Sapienza University of Rome, Italy (in Italian)
Wardrop J (1952) Some theoretical aspects of road traffic research. Proc Inst Civil Eng 1:325–378
Weatherill G, Crowley H, Pinho R, Franchin P, Cavalieri F, Iervolino I, Esposito S (2011) A review and preliminary application of methodologies for the generation of earthquake scenarios for spatially distributed systems. SYNER-G deliverable report D2.13. Available online: www.syner-g.eu

Chapter 11
Application to a Network of Hospitals at Regional Scale

Alessio Lupoi, Francesco Cavalieri, and Paolo Franchin

Abstract The seismic performance of a regional Health-Care System (HCS) is investigated. The earthquake effects both on hospitals and on the Road Network (RDN), connecting towns to hospitals, are evaluated and the interaction among them accounted for. Victims move to hospitals until their request for a bed or for a surgical treatment is satisfied, if possible. A novel "dynamic" model for hospitalization is developed and implemented. The road network is modelled in connectivity terms. The vulnerability of hospitals and bridges is expressed by pre-evaluated fragility curves. Seismic hazard is described by a state-of-the-art model. The reliability problem is solved by Monte Carlo simulation. The un-hospitalized victims, the risk that hospitals are unable to provide medical care, the demand of medical care on hospitals, the hospitalization travel time, are among the useful results of the analysis. The methodology is exemplified with reference to a case-study region, with population of 877,000, 20 towns, 5 hospitals and 32 bridges.

11.1 Introduction

This study presents the probabilistic seismic assessment of a Health-Care System at regional scale. The response of hospitals and of the road network connecting them to neighboring towns jointly contribute to satisfy the demand of assistance by the victims of an earthquake. Damaged hospitals and collapsed bridges reduce and/or delay the capability of providing medical care to the victims. In addition, it is well known that the mortality rate is substantially reduced if those injured receive medical care in a short time; in fact, transportation of the victim plays a crucial role.

A. Lupoi (✉) • F. Cavalieri • P. Franchin
Department of Structural and Geotechnical Engineering, Sapienza University of Rome, Via Gramsci 53, 00197 Rome, Italy
e-mail: alessio.lupoi@uniroma1.it; francesco.cavalieri@uniroma1.it; paolo.franchin@uniroma1.it

K. Pitilakis et al. (eds.), *SYNER-G: Systemic Seismic Vulnerability and Risk Assessment of Complex Urban, Utility, Lifeline Systems and Critical Facilities*, Geotechnical, Geological and Earthquake Engineering 31, DOI 10.1007/978-94-017-8835-9_11,

The evaluation of the reduced performances of the vulnerable elements is affected by a large uncertainty, which has to be added to the uncertainty associated to the seismic hazard. Thus, a probabilistic approach to tackle the problem is here used.

The short-term, emergency period after the seismic event (24/48 h) is the reference time-frame of the study. The main goal is to forecast the expected impact in terms of: (a) the number of un-hospitalized victims; (b) the risk of hospitals not being capable to provide the medical care required by the victims; (c) the demand of medical care on hospitals; (d) the travel time to hospitals.

The Emergency managers (e.g. Civil Protection), the hospital managers as well as all the authorities in charge of managing the emergency phase are the stakeholders which may be interested in the results of this study. The proposed methodology may also serve as a tool for planning risk mitigation measures, by considering alternative strategies (new hospitals, field hospitals, retrofitting bridges, etc.) and comparing the corresponding performances of upgraded systems.

This study follows the line of similar works by Nuti and Vanzi (1998) and Franchin et al. (2006).

11.2 Methodology

11.2.1 Hospital Facility and Treatment Capacity

The seismic assessment of a single hospital facility is studied in (Lupoi et al. 2008); results are employed in this study, which extends the analysis to a regional scale. The probabilistic procedure developed in the cited paper is briefly illustrated in this section.

The hospital is described as a system made of three vulnerable components: physical, human and organizational. These components jointly contribute in providing the medical services for the treatment of patients.

The physical component is the building where the medical services are delivered. It is made of structural elements and non-structural elements (architectural elements, basic contents and equipment). While the former are critical to preserve the life-safety of the occupants, the latter are fundamental to preserve hospital functionality.

The human component is the hospital staff: doctors, nurses and in general whoever plays an active role in providing medical care.

The organizational component is the set of standardized procedures established to ensure that medical services are delivered under appropriate conditions.

The study of the hospital's functioning under emergency conditions has led to the identification of a sub-set of medical services that have to stay operative in order to guarantee, after a seismic event, adequate treatment of patients and victims. These are classified as essential medical services and typically are: Emergency department; Operating theatres; Intensive care unit; Diagnostics; Blood bank; Hemodialysis;

Urology; Neonatology; Gynecology/Obstetrics; Pediatrics; Laboratory; Pharmacy. In addition, it has to be considered that the hospital layout (i.e. the location of medical services) during an emergency can be modified with respect to the "standard" one, i.e. the one developed for daily routine.

Previous experiences have shown that surgery is the bottleneck of medical care services after a mass-casualty event that produces trauma victims. Therefore, the number of surgical treatments that can be operated is selected as the seismic performance measure of a hospital. The number of functioning operating theatres is of primarily importance, of course, but other factors have to be also accounted for. In fact, the proposed "hospital treatment capacity" index, *HTC*, is given by the expression:

$$HTC = \alpha \cdot \beta \cdot \frac{\gamma_1 \cdot \gamma_2}{t_m} \tag{11.1}$$

where: α accounts for the efficiency of the emergency plan (organizational component), β accounts for the quality, training and preparation of the operators (human component), t_m is the mean duration of a surgical operation, γ_1 is the number of operating theatres which remain operative after the hazardous event, γ_2 is a Boolean function equal to 1 if the system "survives" and nil otherwise. The survival condition is defined as follows:

(a) the "operational" limit state has to be satisfied for the areas of the building devoted to the essential medical services;
(b) the "safeguard of human life" limit state has to be satisfied for all other areas.

Condition (a) depends on the response of both structural and non-structural elements, while condition (b) depends on the response of structural elements only.

It is noted that all components "contribute" to the determination of the *HTC* index. At the current stage of development, the influence of the organizational and human components on *HTC* is empirically estimated by expert judgment. The quality of the emergency plan provides the elements for the estimate of the coefficient α; its value typically ranges from 0.5, for very poor emergency plans, up to 1 for excellent and complete ones. The exam of the available resources and of the operators' skill to put in practice the emergency plan provides the elements for the estimate of the β factor; its value may range from 0.5, for poorly trained and understaffed operators, up to 1, for well-trained and adequately-staffed operators.

The γ_1 and γ_2 factors are instead evaluated analytically on the basis of engineering methods. The number of functioning operating theatres, γ_1, is derived from the results of structural analysis, taking into accounts also the effects on the non-structural elements.

The Boolean factor γ_2 is estimated building up a fault-tree of the physical component, which schematically depicts the vulnerable elements (columns, beams, joints, portioning walls, ceiling, windows, furniture, machineries, equipment, pipelines, etc.) and their functional interrelationship. The fault-trees of sub-systems (electrical power, medical gasses, water, elevators, etc.) are appropriately "assembled" to build up the "system" fault-tree of the whole physical component. This is hospital-dependent and has to be customized on a case-by-case basis. A preliminary, thorough examination of the vulnerable elements is recommended in order to reduce as much as possible the branches of the system fault-tree. The fault-tree analysis establishes the relationship between the state of the vulnerable elements and the state of the system.

The fragility curve of the physical component is calculated employing an "advanced" structure-specific approach, which splits in two the reliability problem (Pinto et al. 2004) in order to reduce the computational burden. Uncertainties related to structural properties (strength of materials, amounts of reinforcement, capacity models, etc.) and to the seismic hazard are both taken into account. First, a relationship between the structural response quantities *D* (forces, displacements and deformations) and the ground motion intensity measure, *IM*, is derived by means of a reduced number of numerical analyses carried out for the mean values of the structural random variables. Then, a standard Monte Carlo simulation is performed without carrying out any further structural analysis since, at each run, the response of the structure for the sampled r.v.'s is obtained from the *D*(*IM*) relationship. The state of the system is evaluated at each run of the simulation: if it fails, i.e. $\gamma_2 = 0$, no surgical treatment can be operated, and hence *HTC* is also nil.

The final outcome of the methodology is the fragility curve for the hospital performance index: $HTC = f(IM)$.

11.2.2 Casualty Model and Hospital Treatment Demand

The demand for hospitalization is estimated combining casualty models and epidemiological studies.

Casualty models provide a rapid estimate of the earthquake impact on population for the purpose of response planning and mitigation. Engineers have developed them from limited, anedoctal and historical data. The model by Coburn and Spence (1992), as simplified by Nuti and Vanzi (1998), is here employed. The casualties, *C*, understood as the "Severely-Injured" (*SI*) and the deaths, are estimated as a percentage of the population through the following expression:

$$C(I) = k \cdot (I - I_{\min})^4 \tag{11.2}$$

where *I* is the intensity measure of the seismic event,[1] *k* and I_{min} are model parameters which take into account the occupancy rate and the vulnerability of the building stock, respectively. The model parameters have to be calibrated as function of the environmental conditions. The "Lightly-Injured" (*LI*) people, defined as those not requiring hospitalization, are not included in the estimate given by Eq. (11.2).

[1]In the original model the intensity measure is the Modified Mercalli Intensity (MMI); the relationship in (Wald et al. 1999) is employed to convert MMI into PGA.

In this study an error term having lognormal distribution, unit median and coefficient of variation, ε_{cas}, equal to 0.3, is applied to the expression in Eq. (11.2) in order to account for the large uncertainties that affect the model. The number of casualties is then given by:

$$N_{cas} = N_{SI} + N_{deaths} = C(I) \cdot \varepsilon_{cas} \cdot N_{pop} \tag{11.3}$$

where N_{pop} is the population of the area affected by the earthquake.

The estimate of the people requiring hospitalization, N_{SI}, is of interest for the scope of this study. The proportion of people severely injured and the deaths can be derived from epidemiological studies, which investigate the patterns of disease and injury in human population and provide information of the type and amount of resources needed to treat casualties. The types and numbers of casualties vary with the characteristics of the earthquake, the building stock in the struck area, the demography and also with the time of the day when the earthquake occurs. The "medical severity" of a hazardous event is assessed by two severity indexes:

$$S_1 = N_{deaths}/\,(N_{SI} + N_{LI}) \tag{11.4}$$

which gives an indication of the overall severity of the event (deaths over all injured), and

$$S_2 = N_{SI}/N_{LI} \tag{11.5}$$

which measures the severity of the injuries caused by the event ("Severely-Injured" over "Lightly-Injured"). The larger is S_2, the greater is the demand of medical care. Data from past earthquakes show that S_1 is comprised between 0.1 and 0.5, while S_2 between 0.15 and 0.6. The ratio deaths/"Severely-Injured" is obtained combining Eqs. (11.4) and (11.5):

$$N_{deaths}/N_{SI} = S_1\,(1 + 1/S_2) \tag{11.6}$$

Finally, the number of victims requiring hospitalization, N_{SI}, is obtained by combining Eqs. (11.6) and (11.3):

$$N_{SI} = [S_2/\,(S_1 + S_1 S_2 + S_2)] \cdot N_{cas} \tag{11.7}$$

The Hospital Treatment Demand, *HTD*, is defined as the sub-set of the Severely-Injured that require a surgical treatment:

$$HTD = \zeta \cdot N_{SI} \tag{11.8}$$

The coefficient ζ is defined on a case-by-case basis by expert opinion; it typically ranges between 1/3 and 1/2.

The complement of *HTD* is the number of victims to whom it is sufficient to assign a bed to administer the necessary medical care:

$$\overline{HTD} = (1 - \zeta) \cdot N_{SI} \tag{11.9}$$

11.2.3 Road Network

In this study the function of the road network is to allow the transportation of the injured to hospitals. The primary interest is to identify the portions of the network that may be critical with respect to connectivity as a consequence of the seismic damages eventually experienced by its vulnerable elements.

The analysis is carried out in terms of pure connectivity, i.e. traffic flows are not modeled. This is coherent with the time-frame of the study, which is limited to rescue operations in the aftermath of the seismic event. This approach requires a simple description of the network based on basic graph theory. The road network is represented as a graph consisting of n nodes or vertices, connected by n_a arcs, or links or edges. The relationship between nodes and arcs is described by the adjacency matrix $B = [b_{ij}]$, which is an $n \times n$ Boolean square matrix, whose terms are either 0, when no connection exists between nodes i and j, or 1 when a connection exists. A free-flow travel speed is assigned to each arc of the graph.

Towns within the region are "concentrated" at the graph's nodes; each node is called TAZ (Traffic Analysis Zone). Hospitals are "housed" within towns. A zero-distance/zero-travel time connection is assumed for a hospital and its "hosting" town.

It is assumed that the highways free-flow speed does not change after a seismic event, while a 50 % reduction in speed is considered for the urban portions of the road network in order to account for potential road blockages.

11.2.4 Transportation and Medical Treatment Model of the Victims

Transportation is assumed to take place by private vehicles on the damaged road network. The selection of the hospital, made by users, is affected by both objective constraints and subjective choices. The closure of a road represents one of the former; the user "familiarity" with a specific facility is one of the latter. This section briefly addresses the proposed model for the transportation of casualties to the hospitals of the region of interest.

The implemented hospitalization algorithm is "dynamic": the process follows a chain of events, which are scheduled in ascending order based on their time of occurrence, elapsed from the earthquake time. Victims are allocated based on their arrival time, i.e. following the "first-come, first-served" criterion.

At the very beginning of the process, for each TAZ an event is created to allow its casualties, if present, to be moved to the closest hospital, i.e. the one which can be reached within the minimum travel time. All events are stored in an event matrix, recording the occurrence time, the origin TAZ, the destination hospital and the number of estimated victims which are going to move.

Before starting to go through all the events, the health-care capacity of all the regional hospitals is checked, both in terms of $\overline{HTD}$ (by counting the number of available beds) and in terms of *HTD* (verifying that the *HTC* of the damaged hospitals is greater than zero). "Unavailable" hospitals are excluded.

Successively, the single events are analyzed in their order of occurrence. For each event, in case the destination hospital is able to receive all incoming victims, its capacity is reduced and the next scheduled event is taken into consideration. On the contrary, if the destination hospital is severely damaged (not operative, $\gamma_2 = 0$) or has its capacity saturated, separately for the two types of victims (in terms of number of available beds or number of functioning operating theatres), all or a portion of the victims cannot be hospitalized and, hence, these are forced to move to the closest hospital facility (which has not been visited yet). In this case, in the event matrix is added a further event, with an occurrence time given by the travel time needed to reach the destination hospital at hand plus the travel time to the next hospital. The event matrix is then sorted again.

The iterative process is concluded either when all the casualties are hospitalized or when all the functioning hospitals in the region are saturated (all available beds are assigned or $HTC \leq HTD$).

An alternative modeling choice could be to assume that the injured victims that do not need a surgical treatment, i.e. $\overline{HTD}$, always receive medical assistance at the first operative hospital reached. This assumption of "infinite" beds capacity is justifiable in view of the emergency procedures' activation, where the number of beds may be doubled with respect to the standard, "every-day" condition (eventually by field hospitals).

11.2.5 System's Performance Indicators

A number of results which are useful primarily for hospital disaster managers and civil protection can be inferred from this study. We have selected the following ones: (a) the mean annual frequency of un-hospitalized victims, subdivided into those requiring a surgical treatment and those requiring a bed, as performance measure of the regional health-care system; (b) the risk that a hospital is not capable of providing the required surgical treatment ($HTC < HTD$), as performance indicator for a single facility; (c) the (expected) demand of medical care on each hospital, an essential information to get prepared and to cope with the emergency; (d) the (expected) number of un-hospitalized victims subdivided by towns, a useful information for policy makers; (e) the (expected) maximum travel time for hospitalization, an indicator of the adequateness of the road network and an indication for future upgrade and/or seismic retrofit. All the above information may be useful for the allocation of resources.

11.3 Treatment of Uncertainties

The regional system under assessment is affected by many and large uncertainties of different types, from those related to the regional seismic activity and the (corresponding) local intensity at each site, to those related to the physical damage state as a function of local intensity, to the uncertainty on the parameters (or even the form) of the fragility models employed.

Uncertainty on the seismic hazard is modeled through two models, the event model and the local intensity model (Franchin and Cavalieri 2013). The event model starts with a continuous variable M for the event magnitude, continues with a discrete random variable Z for the active zone, with as many states as the number of seismo-genetic zones, and ends with a random variable L for the epicenter location within the active source. Distributions vary according to the adopted sampling scheme, but that of Z is conditional on the sampled value of M, and that of L is conditional on the sample zone Z.

The local intensity measure at the sites of vulnerable components is described with a vector of *IMs* that are needed as an input to the corresponding fragility model. A scalar random field of a so-called "primary *IM*", e.g. peak ground acceleration (PGA) on rock (no amplification yet), is first sampled as a function of the sampled M and L on a regular grid covering the study region, employing a ground motion prediction equation (GMPE) with inter- and intra-event error terms η and ε. In the application to follow the employed GMPE is that by Akkar and Bommer (2010). Intra-event residuals ε are modeled as a spatially correlated random field (Jayaram and Baker 2009) by means of an exponential auto-correlation function derived for Italian events and consistent with the Akkar and Bommer GMPE in (Esposito et al. 2010). The need for sampling on a regular grid first arises to avoid singularity problems in the covariance matrix of intra-event residuals, since sites usually occur in clusters with very similar source-to-site distances. The primary *IM* is then interpolated to all sites and "secondary *IMs*" (all other components in the intensity vector at a site) are sampled from their distribution conditional on the primary *IM* value (postulating joint lognormality of the *IMs*, see e.g. Bazzurro and Cornell 2002, and using inter-IM correlation values from Baker and Cornell 2006).

The uncertainty related to the vulnerable components, i.e. hospitals and bridges, is accounted for by means of pre-evaluated fragility functions. Its derivation is out of the scope of this study; therefore these are taken from previous studies. The derivation of the fragility curve for a hospital is illustrated in Sect. 11.2.1.

The uncertainty in the estimation of the victims is accounted for by the random variable ε_{cas} as described in Sect. 11.2.2.

The reliability problem is solved employing a standard simulation-based method. A simulation run is carried out as part of either a plain (robust) Monte Carlo simulation or a more (computationally) effective importance sampling scheme (e.g. Jayaram and Baker 2010). The state of all components is sampled as a function of the input intensity measures; once this is known, the functional analysis of the regional system is carried out to determine its performance. Interactions among components are considered at this stage such as, for instance, detour to reach hospitals from area districts due to closure of damaged portions of roads.

11.4 Example Application

11.4.1 The Case Study Area

A hypothetical region composed of towns, hospitals and a road network is shown in Fig. 11.1. The architecture is developed from the application example in Kang et al. (2008), with some modifications and additions in order to form an infrastructure that is subjected to a distributed seismic hazard and in which the road network and

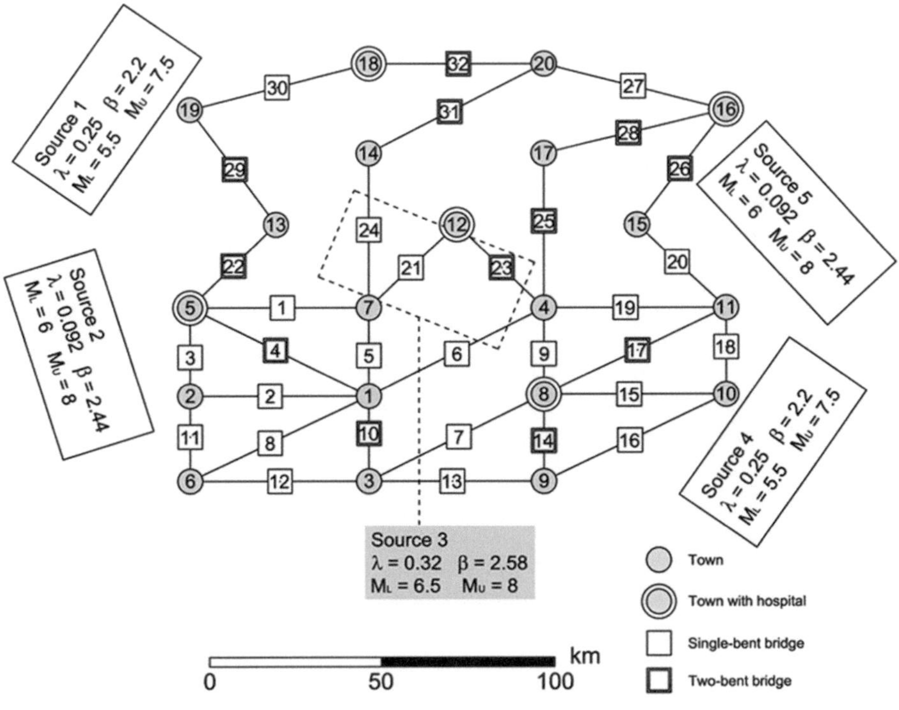

Fig. 11.1 The regional study area

Table 11.1 Number of beds, surgical treatments per day and vulnerability for the five hospitals in the region

Hosp. #	TAZ #	Total beds	Occup. beds	Avail. beds	Max HTC per day (no damage)	HTC curve #
1	5	780	624	156	77	3
2	8	724	579	145	77	3
3	12	690	552	138	77	1
4	16	668	534	134	77	2
5	18	646	517	129	77	2

hospital care system (RDN/HCS) interaction is taken into account. The seismogenetic area sources that can generate events affecting the region are also shown in Fig. 11.1, together with their corresponding activity parameters for the truncated Gutenberg-Richter recurrence law: lower and upper magnitude limits M_L and M_U, magnitude slope β, and mean annual rate of events with $M > M_L$ in the source λ.

The transportation network connects 20 towns by highways with 32 bridges. A pure connectivity approach is employed, i.e. no traffic flows are computed in the damaged network. For simplicity, given the illustrative character of the application, it is assumed that no other roads aside from the highways exist between cities and that the bridges are the only vulnerable components, whose earthquake-induced damage may cause paths to be disconnected.

The 20 towns have populations ranging from 22,000 to 70,000 inhabitants, for a regional population equal to 877,000. The towns centroids (TAZs) are taken as the RDN nodes.

The HCS comprises five hospitals, located in towns #5, 8, 12, 16 and 18. The total number of beds is taken as 0.4 % of population; an occupancy, pre-earthquake rate equal to 80 % is assumed. Hospitals relevant data are summarized in Table 11.1.

The fragility curves employed for the hospitals in the region are relative to an existing RC hospital located in Italy (Lupoi et al. 2008). These have been computed as described in Sect. 11.2.1, introducing uncertainties in both structural and non-structural elements. The factors α and β in Eq. (11.1) have been taken equal to 1 and 0.8, respectively. The reference hospital is equipped with eight operating theatres; the mean duration of a surgical treatment has been taken equal to $t_m = 2$ h. It was assumed that in emergency condition a theatre works 24 h continuously; 24 h is also the maximum time that a severely injured patient can wait for the surgical treatment. The maximum *HTC* for the undamaged configuration of the reference hospital is then: $HTC_{undamaged} = [(1 \times 0.8 \times 1 \times 1)/2] \times 24 \times 8 = 77$ treatments per day.

Figure 11.2 shows three *HTC* fragility curves, expressing the damaged or residual (post-seismic conditions) *HTC* as a function of the ground motion PGA. The displayed curves, corresponding to the mean and the mean minus/plus one standard deviation fragility curves, are indicated as curves #1, #2 and #3, and have been assigned to the five hospitals as indicated in Table 11.1.

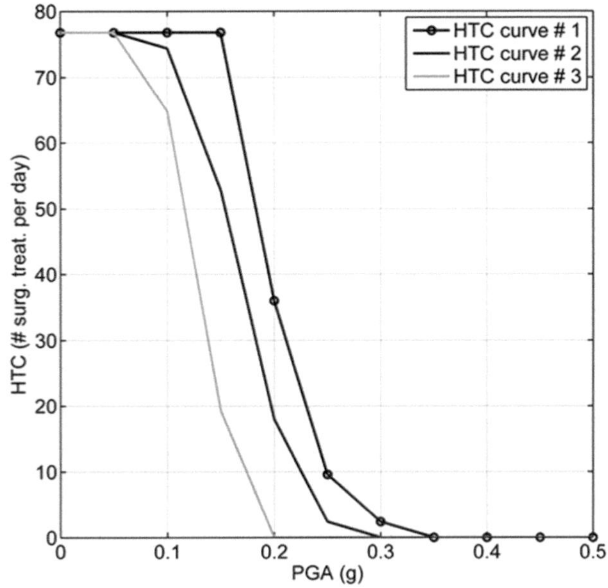

Fig. 11.2 HTC fragility curves for hospitals

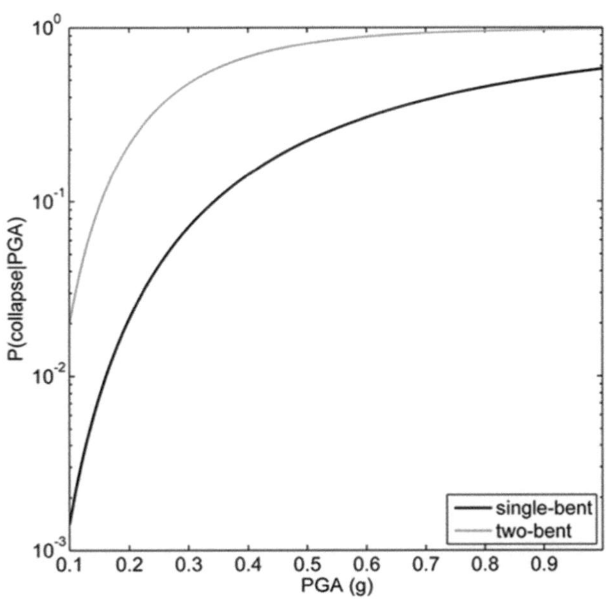

Fig. 11.3 Fragility curve for bridges

Two types of bridge are included in the road network: single-bent and two-bent overpasses. The corresponding fragility curves, expressing the conditional probability of attaining or exceeding the collapse limit state for a given value of PGA, are shown in Fig. 11.3. At each run of the Monte Carlo simulation, the state of a bridge can be "collapsed" or "survived"; no "intermediate" states are considered.

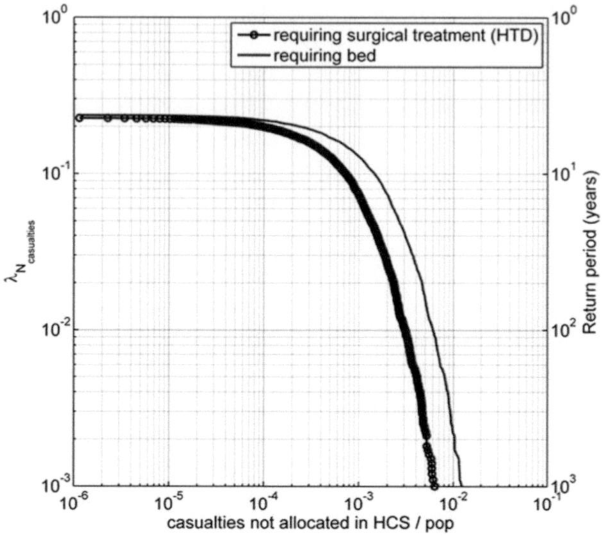

Fig. 11.4 MAF curves of normalized victims (divided in two categories) that are not allocated in hospitals

The casualty model parameters k and I_{min} in Eq. (11.2) are taken equal to 0.05 and 4, respectively; the severity indexes S_1 and S_2 in Eq. (11.7) are taken equal to 0.154 and 0.625, respectively (FEMA 1999).

11.4.2 Simulation Results

A plain Monte Carlo simulation with 10,000 runs is carried out to test the proposed methodology.

The global performance of the regional Health-Care System is measured in terms of the number of victims that cannot receive the medical care. The Mean Annual Frequency (MAF) of exceedance curves for un-hospitalized victims subdivided by HTD and $\overline{HTD}$, normalized to the regional population, are shown in Fig. 11.4. The same curves can also be referred in terms of return period. This latter is commonly used to better communicate the actual risk to stakeholders. For example, the return period of an event where 0.1 % of the regional population cannot receive the (needed) surgical treatment is about 40 years.

The performance of the hospitals in the region is expressed in terms of the probability of not being able to provide the required surgical treatments to victims if an earthquake strikes the region (i.e. the risk), as shown by the bar plot in Fig. 11.5. The risk is higher for the hospitals located in TAZ #8 and #12, because their "tributary" area is greater than the one of the three other hospitals and also because of the proximity to the central seismic source #3.

The (expected) demands of medical care on hospitals are indicated in the area plot of Fig. 11.6. This is the basic information which is needed by hospital disaster

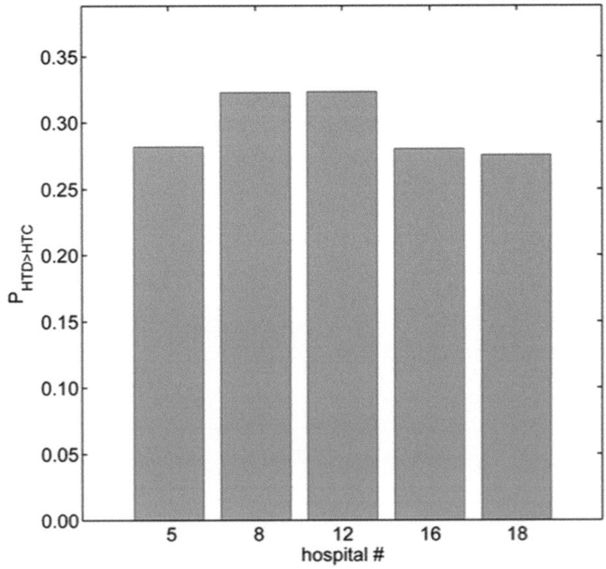

Fig. 11.5 P(HTD ≥ HTC), for the five hospitals

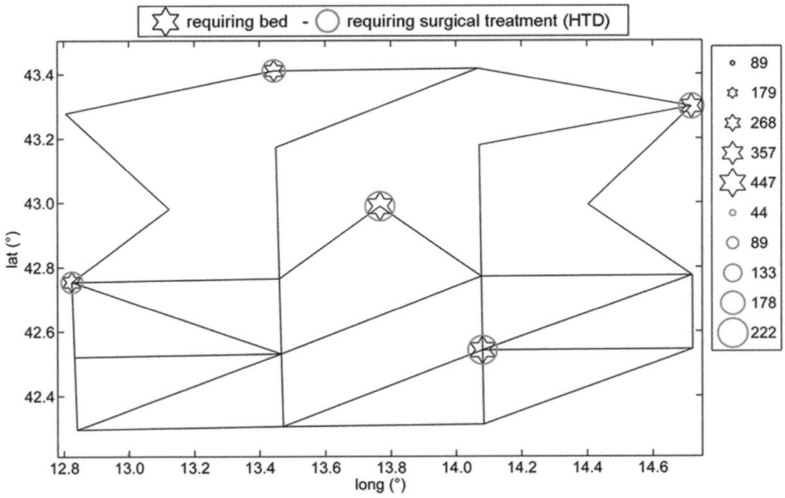

Fig. 11.6 (Expected) demand of medical care on hospitals

managers and civil protections to set up adequate emergency strategies. The higher demand is registered for the central hospital in TAZ #12 for the same reason given above: the (expected) number of victims requiring a surgical treatment that reach the hospital in TAZ #12 is 222. It is noted that this number includes also those that can not be hospitalized in this hospital for "saturation" of its capacity (damaged *HTC*) and have to move to the closest one, the selection of the latter

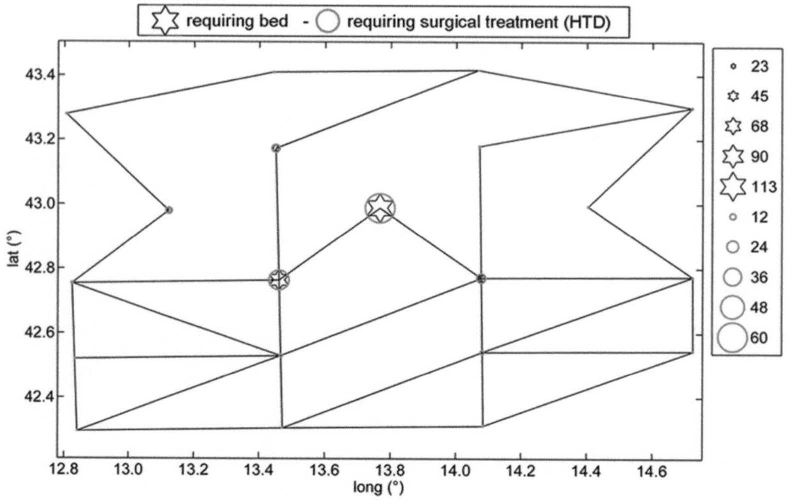

Fig. 11.7 (Expected) victims not able to receive medical care

depending on the state of damaged road network. The hospitalization process of the victims takes place within the first 24 h after the event, or in an even shorter period.

The (expected) number of victims not able to receive medical care is indicated in the area plot of Fig. 11.7; victims are referred to their town of origin. These data give indication of the critical (worst-served) towns, information that can be useful to policy makers for the planning of new facilities and/or for the retrofit of existing ones.

Finally, the (expected) maximum hospitalization travel time for the investigated region is about 70 min. The moving average μ and moving standard deviation σ are computed at each simulation run. Corresponding curves of μ and $\mu \pm \sigma$ are shown in Fig. 11.8. The mean of the indicator becomes stable after about 1,000 runs; this justifies the adopted number of runs.

11.5 Conclusions

A methodology for the seismic assessment of a regional Health-Care System is illustrated in this study. The system is composed of hospitals, towns and a road network. The road network is deputed to connect towns to hospitals allowing the transportation of the injured people.

A probabilistic approach has been employed to model the large uncertainties that affect the problem. The vulnerability of the main components of the system,

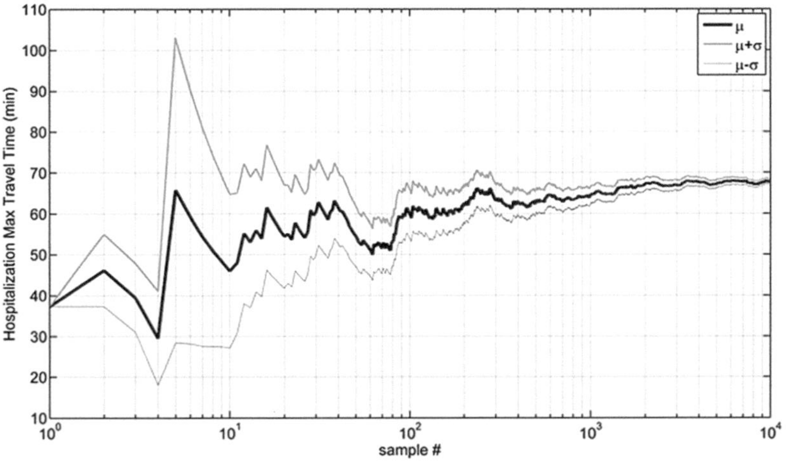

Fig. 11.8 Evolution of maximum travel time for hospitalization

i.e. hospitals and bridges, is accounted for as well as their interaction. In particular, the hospitals' capacity and the bridges' physical damage are represented by fragility curves. Uncertainties in the evaluation of the casualties are also introduced. A state-of-the-art model to compute the seismic hazard is also employed. Finally, a "dynamic" model for the hospitalization of the victims has been developed and implemented: the path of the victims from the town of origin to the hospital of destination through the road network is considered. Hospitalization, for each homogenous type of victims, is accorded on a "first-come, first-served" criterion. This model represents a novelty of the proposed methodology with respect to other similar applications available in the literature.

The reliability problem is solved by Monte Carlo simulation. A number of interesting results are provided by the analysis: the number of un-hospitalized victims expressed either as mean annual frequency of exceedance or as return period; the risk of not being able to provide the required medical care for each hospital of the region; the (expected) demand of medical care on each hospital; the (expected) un-hospitalized victims for each town in the region; the (expected) maximum hospitalization travel time. These results are certainly of great value to emergency managers, policy makers and authorities involved in planning emergency operations and in developing mitigation strategies.

The capability of the proposed methodology is successfully tested through the application to a realistic study region.

Acknowledgements The contribution of Dr Giorgio Lupoi for the development of hospitals' fragility curves is gratefully acknowledged.

References

Akkar S, Bommer JJ (2010) Empirical equations for the prediction of PGA, PGV and spectral accelerations in Europe, the Mediterranean region and the Middle East. Seismol Res Lett 81(2):195–206

Baker JW, Cornell CA (2006) Correlation of response spectral values for multicomponents ground motions. Bull Seismol Soc Am 96(1):215–227

Bazzurro P, Cornell CA (2002) Vector-valued probabilistic seismic hazard analysis (VPSHA). In: 7th U.S. national conference on earthquake engineering, Boston, MA, 21–25 July 2002, Paper no. 61

Coburn A, Spence R (1992) Earthquake protection. Wiley, Chichester

Esposito S, Iervolino I, Manfredi G (2010) PGA semi-empirical correlation models based on European data. In: 14th European conference on earthquake engineering, Ohrid, FYROM

FEMA (1999) Earthquake loss estimation methodology Hazus99 SR2: Technical Manual. Federal Emergency Management Agency, Washington, DC

Franchin P, Cavalieri F (2013) Seismic vulnerability of a complex interconnected infrastructure. In: Tesfamariam S, Goda K (eds) Handbook of seismic risk analysis and management of civil infrastructure systems. Woodhead Publishing Limited, Cambridge. ISBN 978-0-85709-268-7 (Not yet published)

Franchin P, Lupoi A, Pinto PE (2006) On the role of road networks in reducing human losses after earthquakes. J Earthq Eng 10(2):195–206

Jayaram N, Baker JW (2009) Correlation model of spatially distributed ground motion intensities. J Earthq Eng Struct Dyn 38(15):1687–1708

Jayaram N, Baker JW (2010) Efficient sampling and data reduction techniques for probabilistic seismic lifelines assessment. Earthq Eng Struct Dyn 39(10):1109–1131

Kang W-H, Song J, Gardoni P (2008) Matrix-based system reliability method and applications to bridge networks. Reliab Eng Syst Saf 93(11):1584–1593

Lupoi G, Franchin P, Lupoi A, Pinto PE, Calvi GM (2008) Probabilistic seismic assessment for hospitals and complex-social systems, Rose School technical report 2008/02. IUSS Press, Pavia. ISBN 978-88-6198-017-4

Nuti C, Vanzi I (1998) Assessment of post-earthquake availability of hospital system and upgrading strategies. Earthq Eng Struct Dyn 27(12):1403–1423

Pinto PE, Giannini R, Franchin P (2004) Seismic reliability analysis of structures. IUSS Press, Pavia

Wald DJ, Quitoriano V, Heaton TH, Kanamori H (1999) Relationships between peak ground acceleration, peak ground velocity, and modified Mercalli intensity in California. Earthq Spectra 15(3):557–564

Chapter 12
Application in the Harbor of Thessaloniki

Kalliopi Kakderi, Jacopo Selva, and Kyriazis Pitilakis

Abstract The SYNER-G methodology and tools for the assessment of the systemic vulnerability and performance of harbors are applied in the case of Thessaloniki's port, a major export and transport harbor of Greece and the European Union's closest port to the countries of Southeast Europe. Following the methodological framework for the systemic analysis developed in SYNER-G, waterfront structures, cargo handling equipment, power supply system, roadway system and buildings are examined. The systemic vulnerability methodology and software implementation are first described, followed by the description of the system topology and characteristics and the input for the analysis. Characteristic results of the application are provided and commented. Performance Indicators (PIs) are calculated based on the estimated damages and functionality losses of the different components. Apart of the average performance and the Mean Annual Frequency (MAF) of exceedance of the PIs, the distribution of estimated damages and losses for a specific event is also given through thematic maps. The most critical elements for the functionality of the port system are defined through correlation factors to the system PIs. Finally, the epistemic uncertainty related to the use of different fragility functions and functionality definitions is investigated by performing selected sensitivity analyses.

K. Kakderi (✉) • K. Pitilakis
Department of Civil Engineering, Aristotle University, 54124 Thessaloniki, Greece
e-mail: kkakderi@gmail.com; kpitilak@civil.auth.gr

J. Selva
Istituto Nazionale di Geofisica e Vulcanologia, Via D. Creti 12, 40128 Bologna, Italy
e-mail: jacopo.selva@bo.ingv.it

K. Pitilakis et al. (eds.), *SYNER-G: Systemic Seismic Vulnerability and Risk Assessment of Complex Urban, Utility, Lifeline Systems and Critical Facilities*, Geotechnical, Geological and Earthquake Engineering 31, DOI 10.1007/978-94-017-8835-9_12,

12.1 Introduction

Port transportation systems are vital lifelines whose primary function is to transport cargos and people. They contain a wide variety of facilities for passenger operations and transport, cargo handling and storage, rail and road transport of facility users and cargoes, communication, guidance, maintenance, administration, utilities, and various supporting operations. Ports offer wide-open areas that can be used for emergency or refuge activities after a damaging earthquake. Moreover, they can play an important role during the recovery period, as they contribute to the reconstruction assistance and the transportation of goods for homeless citizens.

In view of the importance of ports in today's society, it is clear that the extended loss of function of major ports could have major regional, national, and even worldwide economic impacts. Harbors comprise complex systems consisting of several lifelines and infrastructures, which interact with each other and with the urban fabric. For the assessment of the complex system performance, contributions of all components, and their interactions, have to be appropriately accounted for.

Current engineering practice for seismic risk reduction of port facilities is typically based on design or retrofit criteria for individual physical components. However, the resilience and continuity of shipping operations at a port after an earthquake depends not only on the performance of these individual components, but on their locations, redundancy, and physical and operational connectivity as well, that is, on the port system as a whole. Within SYNER-G a general methodology and appropriate tools were developed for the assessment of the systemic vulnerability and performance of harbors, simulating port operations and considering also the interactions among port elements.

The objective of the present study is to apply and test the methods and tools developed in SYNER-G through the case study of the port of Thessaloniki. The general methodological framework is described in Chap. 2 and its specificities for the harbor system in Chap. 5. Specific details on the port systemic vulnerability methodology and software implementation are provided in this chapter. Following the methodological framework for the systemic analysis, waterfront structures, cargo handling equipment, power supply system, roadway system and buildings are examined. Appropriate fragility curves are applied for the vulnerability assessment of each element at risk. The seismic hazard is assessed through the shakefield method (Chap. 3), making use of the seismic zones proposed in SHARE project (Giardini et al. 2013). The results are obtained through a Monte Carlo simulation (MCS) (10,000 runs) which samples earthquake events, consequent hazard scenarios and, for each scenario, damages and functional consequences based on the methods and tools developed in SYNER-G.

The overall performance of the port system is expressed through the "moving average" μ and "moving standard deviation" σ (both computed by averaging over simulations, from 1 to the j-th simulation), as well as the Mean Annual Frequency (MAF) of different values of each PI ("performance curve"). The average loss is defined based on the moving average graph, at the end of the simulation (j = 10,000). The earthquake event that corresponds to a return period of the PI equal to $T_R = 500$ years is identified from the "performance curve" and maps with the distribution of damages for such selected events are produced. The correlation of each component to the system PIs is estimated, based on all simulations. This type of analysis, as it is based on the results of each single event, preserves the information about system's topology, its behavior in case of spatial correlated damages (related to single earthquakes) and functional inter-dependencies. Thus, it allows identifying the most critical elements for the functionality of the port system (i.e., the damaged components that more closely control the performance of the harbor).

Several sources of uncertainties are inherent to the analysis, related among others to the seismic hazard and spatial correlation models, and the fragility or the functionality assessment of each component. The epistemic uncertainty related to different fragility functions and functionality definitions is investigated by performing sensitivity analysis with the use of alternative fragility curves and functionality thresholds for the waterfront structures.

12.2 Systemic Vulnerability Methodology

For the assessment of the systemic vulnerability of harbors, it is essential to simulate port operations. Since most of the dry cargo in modern ports is containerized, it has been decided in this application to focus on the operation of container handling. However, bulk cargo is also important from the viewpoint of risk management on economic activities such as industrial and insurance market. Given that, in the aftermath of significant natural disasters such as an earthquake event, a port can operate as a "gate" for delivering the necessary assistance to the city, the importance of this analysis may also go beyond the strictly economic consequences. The passenger movement is also an important element to monitor a depression and recovery process of port function. However, there is not enough data on passenger movement to assess the vulnerability in past earthquake events. From this point of view, it would be difficult to develop simulation models for movement of passengers.

Therefore, only container and bulk cargo movements of ports are simulated. The assumption of discrete type of cargo handling (container or bulk cargo) per terminal is made (each terminal is assumed to be either container or bulk cargo). The elements studied include piers, berths, waterfront and container/cargo handling equipment (cranes). Waterfronts and cranes are the physical components of the harbor. Piers and berths are structural (functional) elements. Groups of several berths compose a pier. Each berth is a part of a waterfront designed to serve one ship, and it consists of a portion of a waterfront served by one or more cranes. The berth length is estimated based on the pier's operational depth. To quantify the capacity of berths, the capacity of cranes (lifts per hour/tons per hour) is considered in the evaluation. The main Performance Indicator (PI) used is the *total cargo/containers handled and/or delivered (to the port's gate) in a pre-defined time frame per*

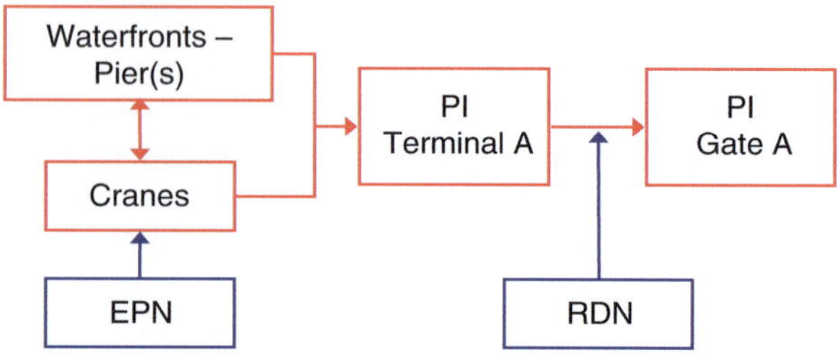

Fig. 12.1 Functionality simulation of port facilities

terminal and for the whole port system. The main interdependency considered within SYNER-G is between the cargo handling equipment and the Electric Power Network (EPN), in particular for the electric power supply to cranes. If a crane node is not fed by the reference EPN node (electric supply station) with power and the crane does not have a back-up power supply, then the crane itself is considered out of service. Road (RDN) closures are also another important dependency, since the delivering process of cargo/containers from the terminals to the port gates could be hampered.

The functionality of the harbor is assessed through several system-level Performance Indicators (PIs), as evaluated starting from the effects of seismic events (Fig. 12.1). The general outline of the method is the following:

(i) A set of shakefields seismic events sampled from the seismic hazard is defined.
(ii) For each event defined in step (i):

  (a) The fields for different intensity measures (shakefields, Chap. 3) within the harbor area are sampled.
  (b) For all components, physical damages are sampled from their probability of occurrence, as assessed through fragility curves and the modeled intensity measures (step ii, a). In case of components sensitive to both ground shaking (PGA, PSA) and ground failure (PGD), like cranes, multiple IMs and damage occurrences are sampled independently, and the results are combined through a Fault Tree Analysis (OR gate).
  (c) Based on the sampled physical damages for each event (step ii, b), the functionality state of each component is assessed, taking also into account system inter- and intra-dependencies.
  (d) For all systems, the PIs are evaluated based on functionality states of their components (step ii, c) and the systemic analysis. The "moving average" (average over all simulated events) is then computed.

(iii) The results of the simulation are estimated. In particular:

  (a) The mean annual frequency of exceedance (MAF) curve ("performance curve") for all PIs, based on the annual rates of seismic events (step i) and the evaluated PIs (step ii, d).
  (b) The rates of functionality (or damages) for each component, based on the results of steps ii, c (or ii, b).

(c) The correlation between functionality states (or damages) and PIs, based on the results of steps ii, c (or ii, b) and ii, d.
(d) Damages, functionality states and PIs are defined for specific events (selected through the MAF curves) corresponding to predefined return periods (step iii, a).

The set of events defined in step (i) must be large enough to obtain reasonably stable results. During the simulation, the process of convergence toward stable results is visually checked from the "moving average" of each PI (step ii, d).

In the followings, the PIs selected for the harbor system and used in this application (Chap. 5) are described in detail.

12.2.1 Container Terminal

12.2.1.1 Terminal (Container Handling)

The terminal performance is measured in terms of:

TCoH = total number of containers handled (loaded and unloaded) per day, in Twenty-foot Equivalent Units (TEU)

For the harbor, the sum of the PIs relative to all container terminals is considered. The berth (one ship) length is estimated based on the pier's operational depth, inverting the following regression, which gives the depth of the waterfront as a function of the ship overall length (Pachakis and Kiremidjian 2005):

$$Draft = \begin{cases} -0.100 + 0.056 \cdot LOA, & for \quad LOA \leq 200 \text{ m} \\ 7.668 + 0.018 \cdot LOA, & for \quad LOA > 200 \text{ m} \end{cases} \tag{12.1}$$

where Draft represents the depth of the waterfront, and LOA is the berth (ship) length.

In practice, for each waterfront, the minimum required berth length is estimated from Eq. 12.1. The waterfront is divided into the maximum possible number of berth(s) with length longer than the minimum length required; then, each crane is assigned to its closest berth.

For each crane, a demand node of the electric power system (EPN) is defined. This demand node is connected to an EPN substation through non-vulnerable lines. In case of failure of power supply, cranes can work with their back-up power supply, if available. The functionality of the demand node is generally based on EPN system analysis, and it can be based on either capacity or pure connectivity analysis (Chap. 5). In this application for the port Thessaloniki, a pure connectivity analysis is performed.

To assess the functionality of components, the following rules are set:

- The waterfront-pier (berth) is functional if damage (D) is lower than moderate (for each IMtype).

- The crane is functional if damage (D) is lower than moderate and there is electric power supply (from the electric network or from the back-up supply).
- The berth is functional if the waterfront and at least one crane is functional, otherwise its PI is set to 0.

If the Berth is functional, the PI is set to the sum of the capacities relative to the functioning cranes that contains. Note that, in case of more than one crane, they can work simultaneously to load/upload containers from the same ship – the time the ship stays at each berth is then reduced.

- $CraneCapacity_k = r*24$ TEU/day (Twenty-foot Equivalent Units per day)
- Berth: $PI_{bi} = \Sigma_k * CraneCapacity_k$
- Pier: $PI_{pm} = \Sigma i\ PI_{bi}$
- Terminal: $PI_{tr} = \Sigma_m\ PI_{pm}$
- Harbor: $PI_H = \Sigma_r\ PI_{tr}$

where $CraneCapacity_k$ is the capacity of the k-th crane, r is the crane productivity, PI_{bi} is the Performance Indicator of the i-th berth, PI_{pm} is the Performance Indicator of the m-th pier, PI_{tr} is the Performance Indicator of the r-th terminal and PI_H is the Performance Indicator of the harbor. An assumption is made of 24 h shifts.

12.2.1.2 Gate (Container Delivering)

The port performance at the gate is measured in terms of:

TCoM = total number of containers' movements per day, in Twenty-foot Equivalent Units (TEU) for the whole harbor facility

In this case the total number of containers' movements per DAY is equal to the sum of total number of containers handled per DAY (TCoH) in all the container terminals that are connected to the gate through the road system (RDN).

For the assessment of TCoM, in addition to the input parameters reported above, it is necessary to consider the road system that connects each terminal to the harbor's gate, with all its important components (i.e., bridges, overpass, tunnels), and of course the buildings and the storage units inside the harbor that may collapse and block the road system (see Chaps. 5, 7 and 10).

The connectivity between terminals and harbor's gate is based on the *RDN* system analysis (see Chaps. 5, 7 and 10).

12.2.2 Bulk Cargo Terminal

12.2.2.1 Terminal (Bulk Cargo Handling)

The terminal performance is measured in terms of:

TCaH = total cargo handled (loaded and unloaded) per day, in tones

For the harbor, the sum of all container terminals is considered.

For the cargo, the same methodology presented above for the container terminals is used, with the following modifications:

- The crane productivity (r) is given in tones per hour.
- CraneCapacity = r*24 tones/day (an assumption is made of 24 h shifts).

12.2.2.2 Gate (Bulk Cargo Delivering)

The port performance at the gate is measured in terms of:

TCaM = total cargo movements per DAY, in tones for the whole harbor facility

In this case, the total cargo movements per DAY are equal to the sum of total cargo handled per DAY in all the bulk cargo terminals that are connected to the gate through the road system. The methodology to assess TCaM, and the required additional parameters, is analogous to the ones described above for TCoM.

12.3 Software Implementation

12.3.1 HBR Class

The HBR (harbor) class is the composition of *HRBStructures*, *HRBEdges* (or *HRBSides*) and *HRBNodes* classes, that are all abstract classes. *HRBStructures* is the generalization of terminals, piers and berths. *HRBNodes* is the generalization of cranes, while *HRBedges* (*HRBsides*) is the generalization of waterfronts.

Structures have a logical hierarchy. Each terminal includes one or more piers, which are made up of one or more berths. Cranes and waterfronts belong to one specific pier and terminal, information that is provided by the user. The position and the configuration of berths are automatically assigned by the software. Following the naming convention set up for other systems, the harbor's elements are classified in the following classes (Fig. 12.2):

- *Harbor*: part of the "network" class
- *HRBnode*: which includes Cranes
- *HRBside*: which includes Waterfronts
- *HRBstructure*: which includes Terminals, Piers, and Berths (internal class – not explicit to the user)

The general properties of the HBR class are summarized in Chap. 5. These include several pointers, harbor system global properties, subclass characteristics that include the main features of sides, nodes and structures, properties that record the state of the HBR for each event and properties required to assess its global performance of the HBR at the end of the simulation. The definition of HBR class includes functions to assess PIs (and relative statistics), to assess damages

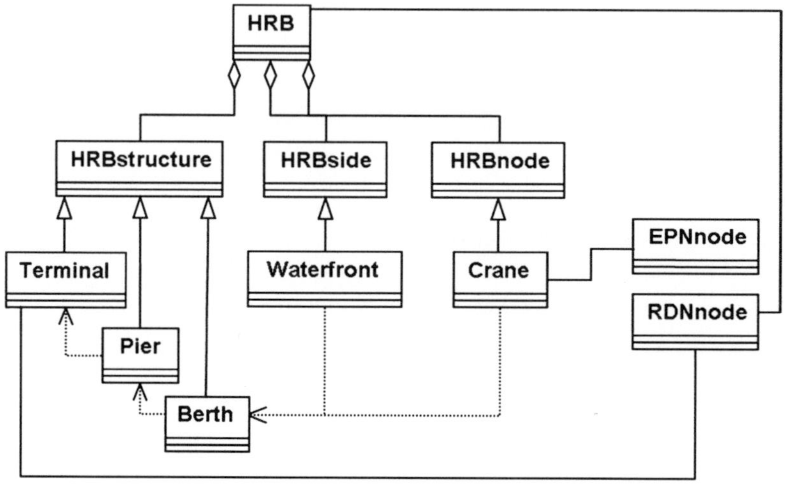

Fig. 12.2 Class diagram for the harbor classes

and functionalities, to retrieve the EPN and RDN functionality states, to locate and characterize Berths, and to plot the system configuration and states (Chap. 5). The average port performance and the Mean Annual Frequency (MAF) of exceedance ("performance curve") of the PIs, are the main outputs of the simulation. Apart from these, the analysis provides also the distribution of estimated damages and losses for specific events (either pre-selected, or selected from pre-defined values of annual frequency of exceedance for the PIs), the average damages/losses (averaged over the simulated events) and the correlation between each element's functionality and the port's PIs (in order to identify the elements that tend to control the port's performance).

12.3.2 HBR Subclasses: Node, Side and Structure

The *HRBnode* and *HRBside* classes include positioning and descriptive properties, general characteristics (e.g., crane capacity, waterfront depth, etc.), reference structures' IDs (pier and terminal), and methods assessing, for each event (scenario), physical damages and functional consequences. The performance of these classes is stored in the states collection, which includes physical and functional states for each event (scenario).

The *HRBstructure* class includes positioning and descriptive properties, and links to *HRBlinks* and *HRBnodes* that each structure includes. The Berth subclass includes reference IDs to the terminal and pier in which it is located. The Pier subclass includes a reference ID to the terminal to which it owns. The Terminal subclass includes a link to a RDN node, which indicates the starting position of the road leading to the harbor gate. The performance of these classes is stored in a states collection (as above), which includes PIs estimation and statistics at each subclass for each event (scenario).

12.4 The Case Study

12.4.1 General Description

The port of Thessaloniki is the nodal point for the transport of goods coming from a large geographic inland, as it is located in a very strategic (geographically, politically and economically) location. It is a major export and transport harbor of Greece and is European Union's closest port to the countries of Southeast Europe, as well as to the countries of the Black Sea and East Mediterranean. It covers an area of 1,550,000 m^2 and trades approximately 16,000,000 tons of cargo annually, having a capacity of 370,000 containers and 6 piers with 6,500 m length (Fig. 12.3). In collaboration with the port authority (Thessaloniki Port Authority, THPA), various data was collected and implemented in GIS format for the construction, typological and functional characteristics of port facilities, including cargo and handling equipment, waterfront structures, electric power (transmission and distribution lines, substations), potable and waste water (pipelines), telecommunication (lines and stations), railway (tracks) and roadway (roads and bridge) systems as well as buildings and critical facilities.

The various components and systems existing inside the port facilities are illustrated in Figs. 12.4 and 12.5.

Fig. 12.3 Thessaloniki port

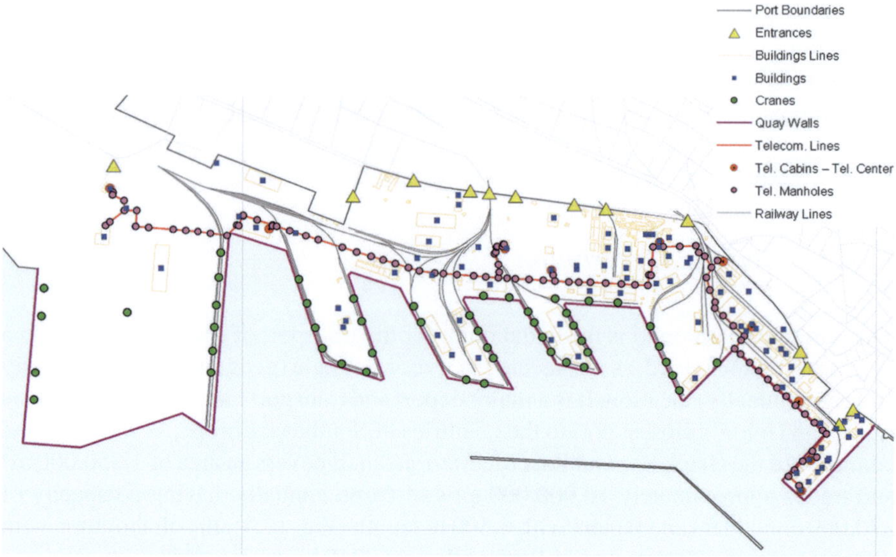

Fig. 12.4 Building facilities, waterfront structures, cargo handling equipment, telecommunication system and railway network of Thessaloniki's port

Fig. 12.5 Water supply, waste-water, fire-fighting, electric power and fuel supply systems of Thessaloniki's port

12.4.2 Seismic Hazard

The seismic hazard is based on the shakefield method (Chap. 3). Seismic zones are set as for the Thessaloniki case study, described in Chap. 7. Also the geotechnical mapping is part of the one given for the Thessaloniki case study (Sect. 7.2). Adopting the Monte Carlo simulation scheme, 10,000 runs are here carried out.

In Thessaloniki's port, soil formations are characterized by very high liquefaction susceptibility, mainly due to loose, saturated, silty-sandy soils that prevail at the area. In previous studies (SRMLIFE 2007), the liquefaction induced permanent horizontal and vertical ground displacements (lateral spreading and settlements), have been evaluated for three seismic scenarios with return periods T_R of 100, 475 and 1,000 years, using empirical and analytical procedures (Seed et al. 2003; Youd et al. 2001, EC8, CEN 2004; Ishihara and Yoshimine 1992; Elgamal et al. 2001). Displacement values range between 0 and 30 cm for settlements and 0 and 6 cm for lateral spreading for the 475 years scenario.

12.4.3 System Topology and Characteristics

Following the methodological framework for the systemic analysis, waterfront structures, cargo handling equipment, power supply system, roadway system and buildings are examined.

Waterfront structures, of 6.5 km length, include concrete gravity quay walls with simple surface foundation and non-anchored components. The majority is block type gravity walls, while the new, actually under construction, part of Thessaloniki's port includes caisson type structures. Backfill soils and rubble foundation include material aggregates with appropriate grain size distributions. Waterfront structures are defined with 17 sides and 24 nodes (pier-nodes).

Cargo handling equipment has non-anchored components without back-up power supply. 48 crane-nodes are considered in the analysis.

For the systemic analysis, two Terminals are considered; one container Terminal (6th pier) and one cargo Terminal (piers 2, 3, 4 and 5).

The electric power supply to the cranes is assumed to be provided from a demand node (substation) through non-vulnerable lines. These demand nodes are the distribution substations present inside the port facilities. They can be classified as low-voltage substations, with non-anchored components. Their functionality is determined from connectivity analysis of Thessaloniki's EPN system (Chap. 7). The geographical representation of Thessaloniki's port waterfronts, cranes and electric power supply system is illustrated in Fig. 12.6.

The majority of the building and storage facilities are also considered in the analyses. In particular, 88 building structures are allocated in 4 building blocks (BC).

The internal roadway network is rather simple with internal roads connecting the port gates to the terminals gates (Fig. 12.7).

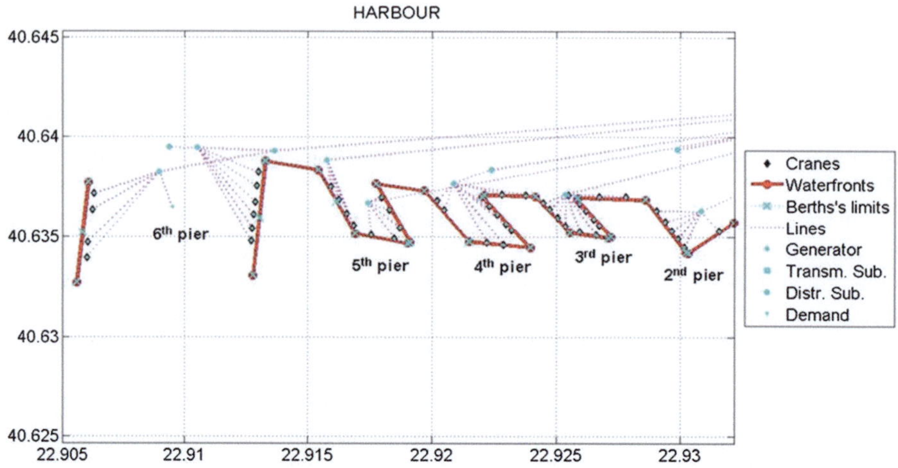

Fig. 12.6 Geographical representation of Thessaloniki's port waterfronts, cranes and electric power supply system

Fig. 12.7 Internal road network of Thessaloniki's port

12.4.4 Description of the Input

12.4.4.1 Harbor Components

Harbor components are comprised of 72 nodes and 17 sides. The nodes are subdivided in pier-edges (non-vulnerable) and (non-anchored) cranes (vulnerable). Edges include only (gravity type) waterfronts. The fragility models used for cranes and waterfronts are expressed in terms of permanent ground deformation (PGD) and peak ground acceleration (PGA) (Table 12.1).

Table 12.1 Fragility functions used for Thessaloniki's port analysis

Component	Intensity measure	Fragility function
Waterfronts	PGD	HAZUS (NIBS 2004)
	PGA	Kakderi and Pitilakis (2010)
Cranes/cargo handling equipment	PGA, PGD	HAZUS (NIBS 2004)
Electric power substations (distribution)	PGA	HAZUS (NIBS 2004)
Electric power substations (transmission)	PGA	SRM-LIFE (2007)
Roads	PGD	HAZUS (NIBS 2004)
R/C and URM buildings	PGA	Kappos et al. (2006)
Steel buildings	PGA	HAZUS (NIBS 2004)

General information for the analysis includes the number of terminals and their characteristics (IDs, type, gate and exit nodes of each terminal). Also, the possibility to activate/deactivate the inter-dependencies with electric power network (EPN) and roadway network (RDN) is provided. The fragility curves and functionality definition of the components are then determined.

The information provided for each node comprises localization, site properties, functional and vulnerability information, crane characteristics and possible links to other system(s) (interdependencies). Localization is given in terms of latitude and longitude in degrees and altitude above sea level in meters. The site properties are specified in terms of V_{s30}, site class according to EC8, (CEN 2004) depth to groundwater, liquefaction and landslides susceptibility class and yield acceleration. Functional information for the harbor nodes is the type of node (either pier-edge or crane-non-anchored). For vulnerable nodes, the Intensity Measure(s) IM(s) of the corresponding fragility model are also given. Specific crane information includes crane capacity, the existence or not of back-up power, the terminal and pier's ID. The interdependency of cranes with the electric power system is provided through links to EPN nodes.

For sides (edges), each edge is determined through its start and end nodes. The site properties are specified in the same way as for nodes. Functional information includes the edge typology (gravity-waterfronts). As for the nodes, each edge may be either vulnerable or not vulnerable. When it is vulnerable, the IM(s) for the corresponding fragility model is defined. Specific waterfront information includes the operational depth, terminal and pier IDs. Pier operational length is computed from the coordinates.

12.4.4.2 Electric Power Network

In total 1 generator, 8 transmission substations, 17 distributions substations and 74 non-vulnerable (assumption) lines are simulated. For the substations, their type and IM is also provided. 48 demand nodes are defined for the electric power supply to cranes.

12.4.4.3 Roadway Network and Buildings

Both vulnerability analysis for the road network and the possible road blockage estimation from collapsed buildings are performed. For non-vulnerable road nodes their type can be either Traffic Analysis Zones "TAZ" (TAZ type is also provided) or intersection (see Chap. 5). Road segments are all classified as principal two-way roads and their capacity is set to 500 vph, while the free-flow speed to 50 km/h. The Intensity Measure is PGD, and it is estimated together with the site and liquefaction susceptibility classes, the landslide susceptibility class and the yield acceleration. Road width, building-road distance, hierarchy and adjacent buildings are also necessary input for the road blockage analysis.

For each Building Block (BC) the total number of buildings inside the block and the percentage of each building typology are given.

12.4.4.4 Fragility Assessment

The fragility curves used for the vulnerability assessment of the components are shown in Table 12.1.

12.5 Results

12.5.1 Main Application Results

The analysis results obtained from a plain MCS of 10,000 runs is presented in the following figures. The chosen number of runs has been shown to yield stable estimates for the considered PIs.

All PIs are normalized to the respective value referring to normal (non-seismic) conditions. For the container terminal this value is equal to $PIa_{max} = 1{,}032$ TEUs per day. For the cargo terminal the value for non-seismic conditions is equal to $PIo_{max} = 43{,}512$ tones per day. These values refer to the maximum capacity of the port, since they are estimated assuming that all cranes are working at their full capacity 24 h per day.

Figure 12.8 shows the moving average (μ) curves for TCoH and TCaH, as well as the $\mu + \sigma$ and $\mu - \sigma$ curves for the two PIs. The figures indicate that the expected loss given the occurrence of an earthquake is higher for TCoH than for TCaH. The Mean Annual Frequency (MAF) of exceedance values for all PIs are given in terms of normalized performance loss ($1 - PI/PI_{max}$) in Fig. 12.9 which shows the MAF of exceedance curves ("performance curve") for TCoH and TCaH. For performance loss values below 20 % TCaH yields higher values of exceedance frequency, while for performance loss over 20 % TCoH yields higher values of exceedance frequency.

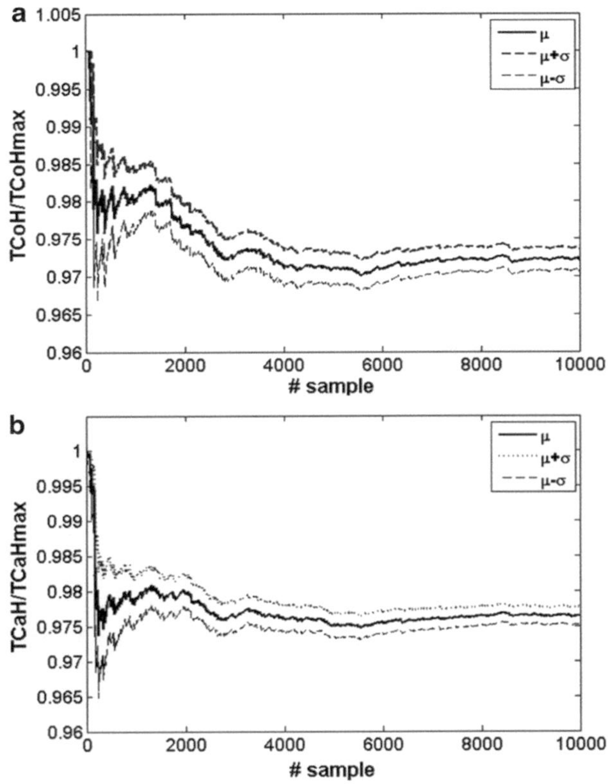

Fig. 12.8 Moving average μ, μ + σ, μ − σ curves for TCoH (**a**) and TCaH (**b**)

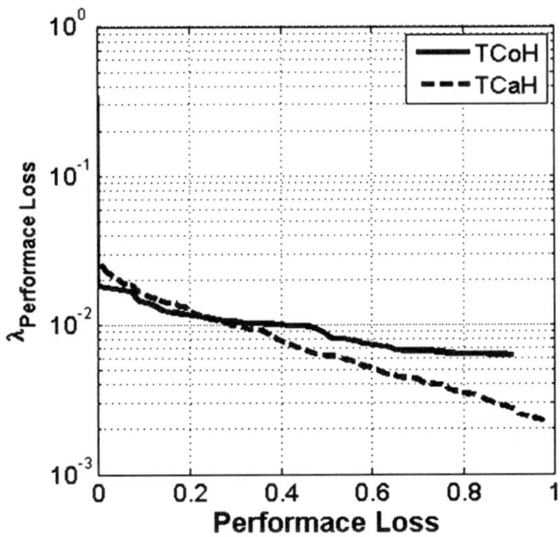

Fig. 12.9 MAF curves for TCoH and TCaH performance loss

Fig. 12.10 MAF curves for TCoH and TCoM performance loss

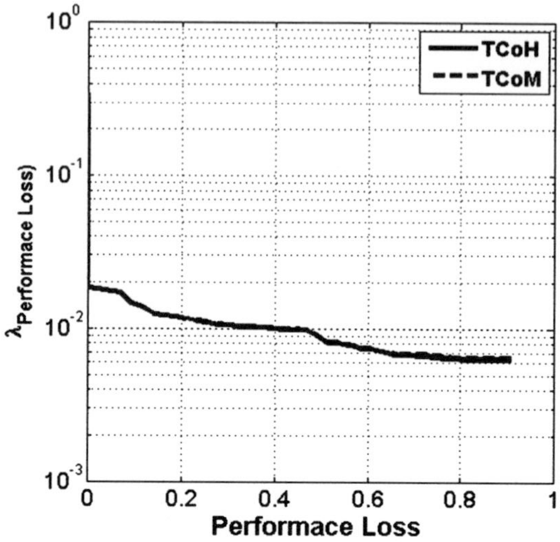

The comparison of the estimated MAF of exceedance curves (in terms of normalized performance loss) for TCoH and TCoM (Fig. 12.10) shows no difference, meaning that no road closures are observed. We recall here that the only difference between these two PIs is the possibility to deliver containers from the pier to the gate. Thus, the interaction with building collapses and consequent road closures is not important to the port's overall performance in this particular case study. Equivalent results are obtained comparing TCaH and TCaM. This can be attributed to the small length of roadways considered in the analysis and the building type found in the port are. The fact that no road closures occur is also confirmed when comparing the moving average (mean) curves for TCoM and TCaM with TCoH and TCaH respectively, which practically coincide. Given the fact that no road closures are observed in the present analysis, results are presented hereinafter only for the TCoH and TCaH PIs.

Figure 12.11 compares the estimated MAF of exceedance curves for TCoH and TCaH when all and no interactions are taken into consideration in the analysis. Note that, for this particular application the interaction is effective only between EPN and cranes. The effect of this interaction can be very important for performance loss levels over 10 % for TCoH and 5 % for TCaH. As an example the TCoH performance loss is increased from about 20 % to about 50 % for $\lambda = 0.01$ ($T_R = 100$ years) when interactions are included in the analysis. In the TCaH MAF curves, for performance loss levels of 50–60 %, it seems to be practically no change in the exceedance frequency values, and values with "no interactions" are higher than those corresponding to the "all interactions" case. This is probably due to a residual instability of the results for very low frequency events, showing that longer simulations, with much more than 10,000 events, would be required to better sample the tails of the distributions.

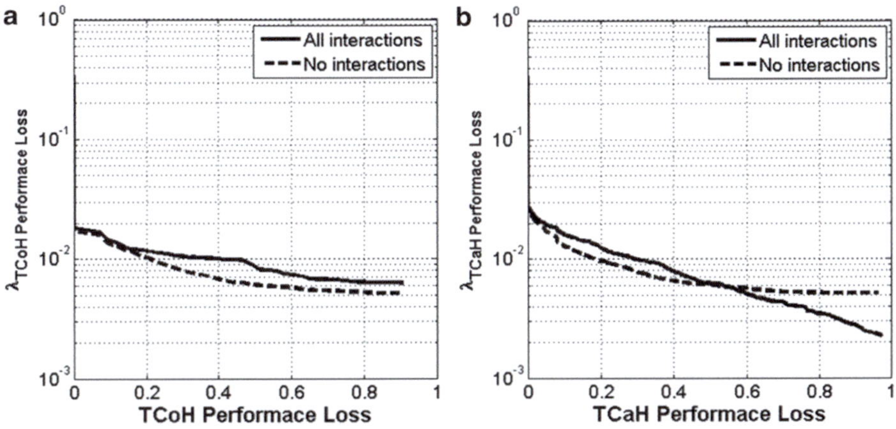

Fig. 12.11 MAF curves for TCoH (**a**) and TCaH (**b**) for Thessaloniki's port, with and without interaction with EPN and building collapses

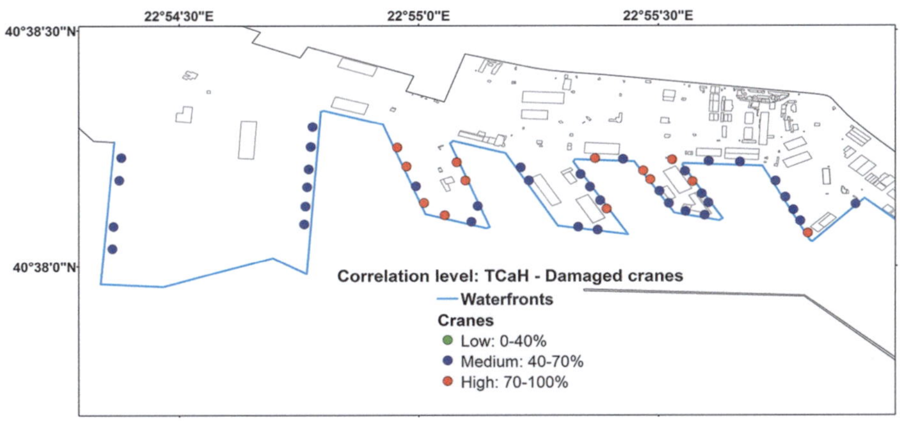

Fig. 12.12 Correlation of damaged cranes to port performance (PI = TCaH)

Figures 12.12 and 12.13 show the level of correlation between the TCaH and the distribution of damages in cranes and non-functionality of electric power distribution substations respectively. In this way the most critical components can be identified in relation with their contribution to the performance loss of the system. All cranes have medium (40–70 %) to high (over 70 %) levels of correlation, indicating their great importance to the functionality of the overall port system. A higher level of correlation is estimated for the EPN distribution substations, with 40 % of the components having values greater than 70 %.

Figures 12.14 and 12.15 show the expected functionality of port components for the events with the highest performance loss for TCoH (100 %) and TCaH (97 %), respectively, among the events with return period of $T_R = 500$ years (i.e. 0.002

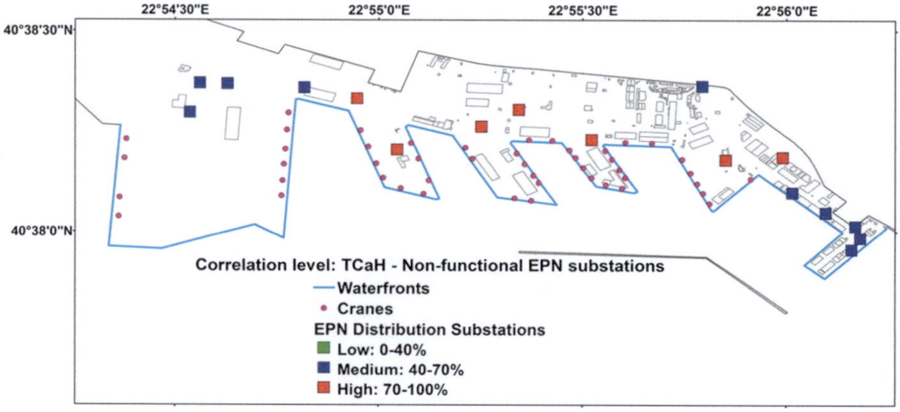

Fig. 12.13 Correlation of non-functional electric power distribution substations to port performance (PI = TCaH)

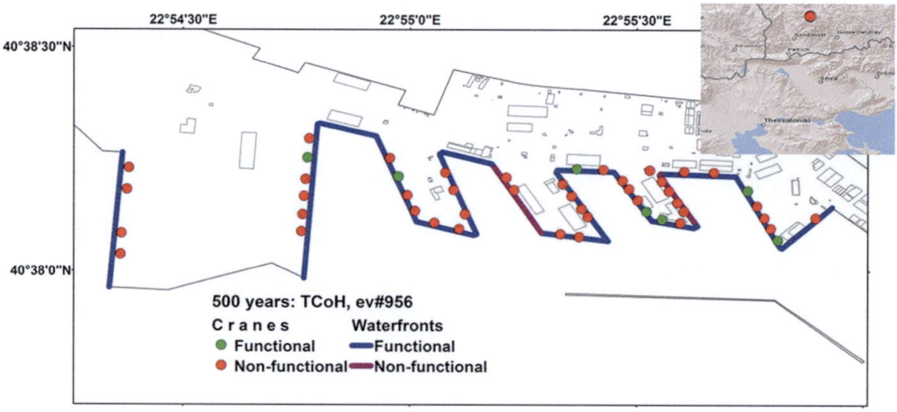

Fig. 12.14 Port components functionality for an event (#956, TCoH loss = 100 %, TCaH loss = 53 %, M = 7.5, R = 135 km) that corresponds to TCoH with T_R = 500 years

probability of exceedance) respectively. For both events, waterfronts structures, with the exception of one component, are functional, but the majority of cranes (85 % and 88 % respectively) are non-functional.

12.5.2 *Uncertainty Issues*

Several sources of uncertainties are inherent in the analysis. They are related among others to the seismic hazard and spatial correlation models, and the fragility or the functionality assessments of each component. The epistemic uncertainty related to different fragility functions and functionality definitions is investigated performing

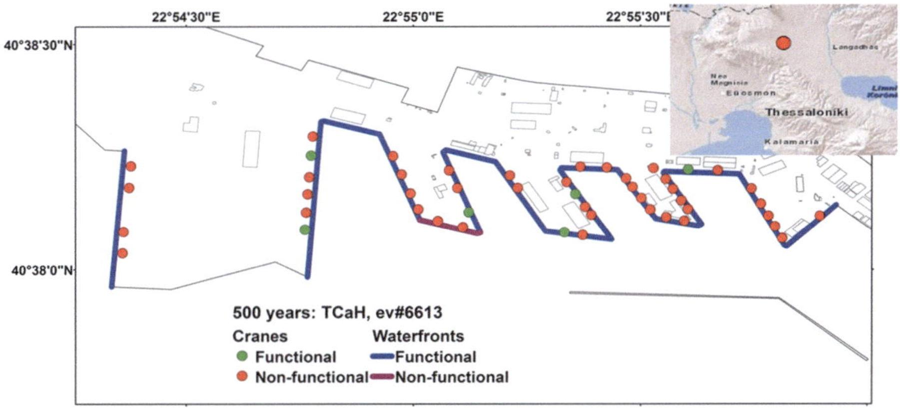

Fig. 12.15 Port components functionality for an event (#6613, TCoH loss = 60 %, TCaH loss = 97 %, M = 5.8, R = 15 km) that corresponds to TCaH with T_R = 500 years

sensitivity analysis with the use of alternative fragility curves and functionality thresholds for the waterfront structures. The vulnerability assessment of cranes could not be performed with alternative functions since HAZUS (NIBS 2004) curves are for the moment the only available in the literature. Also the functionality definition for cranes seems to be the most realistic one, since high levels of occurred damages usually necessitate the withdrawal or even replacement of the component.

In the main analysis, the HAZUS (NIBS 2004) curves are used for waterfronts for the case of ground failure (due to liquefaction), while in case of no liquefaction phenomena the fragility functions proposed by Kakderi and Pitilakis (2010) are adopted ("Fragility 1" case). In the alternative analysis, the fragility functions proposed by Ichii (2003), which take into account the occurrence of liquefaction, are used for the vulnerability assessment of waterfronts ("Fragility 2" case).

In Fig. 12.16, the estimated MAF of exceedance curves (in terms of normalized performance loss) for TCoH and TCaH for both cases of fragility functions of waterfronts are compared; almost no differences are observed. This can be attributed to the small frequency of damage occurrence to the waterfront structures and the fact that the total port performance is mostly prescribed by the cranes functionality.

The functionality of waterfronts depends only on the level of seismic damage. In the basic analysis, the waterfronts were considered fully functional if they sustained minor damages and non-functional for higher levels of damage; this is the "Functionality 1" case. An alternative analysis is performed ("Functionality 2"), where waterfront structures are considered as fully (100 %) functional if they sustain minor damages and partially (50 %) functional if they sustain moderate damages.

Figure 12.17 compares the estimated MAF of exceedance curves (in terms of normalized performance loss) for TCoH and TCaH for the different functionality definitions of waterfront structures. In this case there is some difference in the MAF curves for TCaH with lower values of exceedance frequency for performance loss levels over 65 %. In other words, high levels of performance loss correspond to

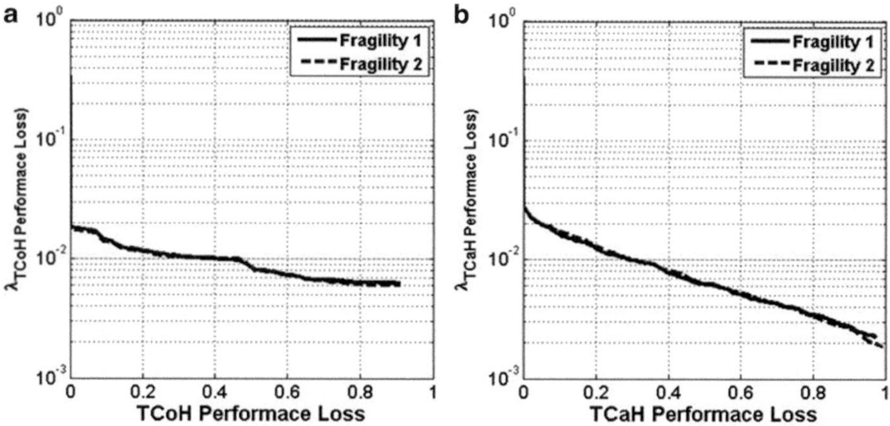

Fig. 12.16 MAF curves for TCoH (**a**) and TCaH (**b**) for Thessaloniki's port using different fragility functions for waterfront structures

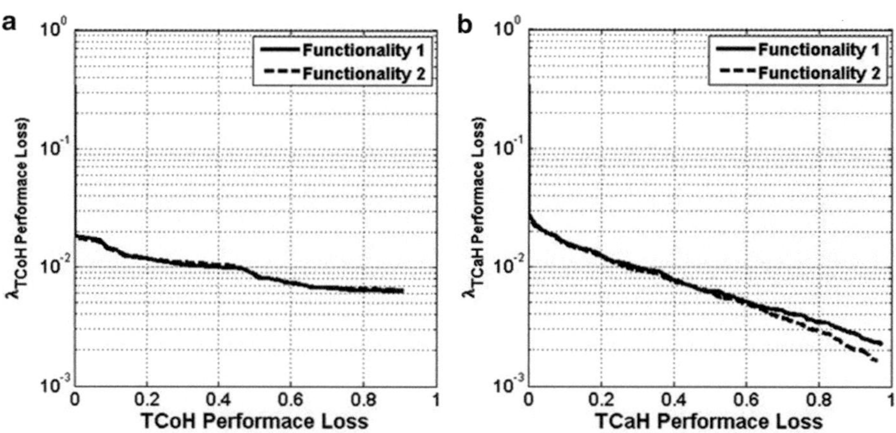

Fig. 12.17 MAF curves for TCoH (**a**) and TCaH (**b**) for Thessaloniki's port using different functionality definitions for waterfront structures

lower probabilities of exceedance (or higher return periods). This is related to the fact that partial functionality of waterfronts is assumed for higher levels of damage, resulting in reduction of the port performance loss.

12.6 Conclusions

The SYNER-G methodology and tools for the assessment of the systemic vulnerability and performance of harbors have been applied in the case of Thessaloniki's port, one of the largest Greek seaports, a major gateway for the Balkan hinterland and southeastern Europe and the second largest container port in Greece.

For the seismic hazard, five seismic zones with $M_{min} = 5.5$ and $M_{max} = 7.5$ are selected based on the results of SHARE European research project (Giardini et al. 2013; www.share-eu.org). Monte Carlo simulation (MCS) has been carried out sampling earthquake events for these zones and computing selected performance indicators (PIs).

Port operations are simulated and system performance is assessed considering specific interdependencies between the components. Port Performance Indicators (PIs) are calculated based on the estimated damages and functionality loss of the different components.

The overall performance is expressed through the moving average μ and moving standard deviation σ (averaged over simulations), as well as the Mean Annual Frequency (MAF) of exceedance of the PIs (performance curves). The average loss is defined based on the moving average graph. Through the MAF graphs the annual probability of exceeding specific levels of loss can be defined and the loss for specific return period of the particular PI can be estimated. The earthquake event(s) that correspond to a particular return period (i.e., 500 year) are identified and maps with the distribution of damages are produced for this event(s). The correlation of each component to the system PIs is also estimated. This type of analysis is based on the results of each single event, and thus it preserves the information about systems' topology and its behavior in case of spatial correlated damages (related to single earthquakes). Thus, it allows identifying the most critical elements for the functionality of the system.

The interactions considered in the analysis are essential for overall risk assessments. For example, it is shown that the performance loss of Thessaloniki's port can be significantly increased due to possible failures of EPN substations that supply power to the cranes without back-up systems.

Several sources of uncertainties are inherent in the analysis. They are related among others to the seismic hazard and spatial correlation models, the fragility assessment or the functionality thresholds of each component. As regards the aleatory uncertainty, a probabilistic (Monte Carlo) approach is performed which samples earthquake events, damages and functionalities based on the methods and tools developed in SYNER-G. In this way, all the characteristics of each event (e.g., spatial correlations) are accounted for and preserved for the systemic analysis. The epistemic uncertainty related to different fragility functions and functionality definitions was investigated performing sensitivity analysis with the use of alternative fragility curves and functionality thresholds for the waterfront structures.

References

CEN (European Committee for Standardization) (2004) Eurocode 8: design of structures for earthquake resistance, Part 1: General rules, seismic actions and rules for buildings. EN 1998-1: 2004. Brussels, Belgium

Elgamal A, Yang Z, Parra E, Ragheb A (2001) Cyclic 1D. UCSD

Giardini D, Woessner J, Danciu L, Crowley H, Cotton F, Grünthal G, Pinho R, Valensise G, Akkar S, Arvidsson R, Basili R, Cameelbeeck T, Campos-Costa A, Douglas J, Demircioglu MB, Erdik M, Fonseca J, Glavatovic B, Lindholm C, Makropoulos K, Meletti C, Musson R, Pitilakis K, Sesetyan K, Stromeyer D, Stucchi M, Rovida A (2013) Seismic Hazard Harmonization in Europe (SHARE). Online data resource, doi:10.12686/SED-00000001-SHARE

Ichii K (2003) Application of performance-based seismic design concept for Caisson-type quay walls. PhD Dissertation, Kyoto University

Ishihara K, Yoshimine M (1992) Evaluation of settlements in sand deposits following liquefaction during earthquakes. Soils Found 32(1):173–188

Kakderi K, Pitilakis K (2010) Seismic analysis and fragility curves of gravity waterfront structures. In: 5th international conference on recent advances in geotechnical earthquake engineering and soil dynamics and symposium in Honour of Professor I. M. Idriss, San Diego, CA. Paper no 6.04a

Kappos A, Panagopoulos G, Panagiotopoulos C, Penelis G (2006) A hybrid method for the vulnerability assessment of R/C and URM buildings. Bull Earthq Eng 4(4):391–413

National Institute of Building Sciences (NIBS) (2004) HAZUS-MH: user's manual and technical manuals. Report prepared for the Federal Emergency Management Agency, Washington, DC

Pachakis D, Kiremidjian AS (2005) Estimation of downtime-related revenue losses in Maritime ports due to earthquakes. Blume Center technical report no 146.

Seed RB, Cetin KO, Moss RES, Kammerer AM, Wu J, Pestana JM (2003) Recent advances in soil liquefaction engineering: a unified and consistent frame work. In: 26th annual ASCE Los Angeles geotechnical spring seminar, keynote presentation, H.M.S. Queen Mary, Long Beach, CA.

SRMLIFE (2007) Development of a global methodology for the vulnerability assessment and risk management of lifelines, infrastructures and critical facilities. Application to the metropolitan area of Thessaloniki. Research project, General Secretariat for Research and Technology, Greece

Youd TL, Idriss IM, Andrus RD, Arango I, Castro G, Christian JT, Dobry R, Liam Finn WD, Harder LF Jr, Hynes ME, Ishihara K, Koester JP, Liao SSC, Marcuson WF, Martin GR, Mitchell JK, Moriwaki Y, Power MS, Robertson PK, Seed RB, Stokoe KH (2001) Liquefaction resistance of soils: summary report from the 1996 NCEER and 1998 NCEER/NSF workshops on evaluation of liquefaction resistance of soils. J Geotech Geoenviron 127(10):817–833

Chapter 13
Recapitulation and Future Challenges

Kyriazis Pitilakis and Bijan Khazai

Abstract This chapter briefly recapitulates the main achievements of SYNER-G project on the systemic seismic vulnerability and risk assessment of buildings, lifelines and infrastructures. Essential needs for future developments and improvements are also summarized.

13.1 Recapitulation

Earthquake engineering has made major steps towards a greater understanding of the effects of earthquakes to human structures and environment as well as to the society and the economy. The vulnerability of structures and society to strong earthquakes is progressively better understood and quantified with the aim to improve preparedness and mitigation. Every new seismic event adds to the knowledge and helps to improve, understand and plan better mitigation measures in the frame of risk assessment and management.

Earthquake risk assessment involves several disciplines where Earthquake Engineering represents the vulnerability side of research into earthquake risk. It is generally expressed through vulnerability functions of physical assets like buildings, infrastructures, facilities and utility systems, which quantify in measurable scale the probability of exceeding a defined level of damage state for a given

K. Pitilakis (✉)
Department of Civil Engineering, Aristotle University, 54124 Thessaloniki, Greece
e-mail: kpitilak@civil.auth.gr

B. Khazai
Geophysical Institute, Karlsruhe Institute of Technology, Hertzstr. 16, 76187 Karlsruhe, Germany
e-mail: khazai@kit.edu

K. Pitilakis et al. (eds.), *SYNER-G: Systemic Seismic Vulnerability and Risk Assessment of Complex Urban, Utility, Lifeline Systems and Critical Facilities*, Geotechnical, Geological and Earthquake Engineering 31, DOI 10.1007/978-94-017-8835-9_13,

intensity measure. A comprehensive state-of-the-art of fragility curves, which is an alternative way to express vulnerability, is provided in the first volume of the work accomplished in SYNER-G published in the same series of Springer editions (Pitilakis et al. 2014). Based on the vulnerability functions of individual elements of a system, the expected physical, human and economic losses related to the damage of this particular element are then estimated. However, the integration of social vulnerability with physical damages, in other words the global impact at city or regional scale, may significantly be increased and diversified if we consider the interactions between elements belonging in the same system and between different systems.

In this regard SYNER-G allowed development of an innovative methodological framework for the assessment of physical as well as socio-economic seismic vulnerability and risk at the urban/regional level. The project paved new ground by proposing methods and simulation software tools to consider inter-element and intra-systems interdependencies, including socio-economic features. The outcomes and results of the project contribute to a better understanding of the vulnerability of systems exposed to seismic risk considering also societal elements at risk belonging to a system (city, region, lifeline network, etc.). In this second volume of SYNER-G, the focus has been on presenting a unified methodology for assessing vulnerability at a systems level considering both interdependencies between elements at risk (physical and non-physical) belonging to different systems and between different systems. Application studies at both the city and regional scale provided the necessary validation of the methodology.

The framework shown here encompasses in an integrated fashion all aspects in the chain, from hazard to the vulnerability assessment of components and systems and to the socio-economic impacts of an earthquake, accounting for most relevant uncertainties within an efficient quantitative simulation scheme, and modeling interactions between the multiple component systems. The unified methodology presented in this volume systematically integrates the most advanced fragility or vulnerability functions to assess the vulnerability of physical assets for the following systems: buildings and building aggregates; utility systems including water, waste water, gas, oil and electric power networks; transportation networks including roadways, railways and harbor systems; and critical facilities such as hospitals.

The assessment of the impact of an earthquake at a composite system of systems like a city is a complex task that requires a comprehensive modeling of the analyzed systems and their interdependencies. The latter is the set of all physical and non-physical systems that make up our society and the infrastructure supporting it. The approach presented in this book and the model framework developed and applied in specific case studies, represent a step in the direction of comprehensiveness. The main goal was that of setting up a framework where multiple interacting systems could be analyzed concurrently and consistently, describing also their inherent and modeling-related uncertainty, in order to evaluate their global state of damage and reach into socio-economic consequences. For practical reasons the focus was put exclusively on the short-term period after the event. Ageing and cascading effects were left for future work and improvements.

An extensive list of systems was considered and analyzed keeping in mind the need for a model that has the flexibility to easily accommodate future extensions. The developed model is object-oriented, a choice proven instrumental to tackle the complexity and size of the considered system of systems. Moreover, some of the principles of the object-oriented design paradigm, such as inheritance and composition, have allowed a high degree of abstraction and hierarchical decomposition in the model, leading to a framework that is easier to maintain and extend. The model encompasses a large number of systems, larger than any integrated model to date in earthquake engineering. The integration of all these systems within the same model and simulation software allows consistent evaluation of demands across all systems starting from the same basic data on buildings, population and activities. It also allows re-evaluation of demands for different post-event conditions.

In a modern city the variety of building typologies and complexity of the aggregate is a major problem to accurately estimate damages and losses. The proposed building sub-model (*class BDG*) is an effective and scalable way of treating the large number of buildings to be considered in a study of regional or urban extension, and it facilitates merging of different data sources. It allows for different levels of granularity in the input data to be considered, through the use of dependency edges between building geo-cells and the reference nodes in all the other systems. Meshing of each component system can be different and appropriate to the level of information about the considered system, with the model taking care of connecting the systems together whenever the analyst does not provide the relevant information. This is the mechanism that allows for demands from tributary cells to be aggregated to demand nodes in each network, and to determine the residual service level of each utility, or degree of residual connectivity/accessibility of each geo-cell. In this respect, the model creates traffic analysis zones for road network analysis automatically.

The model for the evaluation of social metrics also represents a novelty. The basic idea is that population displacement occurs due to several factors besides direct physical damage to buildings. While a collapsed or severely damaged building is obviously non usable, and thus not habitable, partially or even fully usable buildings can be non-habitable due to lack of basic services for example water or gas. Furthermore, the decision to evacuate even a habitable building and to seek or not to seek shelter is influenced by a number of environmental and socio-economic characteristics. The model allows, through the described interaction mechanism, the integration of elements of the built environment system with the relevant societal elements to derive more representative measures of displaced populations.

The consideration of interactions between systems was the central challenge of the project. A general model has been set-up, and a number of interactions have been developed: the interaction for demand and utility loss evaluation between buildings and network-like systems; the interaction between the electric power network and dependent components in all other systems (e.g. pumps in the water supply system or port infrastructures); the interaction of geographical type between buildings and roads in urban environments, with debris from collapsed buildings that can induce additional obstruction/damage to roads in the road blockage model.

The physical modelling of networks is carried out on a detailed level accounting for three main types of solving algorithms:

- *Connectivity analysis* where the damaged components are removed from the network and the network is updated in the adjacency matrix accordingly. This approach is used for all utility networks (water, electricity, gas) and the road transportation system.
- *Capacitive analysis* where graph algorithms and flow equations can be used to optimize capacitive flows from sources (e.g. generators, reservoirs) to sinks (i.e., distribution nodes), based on the damages sustained by the network components (from total destruction to slight damages reducing the capacity). This approach is also used for utility networks.
- *Fault-tree analysis* which aims to evaluate the remaining operating capacity of objects such as health-care facilities. The system is broken down into structural, non-structural or human components, each one of them being connected with logic operators. This method is generally used for the derivation of fragility curves for specific components that comprise a set of sub-components (e.g. healthcare facilities, water treatment plants).

The definition of Performance Indicators, at the component or the system level, is another important contribution of SYNER-G. They depend on the type of analysis that is performed. Connectivity analysis gives access to indices such as the connectivity loss (measure of the reduction of the number of possible paths from sources to sinks). Capacitive modelling yields more elaborate performance indicators at the distribution nodes (e.g. head ratio for water system, voltage ratio for electric buses) or for the entire system (e.g. system serviceability index comparing the customer demand satisfaction before and after the seismic event).

The general methodology has been conceived in order to be general enough to be adequate for each system i.e., buildings, water supply, waste-water, electric power, oil and gas networks, transportation, health care system and harbors. Each system is described based on its structure and taxonomy, on the dependencies it shares with the other systems, on the available methods to describe its systemic vulnerability and, finally, on the existing indicators to evaluate its performance, but also its functionality according to the societal needs. The available methods to describe the different systems that the societies rely on reflect the particularities of each system, and their characteristics. However, in order to integrate all these entities (i.e., lifelines and infrastructures) in a unique framework, a common methodology has been developed which can encompass the diversity of the properties of the systems.

Each system is described according to three main characteristics: (i) the lists of its elements, which is given through the taxonomy of the systems, (ii) the support it provides for the society, which is provided through the system evaluation and the selection of appropriate performance indicators, and (iii) the treatment of interactions with the other interconnected and interdependent systems. Each class of systems is composed of sub-classes that are used to describe the various types of components, based on the geographical extent and their function within the system.

In particular, cell classes are used to define inhabited areas (i.e. buildings system) and contain information on buildings' typologies, population or land use policy; network-like systems (i.e., water supply, waste-water, electric power, gas and road networks) contain two types of sub-classes (edges and nodes), which are further sub-divided in specific classes, according to the role played by the component within the system: network nodes can be stations, pumps, reservoirs, sources, distribution nodes, etc.; finally critical facilities such as components of the health-care system, are modelled as point-like objects. Each of the sub-classes is specified with its characteristic attributes and methods, depending on the type of system considered. For instance, initial properties of the objects may include geographic location, area, length, soil type, typology, associated fragility, capacity, connectivity with other components (for networks), etc. Once the simulation is running, the specific methods update the object properties, such as damage states, losses within each cell or remaining connectivity.

Uncertainty is modeled with a network of random variables. Currently it is entangled with that of the physical system, with random variables being included as further attributes within the classes of the corresponding objects (e.g. magnitude M is an attribute of the event object). At present, the current network, even if embedded in the physical portion of the model, is a rather comprehensive representation of uncertainty in the problem, with a refined and effective seismic hazard model and vulnerability model including epistemic/modeling uncertainty in a hierarchical fashion.

In addition to the physical assets, a unified approach was presented in this volume for modeling socio-economic impacts caused by earthquake damage by integrating social vulnerability into the physical systems modeling approaches with a focus on shelter needs and health impacts. The current state-of-the-art in earthquake engineering produces reasonably accurate estimates of physical damage to buildings and infrastructure systems as well as reasonable estimates of the repair and replacement costs associated with this type of damage. However, poor linkages between damage to physical systems and resultant social and economic consequences remain a significant limitation with existing earthquake loss estimation models. For example, as shown in Fig. 13.1 estimating displaced population should be linked beyond physical damage to buildings typically considered in earthquake loss estimation, and considers various social, economic and climatic factors known to impact the habitability of buildings and perception of households to leave their homes. The shelter model in SYNER-G simulates households' decision-making and considers physical, socio-economic, climatic, spatial and temporal factors in addition to modeled building damage states. Similarly, the health impact model combines a new semi-empirical methodology for casualty estimation with models of health impact vulnerability, and transportation accessibility to obtain a holistic assessment of health impacts in the emergency period after earthquakes.

The integrated health impact model is based on a multi-criteria decision theory (MCDA) framework, which allows the bringing together of parameters influencing the direct social losses represented through estimates of casualties and injuries, with

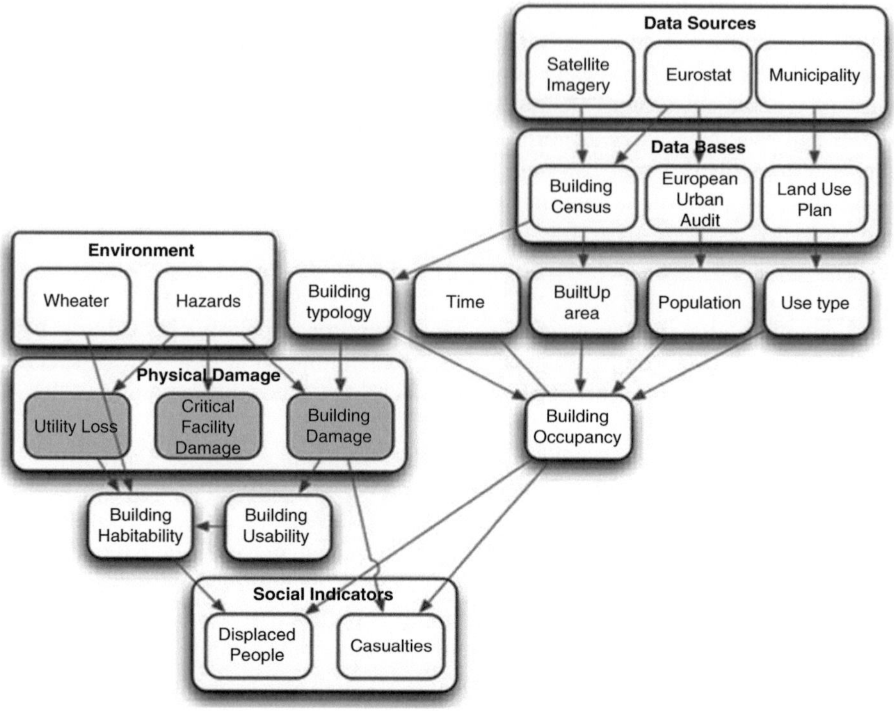

Fig. 13.1 Integrated evaluation of physical and socio-economic indicators in SYNER-G

factors related to overall health impact of the population at risk in an earthquake. The multi-criteria framework is composed of the two main criteria: overall population at risk of mortality in an earthquake (represented by casualties) and an impact factor. Subsequently, the overall health impact for a particular location (i.e., city district, county or country) can be described as the population at risk of mortality, amplified by the set of conditions that can aggravate the health impacts following a disaster which are derived as a weighted index of a set of indicators in four main categories of: social vulnerability, baseline health status, environmental parameters and health-care accessibility.

A key outcome of the SYNER-G project is the development of open-source software and tools that implement the SYNER-G methodology and are capable of dealing with systemic vulnerability for improving the seismic risk assessment and management. The selected city-level case studies for Thessaloniki (Greece) and Vienna (Austria) presented in this volume provide application examples for modeling interactions within and between selected systems and for their implementation within the SYNER-G software. Furthermore, additional case studies such as the gas system of L'Aquila in Italy, an electric power network, a roadway network and a hospital facility in Italy, and also a port facility in Greece, demonstrate the applicability of the methodology at a regional scale.

13.2 Challenges Ahead

SYNER-G is an important step forward for the seismic risk assessment integrating physical damages and socio-economic impact, considering the interactions among different systems and infrastructures at city or regional scale. However there are still several challenges ahead. In the following, an effort is made to present some of the most important ones, always related to the developments accomplished in SYNER-G.

The loss estimation model considers for the moment the short-term period after the event. Future development should allow for the extension of the considered time span including ageing and cascading effects; it is also important to consider the simulation of the recovery process, as well as the changed business and activity pattern, to arrive at the evaluation of overall economic loss and resilience.

The interactions model will probably need further work to widen the scope and the type of interactions considered, taking into account for instance the strength of interactions. In the presented model, the consequence of a failure in the dominant system deterministically follows in the dependent system. This could be weakened, especially when dealing with situations where the input data are incomplete.

The physical modelling of networks is carried out on a detailed level introducing several models. However, the use of other models than connectivity analysis depends entirely on the scale of the problem and interactions considered. While simple connectivity models for all networks may be appropriate in some cases, notably for understanding weaknesses of each network in terms of form, i.e., topology and hence can give very valuable indications on retrofit strategies, the modelling of flow is important to predict the real service level, which may be insufficient even in the presence of a surviving direct connection between a sink and a source. This has an influence on the performance of dependent systems. At present the use of advanced physical modelling of network performance is computationally extremely demanding especially if the problem anticipated concerns a city with all its infrastructures and utility systems included. More effort is still needed in this regard for increasing both the software and hardware capabilities.

Uncertainty is modeled with a network of random variables entangled with that of the physical system, with random variables being included as further attributes within the classes of the corresponding objects. This aspect needs to be improved in order to have the uncertainty network stand alone as a further non-physical network in the model, connected to the physical layers through new dependency edges. A stand-alone network will allow easier introduction of probabilistic analysis methods that will have to deal only with objects on this level. Such a network, used only in a forward simulation, from the marginal variables to the performance metrics, sets the ground for introduction of evidence and Bayesian inference, thus paving the way for a real-time use of the model as a decision support system.

Regarding the seismic hazard part there are pros and cons regarding probabilistic and scenario-approaches to impact, response and recovery estimations. Scenarios allow better and easier analysis of consequences and cascading effects as the conditional probabilities are replaced by if-then considerations. The probabilistic approach associates the frequency of events, conditional probabilities of cascading or induced events and their associated impacts but requires a lot of data, accurate models and can be computationally intensive and results can be difficult to communicate. Reconciling both approaches from a decision perspective represents a major challenge in improved risk management.

The model encodes a fixed, pre-determined sequence for system evaluation. The chosen sequence is very reasonable and grounded in considerations of intrinsic characteristic time each of the considered systems is affected (simply stated, indirect damage in the electric power network propagates faster than in other systems or building collapse occurs and blocks roads before post-event travel is undertaken, etc.). Nonetheless, it is pre-determined and may not apply to all situations. On the other hand, it allows for fixed, though disturbed (with respect to pre-event ones) boundary conditions and the solution of stationary rather than non-stationary flow equations in each system.

Finally, performing more comprehensive risk analyses adds to their complexity and will also necessitate more high quality data, which are rarely available. Acquisition of this type of data and information from various sources is not only very expensive and time consuming, but it may well add extra sources of uncertainties. To this end the rapid acquisition of massive and accurate data in appropriate format needed for the seismic risk assessment is probably one of the most important prerequisites and a major challenge for future research and development.

Solving each one of the challenges outlined above will no doubt make a major contribution, however, achieving a holistic assessment of seismic risk also a better understanding of the links and interrelations to the multi-faceted societal dimensions of risk. Social and economic losses and risks (probability of future losses) of earthquakes are measured with different metrics according to the spatial scale (e.g. national level, regional, city, etc.), time elapsed after the event and need of different groups, organisations, legal entities, etc. who have responsibilities in disaster response, mitigation and preparedness. Thus, the metrics that are used depend on the type of responsibility, the function of the users in the context of disasters, and the services they provide to society. For example, the outputs needed from risk models are quite different for public awareness/education, emergency response planning, post event logistics management, evaluation of mitigation measures or business contingency planning. The scientific and professional earthquake engineering community has thus far focused on human losses, physical losses to buildings and infrastructure using modern approaches such as performance based engineering, which provides probability-based decision variables. The scientific and professional community has also made important strides in terms of the needs of lifeline service providers by computing metrics of downtime due to damaged infrastructure, cascading effects for inter-related networks, and customer response to outages.

However, providing science-based support for diverse requirements in a systematic framework is still a major challenge in earthquake engineering, but is a necessary step towards improved resilience of societies and systems. The impact on population includes more parameters than just the number of deaths, injured persons or lifeline down time, but also it is important to consider the interconnected socio-economic, demographic and cultural context of the populations for which earthquake losses are being modelled. There are a number of factors that limit current modeling efforts in earthquake loss estimation. These include (a) poor quality/inaccessible inventory data with regard to social and economic losses from historic earthquakes; (b) lack of a common and widely accepted reference framework (i.e. definitions, methodology) for the collection of systematic information on social and economic losses; (c) inaccuracies in models of earthquake casualties and displaced persons; (d) lack of quantitative methods to estimate the seismic resilience of communities; (e) poor integration of social, environmental and economic vulnerability in earthquake loss modeling procedures and (f) lack of decision tools and methodologies to capture post-disaster decisions, interactions and changes over time.

In SYNER-G a key thrust has been to go beyond physical risk by linking the outputs of damage to physical assets to the socio-economic dimensions that influence the overall impact within two sectors – shelter and healthcare. The aim has been to introduce improvements in earthquake loss estimation along three interconnected areas: (a) quantitative sub-models (for instance the degree of loss of functionality of water supply given an earthquake), (b) models for the interaction between systems and within systems, and (c) incorporation of social factors for considering drivers of vulnerability and characteristics vulnerable population to possibly suffer larger losses. Future research in earthquake loss estimation must push further along each of these three areas in the development of new models of resilience. Better quantitative models should go beyond estimating the degree of loss of functionality to providing information on the recovery path and time as a consequence of the down time. The interaction models should drive further to consider less tangible but paramount factors such as the influence of institutional and organizational arrangements in dynamic recovery models that can capture post-disaster decisions, interactions and changes over time. Finally science-based implementation schemes of earthquake mitigation measures developed in engineering research and knowledge has to include the social sciences in a more integrated and rigorous way.

Reference

Pitilakis K, Crowley H, Kaynia A (eds) (2014) SYNER-G: typology definition and fragility functions for physical elements at seismic risk, vol 27, Geotechnical, geological and earthquake engineering. Springer, Dordrecht. ISBN 978-94-007-7871-9

Subject Index

K. Pitilakis et al. (eds.), *SYNER-G: Systemic Seismic Vulnerability and Risk Assessment of Complex Urban, Utility, Lifeline Systems and Critical Facilities*, Geotechnical, Geological and Earthquake Engineering 31, DOI 10.1007/978-94-017-8835-9,

T

U

MIX
Papier aus verantwortungsvollen Quellen
Paper from responsible sources
FSC® C105338

If you have any concerns about our products,
you can contact us on
ProductSafety@springernature.com

In case Publisher is established outside the EU,
the EU authorized representative is:
Springer Nature Customer Service Center GmbH
Europaplatz 3, 69115 Heidelberg, Germany

Printed by Libri Plureos GmbH
in Hamburg, Germany